Clinical Aspects of Cyclic Nucleotides

Clinical Aspects of Cyclic Nucleotides

Edited by

Ladislav Volicer, M.D., Ph.D.

Departments of Pharmacology and Medicine
Boston University School of Medicine
Boston, Massachusetts

S P Books Division of
SPECTRUM PUBLICATIONS, INC.
New York

Copyright © 1977 Spectrum Publications, Inc.

All rights reserved. No part of this book may be reproduced in any form, by photostat, microform, retrieval system, or any other means without prior written permission of the copyright holder or his licensee.

SPECTRUM PUBLICATIONS, Inc.
175-20 Wexford Terrace, Jamaica, N.Y. 11432

Library of Congress Cataloging in Publication Data
Main entry under title:

Clinical aspects of cyclic nucleotides.

    Includes papers presented at a symposium held by the Academy of Pharmaceutical Sciences in San Francisco, Apr. 21, 1975.
    Bibliography: p.
    Includes index.
    1. Cyclic nucleotides. 2. Physiology, Pathological. I. Volicer, Ladislav. [DNLM: 1. Nucleotides, Cyclic. QU58 C641]
RB113.C4   616.07'1   76-584-79
ISBN 0-89335-009-5

Contributors

EVA CHABI, B.Sc.
Department of Neurology
Baylor College of Medicine
Houston, Texas

DONALD A. CHAMBERS, Ph.D.
Assistant Professor of Biological
  Chemistry
Department of Dermatology
University of Michigan Medical
  School
Ann Arbor, Michigan

CHU-JENG CHIN, M.D., Ph.D
Associate Professor of Surgery
Division of Cardiovascular and
  Thoracic Surgery
McGill University
Montreal, Quebec, Canada

WAYNE E. CRISS, M.D., Ph.D.
Associate Professor
Department of Biochemistry
Associate Director of Cancer
  Research Center
Howard University School of
  Medicine
Washington, D. C.

MARSHAL P. FICHMAN, M.D.
Associate Clinical Professor of
  Medicine
University of Southern California
  School of Medicine and
  Los Angeles County/USC Medical
  Center
Attending Nephrologist
Cedars-Sinai Medical Center
Los Angeles, California

PAVEL HAMET, M.D., Ph.D.
Assistant Professor in Medicine
University of Montreal
Director, Laboratory of
  Physiopathology of Hormone
  Action
Clinical Research Institute of
  Montreal
Montreal, Quebec, Canada

BRUCE P. HURTER, M.A.
University of Massachusetts
  Medical School
Worcester, Massachusetts

CONTRIBUTORS

LOWELL E. KOPP, Ph.D.
Instructor in Medicine
Roger Williams General Hospital
Brown University
Providence, Rhode Island

ROBERT A. LEVINE, M.D.
Professor of Medicine
Chief, Section of Gastroenterology
State University of New York
Upstate Medical Center
Syracuse, New York

WILLIAM Y. LING, Ph.D.
Assistant Professor
Director of Endocrine Laboratory
Department of Physiology and
  Biophysics
Faculty of Medicine
Dalhousie University
Halifax, Nova Scotia, Canada

CYNTHIA L. MARCELO, Ph.D.
Instructor in Cell Physiology
Department of Dermatology
University of Michigan Medical
  School
Ann Arbor, Michigan

ALEKSANDER A. MATHÉ,
  M.D., Ph.D.
Associate Professor of
  Pharmacology and Psychiatry
Director, Laboratory of Biogenic
  Amines and Allergy
Boston University School of
  Medicine
Boston, Massachusetts

FERID MURAD, M.D., Ph.D.
Professor
Department of Internal Medicine
  and Pharmacology
Director, Division of Clinical
  Pharmacology
University of Virginia
Charlottesville, Virginia

JANET NELL, Ph.D.
Instructor
Department of Neurology
Baylor College of Medicine
Houston, Texas

ARTHUR H. NEUFELD, Ph.D.
Associate Professor
Department of Ophthalmology and
  Visual Science
Yale University School of Medicine
New Haven, Connecticut

SURENDRA K. PURI, Ph.D.
Senior Research Biochemist
Hoechst-Roussel Pharmaceuticals,
  Inc.
Sommerville, New Jersey

EDWIN W. SALZMAN, M.D.
Professor of Surgery
Department of Surgery
Beth Israel Hospital and
  Harvard Medical School
Boston, Massachusetts

L. DANIEL SCHAEFFER,
  Ph.D.
Associate Professor and
Chairman, Department of
  Physiology
School of Dentistry
University of Southern California
Los Angeles, California

GEOFFREY W. G. SHARP,
  Ph.D., D.Sc.
Associate Professor of Physiology
  in Medicine
Biochemical Pharmacology Unit
Massachusetts General Hospital and
Department of Physiology
Harvard Medical School
Boston, Massachusetts

CONTRIBUTORS

RICHARD J. SOHN, Ph.D.
Assistant Professor of Biochemistry
Boston University School of
 Medicine
Boston, Massachusetts

MICHAEL L. STEER, M.D.
Assistant Professor of Surgery
Department of Surgery
Beth Israel Hospital and
 Harvard Medical School
Boston, Massachusetts

ANDREA TAYLOR, M.D.
Clinical Assistant Professor
Department of Internal Medicine
University of Virginia
Charlottesville, Virginia

JOSEPH R. TUCCI, M.D.
Associate Professor of Medicine
Head, Division of Endocrinology
Roger Williams General Hospital
Brown University
Providence, Rhode Island

LADISLAV VOLICER, M.D., Ph.D.
Associate Professor of
 Pharmacology and Assistant
 Professor of Medicine
Boston University School of
 Medicine
Boston, Massachusetts

JOHN J. VOORHEES, M.D.
Professor and Chairman
Department of Dermatology
University of Michigan Medical
 School
Ann Arbor, Michigan

RICHARD WEITZMAN, M.D.
Assistant Professor
Department of Medicine
Harbor General Hospital
University of California, Los
 Angeles, School of Medicine
Torrence, California

K. M. A. WELCH, M.B., Ch.B.
Assistant Professor
Department of Neurology
Baylor College of Medicine
Houston, Texas

Preface

In the past 20 years considerable advances have been made in understanding the role of hormones and mediators in various diseases. As a result of these advances, many drugs that act by modifying the release of mediators or by preventing stimulation of receptors by mediators have been discovered. Until recently, the structure of receptors and the link between receptor stimulation and resulting effects were unknown. The discovery of cyclic nucleotides provided a tool for studying more closely the structure and function of various receptors.

Cyclic nucleotides are intracellular mediators that mediate changes of cell function resulting from receptor stimulation. They are ubiquitous and participate in the mechanism of action of a wide variety of hormones and autacoids. It might, therefore, be expected that cyclic nucleotides would be affected by some diseases and might play an important role in some pathological processes. However, investigation of the role of cyclic nucleotides in clinical situations is very difficult. Cyclic nucleotides are affected by a variety of factors which might mask the relationships between cyclic nucleotides and the disease process. Samples of diseased tissues are seldom available, fixation of the tissue is relatively slow, and, therefore, the cyclic nucleotide levels might change during the fixation procedure. Therefore, many investigators use animal models of pathological processes or study changes of cyclic nucleotides in body fluids which can be readily obtained from clinical subjects.

Despite these difficulties, recent investigations have provided useful clinical information. Determination of cyclic AMP excretion has become a diagnostic test in calcium disorders, and new drugs are being developed which would affect pathological processes through interaction with cyclic nucleotides. Progress of investigation in this field is very fast, and the reports are scattered in numerous specialized journals. The intent of this book is to bring together results of current investigations. Papers included in the symposium on Clinical Aspects of Cyclic Nucleotides held by the Academy of Pharmaceutical Sciences in San Francisco on April 21, 1975, provided the nucleus for the book. Since the time allotted for the symposium limited the number of participants and subjects, the book was expanded by the inclusion of chapters written by additional contributors.

I would like to thank the original panelists of the Symposium and the contributors to this volume for the time and effort required for preparation of their chapters. I would also like to acknowledge the valuable help of Dr. Thomas P. Dousa, Dr. Simmons Lessell, Dr. Peter Polgar, and Dr. Alexander Rutenburg and Dr. Joseph R. Tucci in preparation of the manuscript. Thanks also to the publishers for permission to reproduce figures and tables from their journals.

Ladislav Volicer

March 1976
Boston, Massachusetts

Contents

Preface

1. **Use of Cyclic Nucleotides to Evaluate Calcium Disorders**   1
   by Ferid Murad, Richard Weitzman and Andrea Taylor

2. **Cyclic Nucleotides in Thyroid Disorders and Diabetes Mellitus**   19
   by Joseph R. Tucci and Lowell Kopp

3. **Changes of Cyclic Nucleotides in Normal and Pathological Pregnancy**   69
   by William Y. Ling

4. **Role of Cyclic Nucleotides in the Etiology and Therapy of Polyuric Disorders**   93
   by Marshal P. Fichman

5. **Extracellular Cyclic AMP in Human Hypertension**   169
   by Pavel Hamet

6. **Cyclic Nucleotides in Shock and Trauma**   181
   by Chu-Jeng Chiu

CONTENTS

7. **Role of Cyclic Nucleotides in the Normal Lung and in Bronchial Asthma** ⸺ 193
   by Richard J. Sohn, Aleksander A. Mathé and Ladislav Volicer

8. **Role of Cyclic Nucleotides in Gastroinestinal Diseases** ⸺ 229
   by Robert A. Levine

9. **Adenylate Cyclase and the Stimulatory Effect of Cholera Toxin in the Causation of Diarrhea** ⸺ 263
   by Geoffrey W. G. Sharp

10. **Cyclic AMP in Saliva, Salivary Glands and Gingival Tissues** ⸺ 283
    by L. Daniel Schaeffer

11. **Cyclic Nucleotides in Hemostasis and Thrombosis** ⸺ 295
    by Michael L. Steer and Edwin W. Salzman

12. **The Role of Cyclic AMP in Neurologic and Affective Disorders** ⸺ 327
    by K. M. A. Welch, Janet Nell and Eva Chabi

13. **Role of Cyclic Nucleotides in Drug Addiction and Withdrawal** ⸺ 361
    by L. Volicer, S. K. Puri and B. P. Hurter

14. **Roles for Cyclic Nucleotides in Diseases of the Eye** ⸺ 379
    by Arthur H. Neufeld

15. **The Role of Cyclic Nucleotides in the Pathogenesis of Psoriasis** ⸺ 407
    by Donald A. Chambers, Cynthia L. Marcelo and John J. Voorhees

16. **Cyclic Nucleotide Metabolism in Tumors** ⸺ 429
    by Wayne E. Criss and Ferid Murad

    Index ⸺ 449

Clinical Aspects
of Cyclic Nucleotides

1

Use of Cyclic Nucleotides to Evaluate Calcium Disorders*

FERID MURAD
RICHARD WEITZMAN
ANDREA TAYLOR

In this chapter, we wish to discuss some of our clinical studies and those of others in which cyclic nucleotides have helped in the diagnosis of some calcium disorders and have provided some understanding of their pathophysiology.

Figure 1 summarizes the second messenger concept of adenosine 3',5' monophosphate (cyclic AMP) as proposed by Sutherland and his associates (2). In brief, hormone or drug after receptor binding activates adenylate cyclase that catalyzes the conversion of ATP to cyclic AMP. The cyclic AMP accumulates in the target tissue and regulates various subsequent biochemical or physiologic responses. Greengard (3) has proposed that many, and perhaps all, of the biochemical alterations mediated by cyclic AMP are a result

---

* Supported by a research grant from the USPHS (AM-15316), a Diabetes Endocrinology Research Center grant (AM-17042), and a General Clinical Research Center grant (RR 847). F. Murad is the recipient of a USPHS Research Career Development Award (AM-70456).

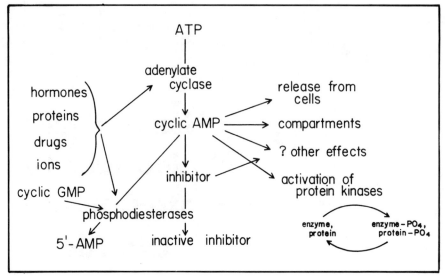

1 Generalized scheme of cyclic AMP metabolism and its effects. From Murad and Weitzman (1).

of the activation of cyclic AMP dependent protein kinases. The latter enzyme(s) catalyzes the transfer of gamma phosphate from ATP to a protein substrate. Some of these substrates may include enzymes, such as phosphorylase kinase, glycogen synthase, hormone-sensitive lipase, etc. The activity of the particular enzyme is altered as a result of its phosphorylation or dephosphorylation. Cyclic AMP may be inactivated by its hydrolysis to 5'-AMP by cyclic nucleotide phosphodiesterases, and these enzymes may be inhibited by a variety of agents (4). Methylxanthines, such as caffeine and theophylline, are commonly used to prevent cyclic AMP hydrolysis in tissues and thereby mimic or enhance a given hormone's response.

Another system consisting of guanylate cyclase and guanosine 3',5'-monophosphate has also been described (5). It is apparent that the formation and metabolism of the two cyclic nucleotides share some common features (Fig. 2). However, work with cyclic GMP has not developed to the same extent as cyclic AMP; thus the reasons for many more question marks in the overall scheme. Little is known about the cellular processes altered by cyclic GMP or the

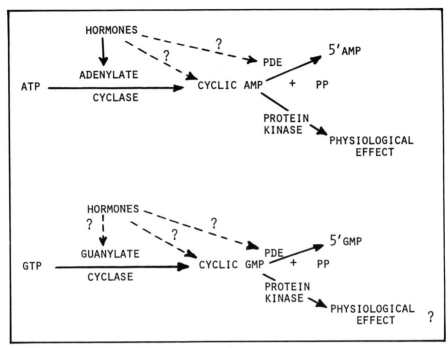

2 Comparisons of cyclic AMP and cyclic GMP metabolism. The analogies in the two systems for synthesis, metabolism and probable mechanism of action are apparent. The question marks designate possible areas for control and/or effects in which data are lacking or inconclusive.

mechanisms by which these occur. The hormones that alter either cyclic AMP or cyclic GMP accumulation in tissues have generally been different. It has also been suspected that the two systems may be interactive, i.e., that one cyclic nucleotide influences the synthesis, metabolism or activity of the other. However, there have been only a few direct observations to support this contention (6–8).

One consequence of cyclic AMP and cyclic GMP accumulation in tissues with hormones is its release into extracellular fluids (9–15). It is not known if this represents an active or passive process, but this could be an important mechanism to regulate intracellular levels of cyclic nucleotides. Levels of cyclic nucleotides in extracellular fluids, particularly plasma and urine, can provide us with an assessment of end organ responses to various hormones and drugs (10–15).

When we began our clinical studies with cyclic nucleotides, we expected that the clinical application of some of the basic information obtained from animal studies should offer the endocrinologist, clinical pharmacologist, and other subspecialists a powerful tool in several areas: (1) diagnosis of some endocrine and metabolic disorders; (2) evaluation of surgical or medical therapy in several diseases or both; (3) evaluation of the pathophysiology of some endocrine and metabolic disorders; and (4) a model with which to design and test new therapeutic agents (10, 14).

The utility of urine specimens for clinical studies with cyclic nucleotides is attributable to the relatively high concentrations of cyclic AMP and cyclic GMP in urine and the ease with which individual or serial specimens can be obtained. Under basal conditions, the cyclic AMP concentration in human urine is of the order of several micromolar. The concentration of cyclic GMP is about one-fifth that of cyclic AMP (9, 15).

Only minor preparation of urine specimens before cyclic nucleotide assay appears necessary (14–18). Some laboratories have collected urine specimens with HC1, acetic acid and thymol, sodium metabisulfite, or chloroform to decrease bacterial growth. Other laboratories, including our own, collect urine samples in refrigerated containers (4°); and at the termination of the collection period aliquots are stored at $-20°$ for assay. We have seen no significant alteration in cyclic AMP or cyclic GMP levels when human urine is stored at $-20°$ for up to several years. However, storage of unacidified urine samples at room temperature for more than 2 to 3 days or at 4° for more than several weeks has resulted in losses in cyclic nucleotides, which we have attributed to bacterial growth and hydrolysis. The ability to assay crude urine samples has obviously permitted processing more samples in clinical screening studies. In addition, the ability to assay cyclic AMP and cyclic GMP content in unpurified urine samples simultaneously with a double-isotope-binding assay (17) or a radioimmunoassay (18) has also decreased technical time considerably.

Although another readily available source for specimens is plasma, studies with plasma have been considerably fewer (9, 10, 12, 14). This is probably due to the relatively low concentrations of cyclic nucleotides in plasma and the need to purify samples before assay with most current methods. Cyclic nucleotides have also been found in other extracellular fluids (see reviews 9 and 10 for references). However, much less work has been done with these other fluids.

The studies of Broadus et al. (19) in normal volunteers under basal conditions indicate that about one-half of the cyclic AMP in urine is derived from plasma load to the kidney and passive glomerular filtration. The remainder of the cyclic AMP in urine is contributed primarily by the proximal nephron under the influence of parathyroid hormone. In contrast, urinary cyclic GMP appears to be derived predominantly from glomerular filtration under basal conditions. In 12 hospitalized normal adult volunteers, we found the mean ($\pm$ SE) renal clearances of cyclic AMP and cyclic GMP to be 258 ± 45 and 83 ± 20 ml/min, respectively. Similar values were reported by Broadus et al. (19). One would predict that reduced glomerular filtration and renal function would decrease cyclic nucleotide excretion; this has been the case (12, 13, 15). We have found that the quantities of cyclic nucleotide excreted are also independent of urine volume (15). This relationship has been observed with wide variations in fluid intake, urine volume, and osmolarity.

Most adults (15) and children (20) demonstrate diurnal variations in cyclic AMP and cyclic GMP excretion. The circadian pattern of urinary cyclic nucleotides has also been observed in hypoparathyroid subjects, suggesting that factors other than fluctuating levels of serum parathyroid hormone are responsible for this phenomenon (15). We have also found circadian patterns in plasma cyclic AMP and cyclic GMP, which could explain the phenomenon observed in urine (unpublished observations). In several patients we examined the circadian pattern in cyclic AMP and cyclic GMP excretion and phase-shifted the patient's sleep cycle so that they were awake during the night and slept during the day (Fig. 3). We found a phase shift in the urinary excretion of cyclic AMP and cyclic GMP, with peak excretion during the awakening hours.

The daily basal excretion rates of cyclic AMP in an individual are generally very stable (10, 15). In many patients who we examined for several days or weeks, we found daily excretion rates of cyclic AMP to vary as little as 10 to 20 percent. However, daily excretion of cyclic GMP may vary as much as 50 to 100 percent for unexplained reasons (10, 14, 15).

In hospitalized adult normal volunteers on a house diet and normal activity, we have observed 24-hr excretion rates of $2.75 \pm$

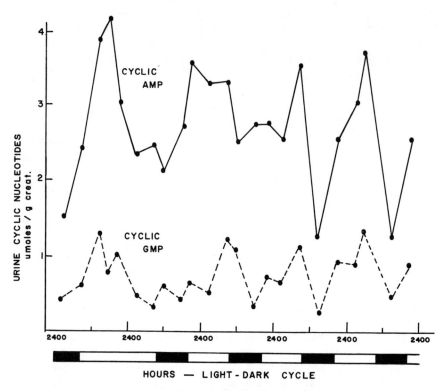

3 Urinary excretion of cyclic AMP and cyclic GMP with alteration of sleep cycle. A normal adult female volunteer remained in bed for a period of 5 days. Her sleep cycle (darkened room) was phase shifted on the second day as indicated. This resulted in a shift of her diurnal pattern of cyclic nucleotide excretion with peak excretion rates during the awakening hours. (From a study done in collaboration with Dr. G. Harbert.)

0.13 μmoles cyclic AMP/gm creatinine and 0.46 ± 0.08 μmoles cyclic GMP/gm creatinine (Table 1). Our values for cyclic AMP excretion rates are somewhat lower with a narrower range than those reported from other laboratories. This may be attributable to our studies with 24-hour urine samples to eliminate effects of diurnal variations and studies in hospitalized patients. In a small series of nonhospitalized normal subjects we found 24-hr excretion rates

Table 1

Daily urinary excretion of cyclic nucleotides in hospitalized and ambulatory normal adult volunteers. Values are means ± S.E. of the number of patients indicated in paraentheses.

	μ moles/g creat/day	
	Cyclic AMP	Cyclic GMP
Hospitalized normals (27)	2.75 ± 0.13	0.46 ± 0.08
Ambulatory normals (6)	3.81 ± 0.31*	0.69 ± 0.12

*Significantly different ($p < 0.01$)

of cyclic AMP to be about 39 percent higher (Table 1). Excretion of cyclic GMP was no different (Table 1).

Patients with reduced renal function and diminished endogenous creatinine clearances, as mentioned earlier, have decreased cyclic nucleotide excretion rates (12, 13, 15); this may be because of decreased glomerular filtration of cyclic nucleotides. Urinary excretion of cyclic AMP and cyclic GMP also changes markedly with age in children and adolescents (Fig. 4) (20). Therefore, there are many factors that can alter basal excretion rates of cyclic nucleotides (see Ref. 1 and 10 for additional discussion). Obviously, the utility of cyclic nucleotides in extracellular fluids in clinical medicine will depend in part on the ability to establish what most of these factors are and to design specific and/or provocative tests that control or minimize such factors.

Liver and kidney appear to be the major sources of plasma cyclic AMP under the influence of glucagon and parathyriod hormone, respectively. Kaminsky et al. (12) found that administration of glucagon in hepatectomized dogs failed to increase plasma or urinary cyclic AMP. Similarly, in nephrectomized patients maintained on dialysis, parathyroid hormone failed to increase plasma cyclic AMP. Studies of arterio-venous differences in cyclic AMP and cyclic GMP in anesthetized dogs from Steiner's laboratory (21) and our own (unpublished observations) suggest that under basal conditions many tissues can contribute small quantities of cyclic nucleotides to plasma.

Incubation and perfusion media from in vitro or in situ studies with a number of tissues contain variable quantities of cyclic AMP.

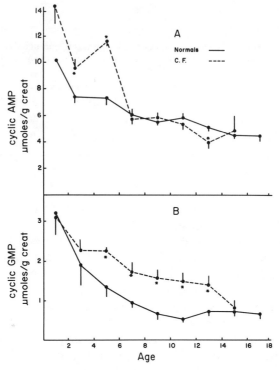

Cyclic nucleotide excretion in normal children and those with cystic fibrosis (C.F.). Values are means ± SE of each age group, and asterisks indicate values that are significantly different ($p < 0.05$) from those of normal children at that age. From Murad et al., (20).

In general, hormonal stimulation of cyclic AMP accumulation in intact cell systems has resulted in accumulation of nucleotide in the medium (6, 22, 23). The rate of this process has varied in different systems, and, except for pigeon erythrocytes, the intracellular concentration has greatly exceeded that of the medium. Release, leakage or transport of cyclic AMP from cells to interstitial fluid, may be a mechanism for communication between cells. However, its importance in plasma, if any, has not been established. The low concentration in plasma (about 10 to 30 nM) and the relatively high exogenous concentrations required for effects in intact cell systems (generally 0.1 to 1 mM) suggest that plasma cyclic AMP is

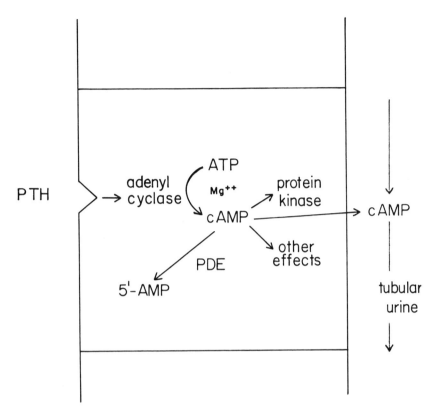

5 Effects of parathyroid hormone on the renal tubule. Approximately one-half of the cyclic AMP in urine is derived from filtration of plasma and one-half from the tubule under the influence of parathyroid hormone.

not important unless mechanisms existed that permitted its concentration in tissues.

The studies of Chase and Aurbach (24) with parathyroid hormone administration demonstrated pronounced rapid increases in urinary cyclic AMP levels. These effects preceded the well-known phosphaturic effect of parathyroid hormone. These studies subsequently led to the demonstration of parathyroid hormone stimulation of renal cortical and bone adenylate cyclases (Fig. 5).

Intravenous administration of 200 U of parathyroid hormone

Table 2

Cyclic nucleotide excretion with various calcium disorders.

Values are the mean ± S.E. daily excretion of cyclic nucleotide normalized for urinary creatinine. Normals are hospitalized volunteers. All patients with primary hyperparathyroidism had histologically verified parathyroid adenomas or hyperplasia. Patients with suspected hyperparathyroidism had elevated serum PTH levels and/or hypercalcemia. Patients with osteoporosis or renal stones were normocalcemic; some had a negative neck exploration and/or normal serum PTH levels. Hypoparathyroid patients include those with idiopathic, pseudo or post-surgical hypoparathyroidism. The number of patients in each group is indicated in parentheses; asterisks indicate values that are significantly different from normals.

Diagnosis	Cyclic AMP (u moles/g creatinine)	Cyclic GMP
Normals	2.75 ± 0.13 (27)	0.46 ± 0.08 (23)
Primary Hyperparathyroidism	5.86 ± 0.31 (54)*	0.90 ± 0.15 (12)*
Suspected Hyperparathyroidism	6.73 ± 0.74 (16)*	1.17 ± 0.21 (3)*
"Ectopic PTH-tumor"	8.99 ± 2.36 (4)*	— —
After surgery for hyperparathyroidism	2.79 ± 0.36 (14)	0.62 ± 0.08 (8)
Hypoparathyroidism	1.81 ± 0.19 (20)*	0.53 ± 0.06 (13)
Osteoporosis	3.48 ± 0.28 (10)*	0.71 ± 0.10 (9)*
Nephrolithiasis	2.98 ± 0.24 (25)	0.64 ± 0.12 (18)

may increase urinary cyclic AMP as much as 200-fold within 30 min (10–14, 25). Large doses of parathyroid hormone will also increase urinary excretion of cyclic GMP and plasma levels of cyclic AMP (10, 12, 25). The effects of parathyroid hormone on plasma cyclic AMP are presumably of renal origin, since others have not observed increases in plasma after its administration in nephrectomized patients (12). The mechanism by which parathyroid hormone increases urinary cyclic GMP is unknown. In vitro effects of parathyroid hormone on cyclic GMP levels have not been observed.

Since parathyroid hormone has such a pronounced effect on cyclic nucleotide excretion, we examined 24-hr urines in hospitalized normals and patients with various calcium disorders and normal renal function (Table 2) (10, 14, 15). Hypercalcemic hyperparathyroid and normocalcemic hyperparathyroid patients excrete significantly greater amounts of cyclic AMP and cyclic GMP than do normal subjects (15). Hypoparathyroid patients (idiopathic, postsurgical and pseudohypoparathyroid) excrete less cyclic AMP than normals. On the other hand, patients with nonparathyroid carcinoma and hypercalcemia excreted normal quantities of cyclic AMP and cyclic GMP (15). Dohan et al. (26) have also reported that patients with nonparathyroid carcinoma and hypercalcemia have low cyclic AMP excretion rates.

Of the 54 hyperparathyroid patients with normal or only moderate reductions in endogenous creatinine clearance (greater than 40 ml/min) whom we have examined to date, 7 patients (14%) had normal cyclic AMP excretion (normal ± 2 SD, 2.75 ± 2 × 0.7). After surgery for hyperparathyroidism cyclic AMP and cyclic GMP excretion returned to normal (Table 2). While hypoparathyroid patients as a group (idiopathic, postsurgical, and pseudohypoparathyroid) excreted significantly less cyclic AMP, about 30 percent of these patients had normal excretion rates. Patients with idiopathic osteoporosis had elevated cyclic AMP and cyclic GMP excretion. These observations suggest that some of these patients may have been hyperparathyroid, since serum PTH or neck surgery were not performed on all of these normocalcemia patients to exclude the diagnosis. Patients with renal stones without hyperparathyroidism excreted normal amounts of cyclic AMP and cyclic GMP. A patient with hypercalcemia and sarcoidosis had a low cyclic AMP excretion rate of 1.26 μmoles/gm creatinine (15).

Our studies (10, 14, 15, 25) and studies from many other laboratories (12, 13, 26–29) to date indicate that cyclic AMP excretion rates can be successfully used to evaluate parathyroid function in either normocalcemic or hypercalcemic patients. The test is simple and useful with less than 15 percent false-negative results provided that appropriate controls and 24-hr urines are used and that endogenous creatinine clearance in patients examined is greater than about 40 ml/min.

Some laboratories have also examined cyclic AMP excretion before and after calcium administration to evaluate hypercalcuric patients (30). Patients with primary hyperparathyroidism fail to suppress their urinary excretion of cyclic AMP after calcium as do normal patients or patients with other causes of hypercalcuria (Table 3). Patients with "ectopic PTH-secreting tumors" also have elevated cyclic AMP excretion rates (12), and, thus, this test cannot currently distinguish primary from extopic hyperparathyroidism (Table 2). It also appears to be of little use in patients with primary or secondary hyperparathyroidism with a significant reduction in renal function, since their cyclic AMP excretion is frequently normal or low (15). Thus, values for urinary cyclic AMP must be

Table 3

Effect of calcium on cyclic AMP excretion*

Four patients with primary hyperparathyroidism, 2 patients with hypoparathyroidism and 16 control patients were studied. Urine was collected in sequential 12 hour aliquots. At the beginning of the third collection period calcium (15 mg/kg) was administered intravenously over 4 hours.

	Cyclic AMP excreted (μ moles/12 hr)			
	Period 1	Period 2	Period 3	Period 4
Controls	2.47 ± 0.27	2.13 ± 0.21	1.61 ± 0.17	1.65 ± 0.11
Hyperparathyroid	2.85 ± 0.43	2.62 ± 0.35	3.57 ± 1.40	2.45 ± 0.50
Hypoparathyroid	1.46 ± 0.51	1.53 ± 0.58	1.48 ± 0.36	1.46 ± 0.34

*Studies done in collaboration with Dr. C.Y. Pak

interpreted in view of the clinical setting and in conjunction with other laboratory tests.

Elevated cyclic-AMP–excretion rates are also not specific for hyperparathyroidism. This could be expected from the numerous hormones that are capable of increasing cyclic AMP levels in various tissues. We have found some patients with pheochromocytoma, carcinoid, and inappropriate ADH syndrome to have increased cyclic-AMP–excretion rates. Increased cyclic-AMP–excretion in hyperthyroid patients has been reported previously (31, 32).

Presumably with additional knowledge of factors that influence cyclic nucleotide levels in tissues and extracellular fluids, false-positive and false-negative tests employing cyclic AMP and/or cyclic GMP in parathyroid disorders or other diseases could be avoided. For example, one would predict that the renal clearance of cyclic AMP relative to creatinine or inulin clearance is increased in primary or secondary hyperparathyroidism with or without diminished renal function. However, these studies have not been reported.

The examination of cyclic nucleotide levels and metabolism have also helped clarify the pathophysiology of some diseases. Dr. Albright and his associates described pseudohypoparathyroidism as a genetic defect characterized by end organ resistance to parathyroid hormone (33). The nature of the defect was somewhat clarified by the studies of Chase et al. (11), who found that parathyroid hormone administration resulted in less of an increase in cyclic AMP excretion in these patients than the response observed in nor-

Table 4

Cyclic nucleotide levels in urine and plasma after parathyroid extract.

Normal volunteers or patients with pseudohypoparathyroidism were given 200 Units of parathyroid extract intravenously. Urine and plasma were collected every 30 minutes before and after PTE. The greatest changes were observed within 30 minutes and are presented as percent of control period (mean ± S.E.). Values in parentheses are the number of patients studied and asterisks indicate values that are significantly different from values in normal controls.

	Urine Cyclic AMP	Urine Cyclic GMP	Plasma Cyclic AMP	Urine PO_4/creatinine
Normals (4)	5108 ± 1767	1519 ± 260	626 ± 200	255 ± 52
PHP (6)	352 ± 89*	219 ± 75*	125 ± 10*	285 ± 78

mals and other patients with various calcium disorders. Table 4 summarizes some of our studies with 200 U of parathyroid extract intravenously in four normal and six pseudohypoparathyroid patients (34). Patients with pseudohypoparathyroidism demonstrated small increases in urinary cyclic AMP and cyclic GMP and plasma cyclic AMP with parathyroid hormone. Previously, the diagnosis of pseudohypoparathyroidism had been made in patients by the lack of a phosphaturic and/or hypercalcemic response to parathyroid hormone administration. However, these effects in normals and pseudohypoparathyroid patients are frequently small and indistinguishable, as illustrated by the phosphate data shown in Table 4. Because of the relatively large increase in cyclic AMP excretion observed normally, and the small effect seen with pseudohypoparathyroid patients, the examination of urinary cyclic AMP before and after parathyroid hormone has proved to be a much better diagnostic test for the disorder.

The diminished cyclic AMP excretion with parathyroid hormone administration in pseudohypoparathyroidism could result from (1) decreased renal synthesis; (2) increased renal hydrolysis; or (3) a combination of the two.

We, therefore, examined plasma cyclic AMP and urine cyclic AMP, cyclic GMP and phosphate with intravenous administration of 200 U of parathyroid extract in pseudohypoparathyroid patients before and after administration of phosphodiesterase inhibitors (Table 5) (34). We gave patients either oral chlorpropamide (500

Table 5

Effects of chlorpropamide, aminophylline and parathyroid extract in patients with pseudohypoparathyroidism.

Patients with pseudohypoparathyroidism were given 200 U PTE intravenously as described in Table 3. Some patients were rechallenged with PTE after receiving either 500 mg chlorpropamide orally for 5 days or an infusion of aminophylline (500 mg over 20 minutes followed by 1 mg/min for 2 hours); PTE was administered after 30-60 minutes of aminophylline infusion. Values are presented at peak mean (± S.E.) responses (percent of the basal period). Peak responses were observed within 30 minutes after PTE. The number of patients examined is indicated in parentheses, and asterisked values are significantly different from PTE alone.

	Urine cAMP	Urine cGMP	Plasma cAMP	Urine PO_4/creat.
PTE (5)	373 ± 107	239 ± 89	129 ± 11	208 ± 20
PTE + Chlorprop. (3)	294 ± 75	130 ± 17	103 ± 4*	162 ± 13*
PTE + Aminoph. (3)	289 ± 34	165 ± 63	143 ± 8	334 ± 53*

mg/day) for 5 days or intravenous aminophylline (500 mg over 20 min followed by 1 mg/min for 2 hr) and rechallenged them with intravenous parathyroid extract. In normals and several pseudohypoparathyroid patients the cyclic nucleotide response in plasma or urine was either unaltered or minimally changed with phosphodiesterase inhibitors. The phosphaturic response to parathyroid hormone was decreased with chlorpropamide and increased after aminophylline (Table 5). One of our patients with pseudohypoparathyroidism had a normal or supernormal response to parathyroid extract after either chlorpropamide or aminophylline (Table 6) (34).

These studies indicate that pseudohypoparathyroidism is a syndrome encompassing several different types of genetic defects (11, 34–36.) As originally suggested by Chase et al. (11), some patients may have an abnormal PTH-renal receptor or diminished renal cortical adenylate cyclase to explain the decreased cyclic AMP excretion with the hormone. All of the pseudohypoparathyroid patients that we have examined (Tables 4–6) had a small but detectable increase in urinary cyclic AMP after parathyroid extract, suggesting that the deficiency is not complete. The ability to obtain a normal increase in urinary cyclic nucleotides with parathyroid

Table 6

Effects of chlorpropamide, aminophylline and parathyroid extract in a patient with pseudohypoparathyroidism.

A patient with pseudohypoparathyroidism was studied with the protocol described in Table 4. Values are peak responses and are percent of the basal period.

	Urine cAMP	Urine cGMP	Plasma cAMP	Urine PO_4/creat.
PTE	247	119	109	665
PTE + Chlorprop.	3,520	600	446	293
PTE + Aminoph.	22,172	3,093	508	591

extract after either chlorpropamide or aminophylline in the latter patient (Table 6) indicates that some pseudohypoparathyroid patients probably have normal cyclic nucleotide synthesis in the face of acccelerated cyclic nucleotide hydrolysis (34). Drezner et al. (35) and Rodriquez et al. (36) have also described two pseudohypoparathyroid patients who increase their urinary cyclic AMP with parathyroid extract but did not demonstrate the phosphaturia. These workers concluded that these patients had defects distal to cyclic AMP formation and perhaps at the level of protein kinase (35, 36). Therefore, the phenotypic expression of pseudohypoparathyroidism (failure to increase urinary cyclic AMP or phosphate after parathyroid hormone) may result from one of several different types of genetic defects. The evaluation of cyclic nucleotide metabolism in such patients should provide a mechanism for their subclassification and may also lead to a more rational mode of therapy for their hypocalcemia.

In the past several years, some headway has been made with the incorporation of cyclic nucleotide metabolism in clinical studies. Clearly, cyclic nucleotide levels in urine and plasma have proved to be useful additional diagnostic tools in the evaluation of several calcium and parathyroid disorders. However, it is also quite apparent that we have only begun to scratch the surface in terms of clinical application. With some serious thought, many of us could speculate on additional situations in which they would probably prove useful.

At present, there are numerous examples of simple as well as complex questions in clinical medicine in which our current methodology with cyclic nucleotides can be applied to disorders to obtain additional information regarding metabolic regulation, pathophysiology, diagnosis, and therapy.

REFERENCES

1. Murad, F., and R. Weitzman: Hormonal regulation of cyclic AMP. Sem Drug Treatment *3*, 189, 1973.
2. G. A. Robison, R. W. Butcher, and E. W. Sutherland, Eds., *Cyclic AMP*. Academic Press, New York, 1971.
3. Greengard, P. On the reactivity and mechanism of action of cyclic nucleotides. Ann N. Y. Acad. Sci. *185*, 18, 1971.
4. Butcher, R. W., and E. W. Sutherland: Purification and properties of cyclic 3′, 5′-nucleotide phosphodiesterase and use of this enzyme to characterize adenosine 3′, 5′-phosphate in human urine. J. Biol. Chem. *237*, 1244, 1962.
5. Hardman, J. G.: Other cyclic nucleotides. In G. A. Robison, R. W. Butcher, and E. W. Sutherland, Eds., *Cyclic AMP*. Academic Press, New York, 1971, p. 400.
6. Murad, F., V. Manganiello, and M. Vaughan: Effects of guanosine 3′,5′-monophosphate on glycerol production and accumulation of adenosine 3′,5′-monophosphate during incubation of fat cells. J. Biol. Chem. *245*, 3352, 1970.
7. Beavo, J. A., J. G. Hardman, and E. W. Sutherland: Stimulation of adenosine 3′,5′-monophosphate hydrolysis by guanosine 3′,5′-monophosphate. J. Biol. Chem. *246*, 3841, 1971.
8. Goldberg, N. D., R. F. O'Dea, and M. K. Haddox: Cyclic AMP, Adv. Cyclic Nucleotide Res. *3*, 155, 1973.
9. Broadus, A. E., J. G. Hardman, N. I. Kaminsky, J. H. Ball, E. W. Sutherland, and G. W. Liddle: Extracellular cyclic nucleotides. Ann. N. Y. Acad. Sci. *185*, 50, 1971.
10. Murad, F.: Clinical studies and application of cyclic nucleotides. Adv. Cyclic Nucleotides Res. *3*, 355, 1973.
11. Chase, L. R., G. L. Melson, and G. D. Aurbach: Pseudohypoparathyroidism: Defective excretion of 3′,5′-AMP in response to parathyroid hormone. J. Clin. Invest. *48*, 1832, 1969.
12. Kaminsky, N. I., A. E. Broadus, J. G. Hardman, D. J. Jones, J. H. Ball, E. W. Sutherland, and G. W. Liddle: Effects of parathyroid hormone on plasma and urinary adenosine 3′,5′-monophosphate in man. J. Clin. Invest. *49*, 2387, 1970.
13. Taylor, A. L., B. B. Davis, G. Pawlson, J. B. Josimovich, and D. H. Mintz: Factors influencing the urinary excretion of 3′,5′-adenosine monophosphate in humans. J. Clin. Endocrinol. *30*, 316, 1970.
14. Murad, F.: Clinical applications of cyclic nucleotide levels. Proc. 5th Int. Congress Pharmacol. *5*, 233, 1972.

15. Murad, F., and C. Y. Pak: Urinary excretion of adenosine 3',5'-monophosphate and guanosine 3',5'-monophosphate. N. Engl. J. Med. *286*, 1382, 1972.
16. Murad, F., V. Manganiello, and M. Vaughan: A simple, sensitive protein binding assay for guanosine 3',5'-monophosphate. Proc. Nat. Acad. Sci. *68*, 736, 1971.
17. Murad, F., and A. G. Gilman: Adenosine 3',5'-monophosphate and guanosine 3',5'-monophosphate: A simultaneous protein binding assay. Biochim. Biophys. Acta *252*, 397, 1971.
18. Wehmann, R. E., L. Blonde, and A. L. Steiner: Simultaneous radioimmunoassay for the measurement of adenosine 3',5'-monophosphate and guanosine 3',5'-monophosphate. Endocrinology *90*, 330, 1972.
19. Broadus, A. E., N. I. Kaminsky, J. G. Hardman, E. W. Sutherland, and G. W. Liddle: Kinetic parameters and renal clearances of plasma adenosine 3',5'-monophosph and guanosine 3',5'-monophosphate in man. J. Clin. Invest. *49*, 2222, 1970.
20. Murad, F., W. Moss, R. Selden, and A. Johanson: Urinary excretion of cyclic AMP and cyclic GMP in normal children and those with cystic fibrosis. J. Clin. Endocrinol. *40*, 552, 1975.
21. Wehmann, R. E., L. Blonde, and A. L. Steiner: Sources of cyclic nucleotides in plasma. J. Clin. Invest. *53*, 173, 1974.
22. Davoren, P., and E. W. Sutherland: The effects of L-epinephrine and other agents on the synthesis and release of adenosine 3',5'-phosphate by whole pigeon erythrocytes. J. Biol. Chem. *238*, 3009, 1963.
23. Manganiello, V., F. Murad, and M. Vaughan: Effects of lipolytic and antilipolytic agents on cyclic 3',5'-adenosine monophosphate in fat cells. J. Biol. Chem. *246*, 2195, 1971.
24. Chase, L. R., and G. D. Aurbach: Parathyroid function and the excretion of 3',5'-adenylic acid. Proc. Nat. Acad. Sci. *58*, 518, 1967.
25. Murad, F., and R. Weitzman: Effects of parathyroid hormone and calcitonin on cyclic AMP metabolism. Prov. IV Int. Congress Endocrinol. *273*, 468, 1972.
26. Dohan, P. H., K. Yamashita, P. R. Larsen, B. Davis, L. Deftos, and J. B. Field: Evaluation of urinary cyclic 3',5'-adenosine monophosphate excretion in the differential diagnosis of hypercalcemia. J. Clin. Endocrinol. *35*, 775, 1972.
27. Tsang, C. P., D. C. Lehotay, and B. E. Murphy: Competitive binding assay for adenosine 3',5'-monophosphate employing a bovine adrenal protein: Application to urine, plasma and tissues. J. Clin. Endocrinol. *35*, 809, 1972.
28. Neelon, F. A., B. M. Birch, M. Drezner, and H. E. Lebovitz: Urinary cyclic adenosine monophosphate as an aid in the diagnosis of hyperparathyroidism. Lancet *1*, 631, 1973.
29. Pak, C. Y. C., M. Ohata, E. C. Lawrence, and W. Snyder: The hypercalcurias: Causes, parathyroid function and diagnosis criteria. J. Clin. Invest. *54*, 387, 1974.
30. Pak, C. Y. C., R. Kaplan, H. Bone, J. Townsend, and O. Waters: A simple test for the diagnosis of absorptive, resorptive and renal hypercalcurias. N. Engl. J. Med. *292*, 497, 1975.
31. Rosen, O. M.: Urinary cyclic AMP in Grave's disease. N. Engl. J. Med. *287*, 670, 1972.
32. Lin, T., L. E. Kopp, and J. R. Tucci: Urinary excretion of cyclic 3',5'-

adenosine monophosphate in hyperthyroidism. J. Clin. Endocrinol. *36*, 1033, 1973.
33. Allbright, F., C. H. Burnett, P. H. Smith, and W. Parson: Pseudohypoparathyroidism—an example of the "Seabright-Bantam syndrome." Endocrinology *30*, 922, 1942.
34. Weitzman, R., F. Murad, and J. Owen: Cyclic nucleotide metabolism in pseudohypoparathyroidism. Proc. 55th Ann. Meeting Endocrine Soc., Chicago, June 1973.
35. Drezner, M., F. A. Neelon, and H. E. Lebovitz: Pseudohypoparathyroidism type II: A possible defect in the reception of the cyclic AMP signal. N. Engl. J. Med. *289*, 1056, 1973.
36. Rodriquez, H. J., H. Villarreal, S. Klahr, and E. Slatopolsky: Pseudohypoparathyroidism type II: Restoration of normal hormone by calcium administration. J. Clin. Endocrinol. *39*, 693, 1974.

2

Cyclic Nucleotides in Thyroid Disorders and Diabetes Mellitus

JOSEPH R. TUCCI
LOWELL KOPP

There is considerable evidence to support the concept that adenosine 3′,5′-monophosphate (cyclic AMP) has an important and vital role in a wide variety of physiologic processes. Since the discovery of cyclic AMP by Sutherland and Rall (1), the formulation of the second messenger concept (2) and the discovery of cyclic AMP in human (3) and animal urine (4), the action of many hormones has been shown to be mediated by the adenylate cyclase-cyclic AMP system (5, 6). The function of cyclic AMP extends from the peripheral effects of various amine and peptide hormones to the regulatory effects of hypothalamus and pituitary gland on the various endocrine glands themselves and to regulation of secretion of hormones such as glucagon and insulin, which are controlled by peripheral substances and hormones rather than pituitary tropic hormones.

THE THYROID GLAND

Hypothalamic–Pituitary–Thyroid Axis

It is now evident that cyclic AMP plays a critical role in the regulation of thyroid function. The effect of thyroid-stimulating hormone (TSH) releasing factor (TRF) upon pituitary TSH synthesis and release is mediated by an increase in pituitary cyclic AMP (7–9), and, in turn, the effect of TSH upon the thyroid is also mediated by cyclic AMP (9, 10). TRF is a tripeptide amide, pyroglutamyl-histidyl-proline amide, produced in the hypothalamus. Interestingly, TRF synthetase is stimulated by thyroxine (T4), while in the absence of thyroxine there is a decrease in the concentration of this enzyme. Such findings suggest that thyroxine may exert a positive feedback effect on TRF (11). TRF is secreted into the hypophyseal portal venous system which carries blood from the hypothalamus to the anterior pituitary gland, where TRF interacts with highly specific cell membrane receptors of the thyrotrope cells (7, 8). This interaction results sequentially in activation of adenylate cyclase, increased production of cyclic AMP, followed by activation of protein kinase, phosphorylation of proteins, and, finally, enhanced secretion of TSH (7, 8).

TSH, a glycoprotein, stimulates the secretion of thyroxine and triiodothyronine (T3) by the thyroid gland (9). The initial step in thyroid stimulation is an interaction between TSH and thyroid follicular cells, in which TSH binds to receptors on the outer surface of the plasma membrane of the thyroid cell. This is followed rapidly by stimulation of adenylate cyclase acitvity, resulting in an increase in intracellular cyclic AMP concentrations (9, 10, 12–16). The cyclic AMP that is generated then initiates a series of biochemical events, including activation of protein kinase and phosphorylation of proteins, which result in the stimulation of thyroid hormone secretion. Cyclic AMP-dependent protein kinases have been demonstrated in thyroid tissue (10). Most of the biochemical and morphologic effects of TSH on the thyroid are mediated by the cyclic AMP-adenylate cyclase system and can be partially reproduced in in vitro systems by cyclic AMP, or the dibutyryl derivative in in vivo and in vitro systems (9, 10, 12–16).

In studies performed in bovine thyroid glands, Yamashita and Field found that most of the TSH-sensitive adenylate cyclase activity was present in the plasma membrane fractions (17), and these observations have recently been confirmed by Marshall et al. (18). It also appears from various studies that other compounds including isoproterenol, PGE_1, and fluoride can stimulate various stages of thyroid hormone synthesis (18–21). Early studies by Burke demonstrated that both phentolamine and propranolol at $10^{-6}M$ concentrations greatly inhibited the adenylate cyclase responses of sheep thyroid mitochondrial fractions to TSH but did not affect the stimulatory effect of sodium fluoride (19). These data suggested an interaction of TSH and sodium fluoride at different sites on the adenylate cyclase system in the thyroid. However, studies in calf thyroid showed that the effect of isoproterenol, but not that of TSH, was prevented by propranolol (22). In other in vitro studies, the inhibitory effect of propranolol at concentrations of $10^{-4}M$ and $10^{-3}M$ on TSH stimulation of adenylate cyclase activity was shown to be due not to beta adrenergic blockade but to its "quinidinelike" properties (18) as earlier suggested by Levey (23). On the basis of such in vitro observations, Marshall et al. felt that the inhibitory effect of propranolol on TSH stimulation of the thyroid might be noted in vivo only when the dose greatly exceeded that which was necessary to produce beta adrenergic blockade (18).

Recently, in an effort to elucidate the mechanism whereby iodide inhibits thyroid function, Rapoport et al. studied the effect of dietary iodide on the thyroid adenylate cyclase and cyclic AMP response to TSH in the hypophysectomized rat (24). They found that basal levels of cyclic AMP and adenylate cyclase in the thyroids of animals on iodide enriched and on iodide deficient diets were similar, but the increase in cyclic AMP concentration and in adenylate cyclase following TSH administration were signficantly less in the animals fed the iodide-enriched diet. Phosphodiesterase activity did not differ in these two groups of animals. Their results indicate that iodide diminishes the cyclic AMP response to TSH by a direct action on adenylate cyclase in the thyroid gland.

Another substance that has TSH-like effects on thyroid tissue is long-acting thyroid stimulator (LATS) (25, 26). This abnormal substance is a 7S gammaglobulin, which is present in the serum of many patients with Graves' disease. It is thought to be an autoantibody to antigen derived from the thyroid gland and has been implicated in the pathogenesis of Graves' disease. In vitro studies have shown that LATS can reproduce many of the metabolic effects of TSH on thyroid tissue (27–32). Kaneko et al. have reported that LATS increases cyclic AMP levels in dog thyroid slices (33). Levey and Pastan have also reported that LATS stimulates adenylate cyclase activity in bovine and canine thyroid homogenates (34). Yamashita and Field have reported LATS stimulation of adenylate cyclase activity in thyroid plasma membranes (35), and Kaneko et al. have found that LATS increases tritiated adenine incorporation into tritium-labeled cyclic AMP in dog thyroid slices (33). Yamashita and Field noted that the stimulatory effect of LATS was much less than that seen with maximal amounts of TSH and suggested that the effect of LATS might be more related to a change in membrane configuration than to an effect on specific binding sites (35).

Solomon et al. have suggested that abnormalities in adenylate cyclase, intracellular cyclic AMP binding, and/or protein kinase activities might be responsible for the hyperfunctioning thyroid gland found in Graves' disease (36, 37). Recently, however, Orgiazzi et al. in in vitro studies reported that these cellular functions were normal in the thyroid glands taken from patients with Graves' disease (38). Their results were in agreement with other in vitro studies demonstrating normal intracellular responses to TSH in Graves' disease (39–42).

Most data relating to pituitary–thyroid function have been consistent with a negative feedback control, which operates primarily at the level of the anterior pituitary gland rather than at a hypothalamic level (43). Strong evidence in support of such a feedback in the anterior pituitary includes the diminished or absent TSH response to TRF administration following pretreatment with thyroid hormone (43–49). The mechanism of the negative-feedback effect appears to involve thyroid hormone stimulation of DNA- and RNA-dependent synthesis of proteins that exert an inhibitory effect on TSH secretion (48). The role of cyclic nucleotides in this regard is unclear at the present time.

Data consistent with feedback inhibition by thyroid hormones on the thyroid have suggested the need for reassessment of the manner in which thyroid function is controlled in man and in experimental animals. A number of investigators have shown that T4 and T3 act directly on the thyroid gland to inhibit its response to TSH (50, 51). Also, the magnitude of the thyroidal response to exogenous TSH has been shown by some to be greater in normal than in thyrotoxic subjects (52, 53). In vitro studies by Takasu et al. have demonstrated that exogenous TSH can stimulate adenylate cyclase activity in plasma membranes and thyroid slices obtained from thyrotoxic patients and control subjects (54). The magnitude of the response, however, was less in thyroid tissue obtained from the thyrotoxic patients, and five times more TSH was needed to produce a significant effect in such tissue. Small or large doses of T4 and T3 diminished the TSH-induced increase in the adenylate cyclase activity of normal human thyroid tissue, while only large doses of T4 and T3 depressed TSH-stimulated increases in cyclic AMP in thyroid tissue from thyrotoxic patients. The possibility that these effects were related to an increase in phosphodiesterase activity was excluded. In the absence of TSH, T4 and T3 slightly depressed the basal level of adenylate cyclase in the thyroid of normal subjects, while basal activity was not depressed by large doses of T4 and T3 in the thyroids from thyrotoxic patients. Takasu et al. concluded that thyroid hormones play an important role in the control of thyroid function by modifying the thyroidal response to TSH.

For many years, clinicians have been aware of the possible role of the central nervous system and of the sympathetic nervous system in the development or precipitation of Graves' disease. Both stimulatory and inhibitory effects of the sympathetic nervous system and catecholamines on thyroid function have been reported (55). Recently, there has been a resurgence of interest in the effects of catecholamines on the thyroid gland. Catecholamine stimulation of thyroid functions including effects on iodine metabolism, glucose oxidation, adenylate cyclase stimulation, and thyroid hormone secretion has been described in animal and human thyroid tissue (55, 56). Studies showing an effect of catecholamines on the thyroid gland have been consistent with the presence of adrenergic receptors in the plasma membrane of thyroid follicular cells (56). Burke has shown that alpha and beta adrenergic blocking agents, such as phentolamine and propranolol, respectively, greatly inhibited TSH effects on thyroid adenylate cyclase activity (19). In some studies, inhibition with phentolamine has suggested catecholamine stimulation of thyroid function via alpha adrenergic receptors (57, 58) whereas in others, performed in bovine thyroid, antagonism with propranolol has suggested the presence of a thyroidal beta adrenergic receptor system (23, 55, 59). However, findings from studies in the calf thyroid by Spaulding and Burrow suggested that isoproterenol and TSH were acting on different receptor sites, since propranolol was effective in inhibiting the response to isoproterenol but not to TSH (22). These observations and the recent demonstration of a sympathetic adrenergic nerve supply to thyroid blood vessels and thyroid follicles in man (60) have raised the distinct possibility that catecholamines have a direct action on the thyroid follicular cells and on thyroid hormone secretion. Thus far, however, the situation in man remains to be clarified. Catecholamines are not as potent as TSH in their effects on the thyroid (61), and adrenergic blocking agents generally have little or no effect on thyroid function as measured by radiolabeled iodide uptake studies and circulating thyroid hormone levels (62). Woolf et al. found in five normal subjects that basal TSH levels and the TSH response to TRF administration were unaffected by acute adrenergic manipulations (63). On the other hand, chronic L-dopa therapy has been reported to diminish the pituitary TSH response to TRF (64), and in hypothyroid patients L-dopa administration has been reported to acutely lower TSH levels (65).

Guanosine 3',5'-monophosphate (cyclic GMP) is the only other 3',5'-mononucleotide normally found in nature; it appears to be present in most mammalian tissues (66). However, little is known of the role, if any, of cyclic GMP in the functioning of the hypothalamic-pituitary-thyroid axis, in intrathyroidal metabolism and in the effects of T4 and T3 peripherally, and in the factors which may regulate its metabolism. Recently, guanylate cyclase activity has been demonstrated in rat thyroid (67). In dog thyroid slices, the content of cyclic GMP was found to be less than 10 percent of that of cyclic AMP (68). Cyclic GMP levels were unaffected by TSH but were significantly increased by acetylcholine and sodium fluoride (68, 69). In immunofluorescent studies of canine thyroid tissue, cyclic GMP was located primarily in the follicular cell membrane, whereas cyclic AMP was found throughout the cytoplasm (69). TSH had no effect on staining of cyclic GMP, and acetylcholine had no effect on staining of cyclic AMP. On the basis of these studies, Fallon et al. suggested that cyclic GMP may be involved in the iodination of thyroglobulin. Van Sande et al. have recently reported that agents which increased cyclic GMP levels in dog thyroid slices inhibited TSH-induced increases in cyclic AMP (70). However, Macchia and Varonne have obtained data suggesting that cyclic GMP may mediate some of the effects of TSH on the thyroid gland (71). De Nayer has shown that high concentrations of cyclic GMP activate amino acid incorporation into proteins in a bovine thyroid acellular system (72), and Pisarev et al. have shown in in vivo experiments that cyclic GMP stimulates protein synthesis and growth in the rat thyroid (73).

Peripheral Effects of Thyroid Hormones

The diverse metabolic changes and clinical effects which are seen in almost all patients with an excess or deficiency of thyroid hormone may be quite profound. Some of the effects of these hormones are direct ones, while others are indirect, and through their permissive effects, these hormones may modify the response of a particular tissue to other substances or hormones (74–77). Although many of the actions of T4 and T3 are thought to be related to stimulation of protein synthesis, some of the peripheral effects of thyroxine and triiodothyronine have now been shown to involve the adenylate cyclase-cyclic AMP system (76–78). Consequently, it is not unreasonable to expect that a surfeit or deficit of these hormones may be associated with appreciable alterations in cyclic AMP metabolism.

As a number of investigators have noted, many of the cardiovascular manifestations seen in patients with hyperthyroidism resemble those produced by an excess of catecholamines or by stimulation of the sympathetic nervous system, while many of the features seen in hypothyroid patients have suggested diminished sympathetic nervous system activity (74, 76). Also, there has been considerable documentation in animals and in man of an augmented peripheral vascular response to a given dose of catecholamines in hyperthyroidism and, conversely, a diminished response in hypothyroidism (74, 77). Others, however, have found no appreciable difference in the cardiovascular response to epinephrine in hypothyroid patients before and after thyroid treatment (79). Recent studies in animals and man have also failed to document any alteration in sensitivity to catecholamines in hyper- and hypothyroid states (74, 80). There is evidence that the cardiovascular manifestations of thyrotoxicosis are mediated both by the sympathetic nervous system, and by direct inotropic and chronotropic effects of thyroid hormones on the myocardium (74, 76). Spaulding and Noth have suggested a direct action of thyroid hormones on the end organs of the sympathoadrenal system (76).

Levey and Epstein (81) and Brodie et al. (82) have reported stimulation of adenylate cyclase by pharmacologic amounts of thyroid hormones in particulate preparations of cat left ventricle. The effect on cyclic AMP was shown to be unrelated to inhibition of phosphodiesterase activity (81). The observation that adenylate cyclase activation was not abolished by DL-propranolol, reserpine, or guanethidine was consistent with a direct effect of thyroid hormone (83, 84). Interestingly, however, unaltered levels of cyclic AMP have been demonstrated in hearts from hyperthyroid rats (85, 86) and cats (87).

Various studies have demonstrated an unaltered responsiveness of adenylate cyclase to catecholamines in the heart of hyperthyroid and euthyroid rats (80, 85) and in particulate preparations of left ventricle from hyperthyroid (87, 88) and hypothyroid cats (89). Levey et al. have also observed that the inotropic effects of norepinephrine are not augmented in papillary muscle obtained from hyperthyroid cats (88). In fact, Buccino et al. found that papillary muscles from hyperthyroid cats were less sensitive to the inotropic effects of norepinephrine than those from euthyroid animals (90).

Sobel et al. found no differences in the myocardial adenylate cyclase and phosphodiesterase activity in myocardial fractions from hyper- and euthyroid cats nor in the response to epinephrine (87). Other studies have shown a twofold increase in myocardial adenylate cyclase activity in the hypothyroid rat (91), unaltered levels in the hyperthyroid rat (85, 86), and significantly depressed levels in the hyperthyroid cat (88). Levey et al. suggested, however, that in vitro findings may not reflect the levels of cyclic nucleotides which occur in vivo (88). In the hypothyroid animals, oral therapy with thyroxine tended to lower heart adenylate cyclase activity towards normal (91). The increase in adenylate cyclase in the hypothyroid animals could not be attributed to an increase in catecholamines or TSH levels. These authors concluded that the cardiac effects of thyroid hormones are not explained by activation of adenylate cyclase.

Studies performed in other tissues have demonstrated results which are at some variance from those relating to the myocardium. Adipose tissue cells have been shown to be responsive to a variety of hormones, and a lipolytic response has been demonstrated with at least six hormones, including the catecholamines and peptide hormones such as TSH. Studies are consistent with a number of adipose cell receptors (92) with mediation of the lipolytic effect through activation of adenylate cyclase (88, 93). Early studies by Deykin and Vaughan demonstrated an increase in the lipolytic response to norepinephrine in hyperthyroid rats (94). Brodie et al. reported increased adenylate cyclase activity in adipose tissue of hyperthyroid animals (82). Subsequently, Krishna et al. found that the increased lipolytic response to norepinephrine in hyperthyroid rats was associated with an increase in basal adenylate cyclase activity and suggested that this might be related to de novo synthesis of adenylate cyclase in adipose tissue (95). However, Caldwell and Fain found no increase in adenylate cyclase activity in fat cell membrane ghosts taken from rats treated with T3 for 18 hr, despite an increased maximal cyclic AMP accumulation in these animals (96). In in vitro studies of the lipoyltic effect of pathologic concentrations of T3, Mandel and Kuehl confirmed that potentiation of norepinephrine and epinephrine lipolysis was associated with an increase in cyclic AMP content of rat epididymal fat tissue and of beef heart (97). This effect did not involve a direct action on beta adrenergic receptors and adenylate cyclase activity, but competitive inhibition of cyclic AMP phosphodiesterase activity. However, these and similar observations by others have been criticized because of the large amounts of thyroid hormone used (98).

A number of studies have shown that fat cells from hypothyroid rats and humans exhibit a diminished lipolytic response to catecholamines and other adipokinetic hormones (99–101), while fat cells from hyperthyroid animals exhibit an increased response to hormones such as glucagon and epinephrine (94, 102). Despite the exquisite sensitivity of hyperthyroid patients to the effect of epinephrine on free fatty acid mobilization, adrenergic blockade in such patients has been shown not to affect free fatty acid levels (103). In in vitro studies, Rosenquist found that the lipolytic effect of norepinephrine was absent in adipose tissue specimens obtained from hypothyroid patients, but was restored to normal with the addition of the alpha adrenergic antagonist, phentolamine (100). His observations and those of others (104) tend to support the hypothesis that the antilipolytic action of alpha adrenergic stimulation is associated with a reduction of intracellular cyclic AMP levels in the fat cell. He suggested that hypothyroidism in man may be associated with enhanced alpha adrenergic receptor activity due to a decrease in beta adrenergic activity. In the hypothyroid rat, however, phentolamine was without effect, suggesting that the diminished sensitivity to norepinephrine was not mediated by alpha adrenergic receptors (105). On the other hand, Armstrong et al. found that, whereas the soluble low affinity form of phosphodiesterase activity was comparable in the normal and hypothyroid states, the particulate high affinity forms of cyclic AMP phosphodiesterase activity were elevated in fat cells from hypothyroid rats, and an increase in cyclic AMP in response to stimulation with epinephrine was not observed, except in the presence of phosphodiesterase inhibitors (98). Their studies suggested that activation of cyclic AMP phosphodiesterase could be responsible for the insensitivity of adipose tissue to epinephrine and glucagon in the hypothyroid state. Basal, unstimulated cyclic AMP levels were similar in cells from normal and hypothyroid animals. Recently, Karlberg et al. reported cyclic AMP levels in skeletal muscle and adipose tissue taken from hypothyroid patients which were comparable to those from normal subjects (106).

There is up to 10 times as much adenylate cyclase activity in brain than in peripheral tissues (107), and it has been suggested that the central nervous system may be a source of extracellular cyclic nucleotides (108). Thus far, however, there is little direct evidence to support this contention. Thyroid hormone deficiency in the neonatal animal or human is associated with appreciable abnormalities in central nervous system development and function. Despite the critical role of thyroid hormones in the early development of the brain, recent studies in the neonatal rat have shown that thyroidectomy has no effect on the developmental increase in brain adenylate cyclase activity (109). In other studies in the neonatal rat, thyroidectomy partially reduced the increase in brain cyclic AMP levels in the first 40 days of life, but the activity of the cyclic AMP-dependent protein kinase was unaffected as were adenylate cyclase and phosphodiesterase activity (110). Also, a deficiency in thyroid hormone did not blunt the norepinephrine-induced increases in cyclic AMP levels in brain slices from animals ranging from three to 25 days old (110). In the neonatal rat, hyperthyroidism failed to alter protein kinase activity in the developing brain suggesting that the cyclic AMP system in the brain is unaffected by changes in thyroid hormone levels (111). Additionally, thyroxine has been shown not to affect adenylate cyclase activity in the mature brain (111). Notwithstanding the above, it is clear that at a clinical level both hyper- and hypothyroidism may have a profound effect on the mental state and cerebral cortical functions. Cerebrospinal fluid cyclic AMP measurements have yet to be reported in patients with hyper- and hypothyrodism.

An excess or deficiency of thyroid hormones is known to have a dramatic effect on growth in the young, but the mechanism of such effects is unclear. Armstrong et al. have speculated that some of the effects of thyroid hormones on growth and development may be related to changes in intracellular cyclic AMP concentrations on the basis of alterations in cyclic AMP phosphodiesterase activity (98). In chicken epiphyseal cartilage tissue, Thanassi and Newcombe found that T3 was the most significant inhibitor of 3',5'-nucleotide phosphodiesterase activity, whereas other less potent thyroid hormones exhibited smaller effects on this enzyme (112). Indeed, their data suggest that the effect of T4 and T3 on growth and development may in part be related to effects on the adenylate cyclase-cyclic AMP system in epiphyseal cartilage.

Hyper- and Hypothyroidism

The first report of altered cyclic nucleotide metabolism in human disease was that of Chase et al. in 1969 (113). These authors found that patients with pseudohypoparathyroidism had no appreciable urinary cyclic AMP response following parathyroid hormone (PTH) administration. Since then, deviations from normal in cyclic nucleotide metabolism have been described in a number of endocrine and nonendocrine disorders, with much of the literature relating to hypo- and hyperparathyroidism and the possible use of cyclic AMP measurements in the differential diagnosis of hypercalcemia (114, 115). With the profound clinical and chemical changes that occur in the majority of patients with hyper- and hypothyroidism, one might anticipate appreciable alterations in cyclic nucleotide metabolism. However, thus far there have been few studies of extracellular cyclic nucleotide levels in animals and in patients with thyroid dysfunction.

Studies by Hardman and associates in the hypophysectomized rat demonstrated a reduction in cyclic GMP excretion to approximately 50 percent of normal along with a small decrease in cyclic AMP excretion (116, 117). They found that cyclic GMP levels could be restored to normal with the administration of a mixture of pituitary hormones including LH, FSH, TSH, prolactin, growth hormone and ACTH. Hydrocortisone, together with large doses of thyroxine, were also effective in this regard while a more nearly physiologic dose of thyroxine together with hydrocortisone was only partially effective. The effect of hypophysectomy on cyclic AMP excretion was reversed by hydrocortisone administration whereas cyclic GMP excretion was unaffected. These investigators also found that thyroparathyroidectomized rats excreted less cyclic GMP and cyclic AMP in urine, and maintenance doses of thyroxine restored such levels to normal (117).

In acute studies in man, Williams et al. administered 100 μgm of T3 by mouth to four subjects and found no significant change in urinary cyclic AMP levels in specimens collected during four consecutive 2-hr periods (118). In 1972, Rosen, in a letter to the editor, indicated that he had measured cyclic AMP excretion in six patients with Graves' disease who were normocalcemic and found that cyclic AMP/creatinine ratios were significantly increased with a mean value of 6.88 ± .56 μmoles/gm creatinine as compared with a mean value of 3.07 ± .29 μmoles/gm creatinine in his normal subjects (119). Subsequently, Estep and associates reported that in 12 thyrotoxic patients, 4 of whom had hypercalcemia, mean control urinary cyclic AMP excretion was significantly elevated (7.4 μmoles/gm creatinine) compared with normal subjects (4.3 μmoles/gm creatinine) (120). In that study, calcium infusions were associated with a fall in urinary cyclic AMP excretion to one-half the control value both in normal subjects and patients with hypothyroidism. In contrast, there was no change in urinary cyclic AMP levels in the thyrotoxic patients. These authors suggested that the hypercalcemia and elevated urinary cyclic AMP excretion in their thyrotoxic patients were not due to an increased sensitivity to parathyroid hormone but were direct or indirect effects of thyroid hormone.

Initial studies reported from this laboratory by Lin et al. in sixteen patients with hyperthyroidism showed variable increases in urinary cyclic AMP excretion but consistent and significant increases in urinary cyclic AMP/creatinine ratios (Fig. 1) (121). In these patients there were significant reductions in creatinine excretion, and it was apparent that the increases in cyclic AMP/creatinine ratios were to a great extent reflecting this change. In the three male hyperthyroid patients, total cyclic AMP excretion (7.29 ± .80 μmoles/day) was significantly increased when compared to values in the male control subjects (5.47 ± .14 μmoles/day). The small number of hyperthyroid male patients, however, suggested the need for caution in interpreting the data. Urinary cyclic AMP excretion in the 13 female hyperthyroid patients, including 1 with thyrotoxicosis factitia, was not significantly greater than that in the normal female subjects.

Bartley et al. in a study of 14 patients with hypercalcemia, reported normal cyclic AMP excretion in their 1 patient with hyperthyroidism (122). In another study, Carter and Heath reported their urinary cyclic AMP findings in normal subjects, and in 22

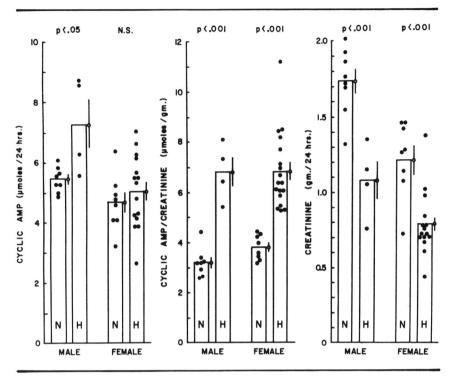

1 Urinary cyclic AMP, creatinine, and cyclic AMP/creatinine ratios in normal (N) and hyperthyroid (H) subjects. The mean ± SEM is shown to the right of each bar. (From T. Lin, L. Kopp, J. R. Tucci: J. Clin. Endocrinol. Metab., 36:1033, 1973.)

normocalcemic patients with hyperthyroidism and 12 patients with hypothyroidism, 3 of whom were hypercalcemic (123). Urinary cyclic AMP/creatinine ratios were raised in the female hyperthyroid patients ($\bar{x}$, 5.4; range, 2.1 to 8.63) as compared with female controls ($\bar{x}$, 3.52; range, 1.88 to 5.96). No comment was made of ratios in the male subjects. Cyclic AMP excretion per 24-hr period was found to be raised in both male and female hyperthyroid patients (females, $\bar{x}$, 5.47 μmoles, range, 2.54 to 8.90; males, $\bar{x}$, 7.54 μmoles, range, 5.88 to 9.33), as compared with controls (females, $\bar{x}$, 3.83 μmoles, range, 2.14 to 5.79; males, $\bar{x}$, 5.49 μmoles, range, 2.95 to 6.99). They noted that in their hyperthyroid patients the

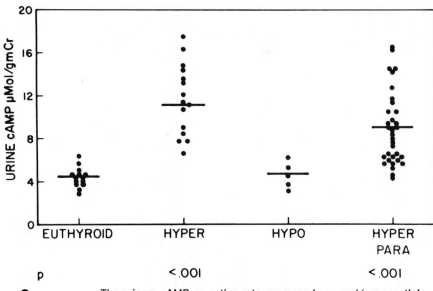

2 The urinary cAMP excretion rate expressed as μmol/gm creatinine, on spot urinary samples is depicted for a series of hyperthyroid, hypothyroid, and primary hyperparathyroid subjects and compared to a euthyroid control group. (From R. B. Guttler, J.W. Shaw, C. L. Otis, J. T. Nicoloff: J. Clin. Endocrinol. Metab. 41:707, 1975.)

finding of a relatively low phosphate excretion index was consistent with the concept of a coexistent relative hypoparathyroidism. This, they felt, made it unlikely that the increases in urinary cyclic AMP excretion in hyperthyroidism were related to renal tubular stimulation of adenylate cyclase by PTH. In their hypothyroid female and male patients, urinary cyclic AMP excretion was diminished as compared with their control subjects (females, $\bar{x}$, 2.66 μmoles/day, range, 1.79 to 3.96 and males, $\bar{x}$, 1.86 μmoles/day, range, 1.63 to 2.10). Cyclic AMP/creatinine ratios were slightly lower in the hypothyroid female patients as compared with the controls ($\bar{x}$, 3.1 μmoles/gm creatinine, range, 1.34 to 4.18). In this brief communication, however, Carter and Heath did not indicate whether any of these differences were significant.

Recently, Guttler et al. reported their urinary cyclic AMP and cyclic AMP/creatinine data in hyper-, eu- and hypothyroid patients (124). They found significantly increased cyclic AMP/creatinine ratios in "spot urines" obtained from 15 hyperthyroid patients, as

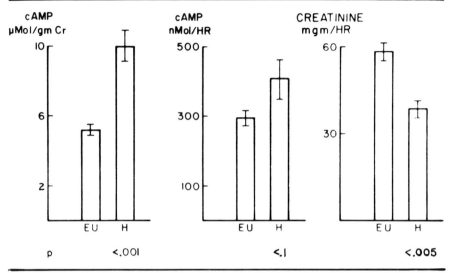

3 The urinary cAMP excretion rate expressed as μmol/gm creatinine, the absolute cAMP excretion in nmol/hr, and the creatinine excretion in mg/hr are depicted in a group of hyperthyroid subjects, and compared to a euthyroid control group. (From R. B. Guttler, J. W. Shaw, C. L. Otis, J. T. Nicoloff: J. Clin. Endocrinol. Metab. 41:707, 1975.)

compared with ratios in the normal subjects, and these elevations were comparable to ratios obtained in 38 patients with surgically proven hyperparathyroidism (Fig. 2). Values in their five hypothyroid patients were comparable to those in the control group. Cyclic AMP excretion and cyclic AMP/creatinine ratios were also measured in single 2-hr timed urine samples in 10 of their euthyroid subjects and in 6 of their hyperthyroid patients. Cyclic AMP excretion was slightly but not significantly increased over normal, whereas cyclic AMP/creatinine ratios were significantly increased (Fig. 3). These increases were attributed to significant reductions in creatinine excretion. However, when 24-hr urine collections were obtained in six of their hyperthyroid patients who were ambulatory, cyclic AMP excretion was found to be significantly increased over normal. A continuous epinephrine infusion at a rate of .05 μg/kg per minute over a 2-hr period resulted in a significantly greater rise in cyclic AMP excretion in their hyperthyroid patients than in the normal subjects, and their hypothyroid patients failed

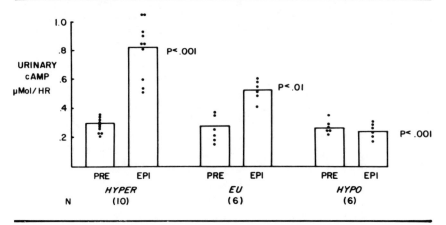

4 The absolute urinary cAMP excretion rate in μmol/hr pre- and post-epinephrine infusion (0.05 μg/kg/hr is shown for a group of 10 hyperthyroid, 6 euthyroid, and 6 hypothyroid subjects). (From R. B. Guttler, J. W. Shaw, C. L. Otis, J. T. Nicoloff: J. Clin. Endocrinol. Metab. 41:707, 1975.)

to exhibit a significant rise in cyclic AMP/creatinine ratios or in cyclic AMP excretion (Fig. 4).

The most recent studies from this laboratory, which now include 38 patients with hyperthyroidism confirm our earlier observations of a variable increase in urinary cyclic AMP excretion.* Urinary cyclic AMP excretion in the male hyperthyroid patients was not significantly greater than that in the normal males (6.36 ± .76 μmoles/day vs 5.82 ± .24 μmoles/day, $p < .37$). Cyclic AMP excretion in the female hyperthyroid patients, however, was greater than that in normals (6.12 ± .41 μmoles/day vs 5.00 ± .26 μmoles/day, $p < .038$). Cyclic AMP/creatinine ratios in both male and female hyperthyroid patients were significantly increased over normal and appear to reflect the tendency to higher cyclic AMP levels together with the decreases in creatinine excretion. The decrease in urinary creatinine in patients with hyperthyroidism has been noted previously by others (125) and is related to an alteration in creatine and creatinine metabolism whereby increased creatine is excreted at the expense of creatinine.

Other studies from this laboratory include urinary cyclic AMP measurements in 24 females and 8 males with hypothyroidism.*

---

* J. Clin. Endocrinol. Metab. In press.

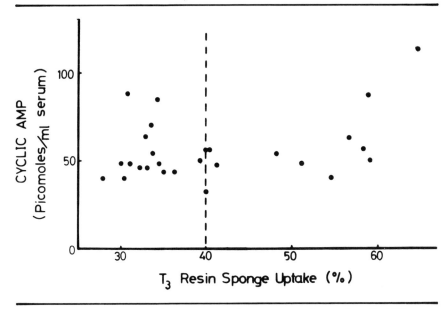

5 Relationship between T_3 resin sponge uptakes and cyclic AMP levels in the sera of normal and hyperthyroid subjects. Vertical dotted line indicates upper normal range of T_3 resin sponge uptake. (From T. Onaya, M. Kotani, T. Yamada, Y. Ochi, J. Clin. Endocrinol. Metab. 36:859, 1973.)

These show qualitatively similar changes as those reported by Carter and Heath (123), but the magnitude of these changes is less. Both hypothyroid male and female patient groups did excrete significantly less cyclic AMP than normals (4.38 ± .50 µmoles/day vs 5.82 ± .24 µmoles/day, $p < 8.1 \times 10^{-3}$ and 4.27 ± .29 µmoles/day vs 5.10 ± .25 µmoles/day, $p < .037$, respectively). Cyclic AMP/creatinine ratios, however, were not significantly different from normal. Cyclic AMP excretion and cyclic AMP/creatinine ratios in the hypothyroid groups were significantly less than in the hyperthyroid patients.

Several groups of investigators have reported plasma or serum cyclic AMP levels in patients with hyper- and hypothyroidism. Onaya et al. measured cyclic AMP levels in the sera of 14 normal subjects and 13 thyrotoxic patients, and these results were evaluated in relation to the resin T3 uptake results (Fig. 5) (39). They found that plasma cyclic AMP levels were comparable in the normal

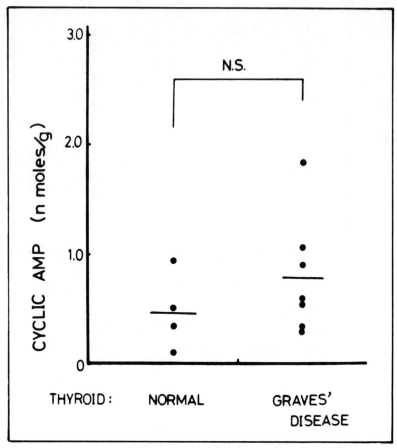

6 Cyclic AMP contents in human thyroid tissues of normal and hyperthyroid subjects. Fresh human thyroid tissues were measured for cyclic AMP. (From T. Onaya, M. Kotoni T. Yamada, Y. Ochi: J. Clin. Endocrinol. Metab. 36:859, 1973.)

subjects and in the hypo- and hyperthyroid patients. Also, they observed that the thyroid cyclic AMP content was slightly but not significantly greater in the patients with Graves' disease as compared with the normal (Fig. 6). In a recent study of the plasma cyclic AMP response to intravenous glucagon administration in euthyroid and in hyper- and hypothyroid patients, Elkeles et al. found no significant differences in fasting plasma cyclic AMP con-

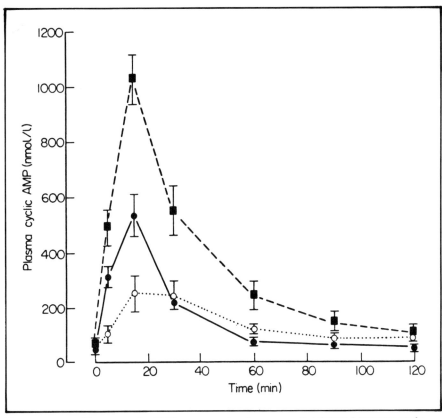

7 Plasma cyclic AMP response to glucagon administered intravenously at 0 min in hyperthyroid (■), euthyroid (●) and hypothyroid (O) subjects. Each point is the mean concentration ± 1 SE. (From R. S. Elkeles, J. H. Lazarus, K. Siddle, A. K. Campbell: Clin. Sci. Mol. Med. 48:27, 1975.)

centrations between any of the groups (Fig. 7) (126). After glucagon administration, plasma cyclic AMP levels rose in all groups but the response was significantly greater in the hyperthyroid group and significantly less in the hypothyroid group as compared with that in the normal subjects.

Guttler et al. reported in abstract form that supine basal plasma cyclic AMP levels were elevated in 12 hyperthyroid patients (26 ± 1.6 nM, $p < .001$) as compared to 10 euthyroid subjects (18

± .7 n*M*) and five hypothyroid patients (18 ± 1.3 n*M*) (127). Augmented cyclic AMP responses were noted in the hyperthyroid patients following administration of glucagon, parathyroid hormone, and epinephrine and a blunted response to PTH and epinephrine in the hypothyroid patients as compared with levels in the euthyroid subjects. Propranolol infusions abolished the plasma cyclic AMP response to epinephrine in the euthyroid subjects but only partially diminished the response in the hyperthyroid patients.

Karlberg et al., on the other hand, have reported significantly increased plasma cyclic AMP levels in eleven patients with hyperthyroidism and significantly decreased levels in 10 patients with hypothyroidism (Fig. 8) (106). Cyclic AMP levels in the hyperthyroid group were about two times higher than those of healthy euthyroid subjects. Values in the hyperthyroid patients ranged from 18 to 41.8 n*M*, with a mean level of 23.6 ± 2.2 n*M*, and in 13 healthy subjects values ranged from 6.3 to 13.6 n*M*, with a mean value of 10.7 ± .6 n*M*. Cyclic AMP levels in the hypothyroid group ranged from 1.5 to 7.9 n*M* with an average of 5.1 ± .7 n*M*. The cyclic AMP content of adipose and muscle tissue was significantly increased in the hyperthyroid patients, but, curiously, was not significantly diminished in the hypothyroid patients (Fig. 9). The authors could give no satisfactory explanation for these latter findings. Interestingly, propranolol was given orally in doses of 160 mg daily to five patients with hyperthyroidism for three weeks resulting in a reduction of plasma and tissue cyclic AMP levels to that of euthyroid subjects (Fig. 10).

These discrepant plasma cyclic AMP results reported in patients with hyperthyroidism from several laboratories are not readily explained. Plasma cyclic AMP levels are known to be easily affected by postural changes and physical activity (128, 129). Karlberg et al. emphasized the importance of strictly basal conditions for plasma measurements and this may have facilitated the discrimination seen between the hyper-, eu-, and hypothyroid groups (106). There was no mention, however, by Onaya et al. of the conditions under which the plasma specimens were obtained (39). In the study performed by Elkeles et al., specimens were obtained after an overnight fast and patients were rested for 20 min before the basal blood samples were drawn (126).

Interpretation of these limited and discordant observations is difficult. If we assume from these studies that plasma cyclic AMP levels are normal or increased in patients with hyperthyroidism,

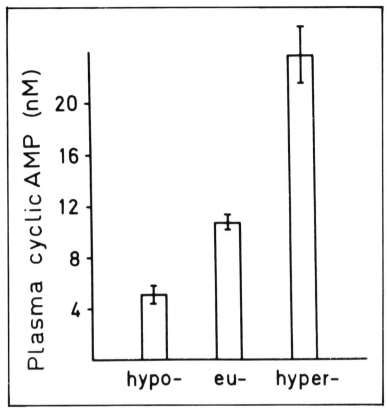

8 Plasma levels of cyclic AMP (nM) in euthyroid (eu-), hypothyroid (hypo-) and hyperthyroid (hyper-) patients. Mean ± SEM. (From B. E. Karlberg, K. G. Henriksson, R. G. G. Andersson: J. Clin. Endocrinol. Metab. 39:96, 1974.)

the urinary findings are in keeping with such an interpretation. If, however, we assume on the basis of the data obtained by Karlberg et al. that plasma cyclic AMP levels are consistently increased, then one must ask why the urinary cyclic AMP levels are not consistently increased in hyperthyroid patients. Several considerations seem relevant. A number of investigators have suggested that many of the findings in hyperthyroidism are consistent with an increase in beta adrenergic tone (74, 77, 106). Karlberg et al. postulated on the basis of increased plasma cyclic AMP levels and decreases with pro-

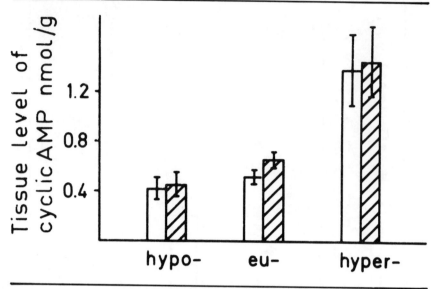

9 Tissue levels of cyclic AMP (nmol/gm wet weight of tissue). Abbreviations as in Fig. 8. Open bars = adipose tissue, hatched bars = skeletal muscle. Mean ± SEM. (From B. E. Karlberg, K. G. Henriksson, R. G. G. Andersson: J. Clin. Endocrinol. Metab. 39:96, 1974.)

pranolol administration that patients with hyperthyroidism do indeed have an increase in beta adrenergic tone (106). They suggested that such a change might involve an increase in the activity of sympathetic neurons, an increase in sensitivity of adenylate cyclase to catecholamines, reduced elimination of catecholamines from sympathetic neurons or an independent action of thyroid hormones and catecholamines on separate receptors with all contributing to the cyclic AMP pool. However, many observations are not in keeping with these notions. For example, total plasma catecholamines are reported to be increased in hypothyroidism and decreased in hyperthyroidism, and catecholamine values inversely correlated with total thyroxine levels in hyperthyroidism (130). Increased plasma norepinephrine and normal epinephrine levels have been found in hypothyroidism with values returning to normal after treatment with thyroxine, while insignificantly decreased norepi-

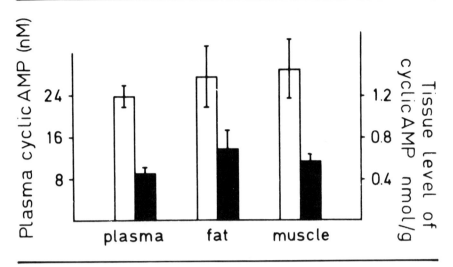

10 Tissue and plasma levels of cyclic AMP in thyrotoxic patients before (open bars) and after (filled bars) treatment with propranolol 40 mg 4 times daily for 3 wk. Mean ± SEM. $n = 5$. (From B. E. Karlberg, K. G. Henriksson, R. G. G. Andersson: J. Clin. Endocrinol. Metab. 39:96, 1974.)

nephrine and slightly elevated epinephrine levels have been reported in hyperthyroidism (131). Urinary catecholamine or catecholamine metabolite excretion has been reported to be normal (132) or low in patients with thyroid hyperfunction (133), and epinephrine excretion normal in thyroid hypofunction along with a small but significant increase in norepinephrine and VMA excretion (134). In addition, decreased dopamine beta hydroxylase activity has been reported in hyperthyroidism (135, 136) and increased dopamine beta hydroxylase activity in hypothyroidism (135).

An apparent increase in adrenergic activity, however, as suggested by Karlberg et al. (106) and Levey (77) would seem to explain the changes in plasma cyclic AMP in hyperthyroidism along with the variable increases in urinary cyclic AMP excretion reported by several groups. In interpreting the urinary findings in hyperthyroidism, one also has to consider the possibility of changes in the renal handling of cyclic AMP as noted by Kaminsky et al.

(137). These authors noted that beta adrenergic stimuli seemed to decrease the nephrogenous fraction of cyclic AMP, in that the increase in urinary cyclic AMP was somewhat less than that which could be calculated from plasma cyclic AMP and inulin clearances.

Another consideration relates to changes in calcium metabolism and parathyroid hormone secretion in hyper- and hypothyroidism. Bouillon and De Morr noted that serum calcium levels were significantly higher in hyperthyroid than in hypothyroid patients, with normal subjects having intermediate values, while serum phosphate levels were increased in hyperthyroid as compared with control and hypothyroid patients (138). The hyperthyroid patients exhibited lower PTH levels than the hypothyroid patients, while normal or increased parathyroid function was noted in myxedema (Fig. 11). In a related study, Castro et al. have found an increase in the sensitivity of hyperthyroid patients to parathyroid hormone and a blunted response in hypothyroid patients (139). As these authors noted, such a change in sensitivity to PTH may explain a functional hypoparathyroidism in hyperthyroidism and, conversely, a functional hyperparathyroidism in hypothyroidism.

In our experience, although cyclic AMP excretion in hyperthyroid patients may be elevated, levels do not reach those found in patients with proven primary hyperparathyroidism. On the other hand, due to the decrease in creatinine excretion in patients with hyperthyroidism, the cyclic AMP/creatinine ratios in such patients approach those found in primary hyperparathyroidism. Conversely, the depressed values in the hypothyroid patients are comparable to those obtained in a small group of patients with hypoparathyroidism.* In the case of parathyroid hormone, it is well documented that this substance stimulates adenylate cyclase activity in renal cortical tubules and approximately 40 percent of urinary cyclic AMP is generally attributable to the stimulatory effect of PTH (108). Therefore, with a decrease or an increase in parathyroid hormone secretion there are concordant changes in nephrogenous production of cyclic AMP and in urinary cyclic AMP excretion (115). In the case of thyroid hormone, however, the exact cause of the changes seen in urinary cyclic AMP in thyroid dysfunction is unknown, and it appears to be a more complex problem. In keeping with the plasma cyclic AMP data obtained by Karlberg and associates (106), one would anticipate that changes in thyroid function would be

---

* Unpublished observations.

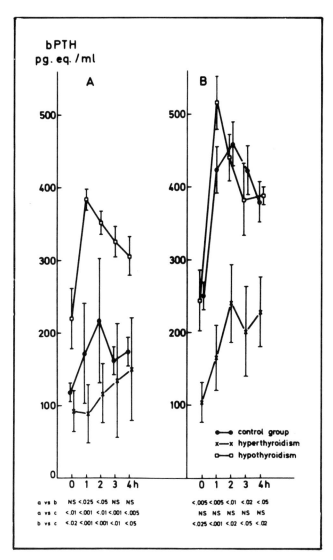

11 Serum parathyroid hormone levels (mean ± SEM) measured with two different antisera (A, B) before, during and after an infusion of disodium EDTA, a) in 13 normal controls (-o-), b) in hyperthyroid patients (-x-) and c) in 5 hypothyroid patients (-□-). The significance of the differences between the groups was assessed by Student's t test. (From R. Bouillon, P. DeMoor: J. Clin. Endocrinol. Metab. 38:999, 1974.)

associated with alterations in the nonnephrogenous component of urinary cyclic AMP. On the other hand, alterations in parathyroid hormone secretion in patients with thyroid dysfunction would be expected to alter the nephrogenous component. Studies in which both plasma and urinary cyclic AMP are measured would help to delineate the net effect of hyper- and hypothyroidism. Also, whereas cyclic nucleotides have been studied in some detail in patients with primary hyperparathyroidism before and after therapy, with the exception of limited conflicting preliminary observations, this has not been the case in patients with hyper- and hypothyroidism (140–142). One would anticipate, however, in keeping with the data obtained in hyperparathyroidism that with appropriate therapy, plasma and urinary cyclic nucleotide levels would return to normal.

In contrast to the variety of clinical disorders in which cyclic AMP has been measured in extracellular fluids, there are few studies of extracellular cyclic GMP levels in man. As far as we know, with the exception of data reported in abstract form in three patients with thyroid dysfunction (143), extracellular cyclic GMP levels have not been reported in patients with hyper- and hypothyroidism. In recent studies from this laboratory in which cyclic AMP was measured in 38 patients with hyperthyroidism and 32 patients with hypothyroidism, urines were also assayed for their cyclic GMP content and results were compared with urinary cyclic GMP measurements in 57 normal subjects.* While cyclic GMP excretion in the hyperthyroid male patients was comparable to that in the normal subjects (.55 ± .09 µmoles/day vs .56 ± .03 µmoles/day, $p < .90$), urinary cyclic GMP levels in the hyperthyroid females were greater than those in the normal females (.68 ± .08 µmoles/day vs .38 ± .02 µmoles/day, $p < 9.1 \times 10^{-4}$). Cyclic GMP/creatinine ratios were significantly increased in both male and female patients as compared with their respective normal groups (5.0 ± .08 µmoles/gm creatinine vs .34 ± .02 µmoles/gm creatinine, $p < 4.1 \times 10^{-3}$; .84 ± .08 µmoles/gm creatinine vs .34 ± .02 µmoles/gm creatinine, $p < 1.1 \times 10^{-7}$). In our hypothyroid male and female patients, urinary cyclic GMP excretion was comparable to that in normal subjects (males, .46 ± .08 µmoles/day vs .56 ± .03 µmoles/day, $p < .17$; females, .38 ± .04 µmoles/day vs .38 ± .02 µmoles/day, $p < .99$). Cyclic GMP/creatinine ratios were similar in the male hypothyroid patients and normals, while levels were marginally increased in the hypothyroid females (males, .31 ± .05 µmoles/gm creatinine vs .34

---

* J. Clin. Endocrinol. Metab. In press.

± .02 µmoles/gm creatinine, $p < .07$; females, .43 ± .04 µmoles/gm creatinine vs .34 ± .02 µmoles/gm creatinine, $p. < .05$). The differences in cyclic GMP excretion and in cyclic GMP/creatinine ratios between the hypothyroid and hyperthyroid female patients were significant ($p < 5.2 \times 10^{-3}$; $p < 1.5 \times 10^{-4}$, respectively). While cyclic GMP excretion was not significantly lower in the hypothyroid male patients as compared with the hyperthyroid male patients ($p < .46$), cyclic GMP/creatinine ratios were significantly lower ($p < .05$).

Whereas cyclic AMP has been shown to mediate the action of a wide variety of hormones, and plasma and urinary measurements of this nucleotide have clinical relevance, the current status of cyclic GMP is rather nebulous, and its relevance to clinical medicine is unknown at the present time. From the data currently available, it appears that the urinary cyclic GMP levels noted in patients with hyper- and hypothyroidism run *pari passu* with urinary cyclic AMP measurements. As with cyclic AMP, the increase in cyclic GMP/creatinine ratios in hyperthyroidism reflects both the consistent change in creatine and creatinine metabolism and the somewhat variable changes in cyclic GMP metabolism. Measurements of plasma cyclic GMP levels would be helpful in delineating the source of such changes. The data do point to the importance of considering the excretion of cyclic nucleotide per 24-hr period as well as the cyclic nucleotide/creatinine ratio since these parameters may not necessarily be concordant and one measurement, taken alone, may be misleading.

DIABETES MELLITUS

Evidence is accumulating that diabetes mellitus is a bihormonal disease involving a relative or absolute deficiency in insulin and a relative or absolute hyperglucagonemia (144). The importance of glucagon in this disorder has only recently begun to be appreciated. Muller et al. have reported that glucagon levels in patients with adult and juvenile forms of diabetes mellitus did not fall despite marked hyperglycemia, in contrast to normal subjects in whom glucagon levels decrease with rising blood glucose levels (145). Glucagon levels in diabetic patients are not restored to normal immediately, that is, within 45 min, with insulin treatment (144) but may return to normal after several hours of such treatment (146). In animals made diabetic, glucagon levels were also found to be elevated despite severe hyperglycemia; they were quickly restored to normal by insulin infusion (147). Glucagon secretion rose early in insulin-dependent diabetics deprived of insulin, and changes in glucagon secretion correlated significantly with the rise in blood ketone concentration (148–150). Somatostatin, a hypothalamic tetradecapeptide (151, 152), which may also be a natural component of D-cells in both gut and pancreas (153), decreased the secretion of both insulin (154, 155) and glucagon (155, 156) and prevented the rise of plasma glucagon, 3-hydroxy butyrate and free fatty acids in insulin-dependent diabetics deprived of insulin (157). These most recent findings suggest that glucagon may be important in the early stages of the development of ketoacidosis.

Cyclic nucleotides, especially cyclic AMP, have a multiplicity of roles to play in the regulation of secretion of insulin and glucagon and in their mechanisms of action. This review will attempt to briefly summarize what is currently known about the roles of these cyclic nucleotides in the mechanisms involved in glucose homeostasis and will discuss the available clinical data on cyclic nucleotides in diabetes mellitus.

Cyclic Nucleotides and Insulin Biosynthesis and Secretion

It is well established that glucose can dramatically increase the rates of insulin or proinsulin biosynthesis in isolated pancreatic islets (158–160). Cyclic AMP may also have a role in the regulation of insulin synthesis. Dibutyryl cyclic AMP, theophylline, and caffeine stimulate insulin and proinsulin synthesis in isolated rat islets (161, 162). Glucagon, cyclic AMP, and dibutyryl cyclic AMP produce similar effects in isolated mouse islets (163). The effect of cyclic AMP on insulin appears to be glucose dependent, since it does not occur in the absence of glucose (162, 163) even when an alternative energy source such as pyruvate is provided (160).

Although it is generally accepted that glucose is the major physiologic stimulus to insulin secretion, cyclic AMP apparently plays a significant role in modulating this effect. It is possible that changes in beta cell cyclic AMP levels are an essential component of the mechanism of glucose-induced stimulation of insulin release, but the evidence concerning this role for cyclic AMP is controversial (see below). Cyclic AMP and dibutyryl cyclic AMP (164–166), and agents which activate adenylate cyclase in a variety of tissues or islets including glucagon (167–171), corticotropin (164, 172, 173), thyrotropin (165), secretin (174, 175), pancreozymin (174), and acetylcholine (173, 176, 177) enhance insulin release. Agents which inhibit cyclic nucleotide phosphodiesterase such as caffeine (178, 179), theophylline (172, 180), and 3-isobutyl-1-methylxanthine (181, 182) also enhance insulin release. Catecholamines inhibit insulin release (176, 183–188). This effect appears to be a result of alpha adrenergic stimulation (176, 188). The inhibitory effect of norepinephrine on insulin secretion is enchanced in the presence of the beta adrenergic blocker, propranolol, and is replaced by a stimulation of insulin secretion in the presence of an alpha adrenergic blocker, phentolamine (176, 180, 189). Inhibition of the insulin response to glucose produced by propranolol may be reversed with the simultaneous administration of aminophylline (190).

The role of cyclic AMP in the mechanism of glucose-induced stimulation of insulin secretion is an issue which remains unresolved. In several studies cyclic AMP levels were elevated in isolated pancreatic islets exposed to high glucose concentrations (191–198), while in other studies, glucose alone did not alter cyclic AMP levels of islets (181, 182) or had only a small transient effect (199). Some investigators found that a high glucose concentration elevated islet cyclic AMP levels when a phosphodiesterase inhibitor was present (191, 192, 199) but others did not confirm this (182, 199). It has been reported that the presence of extracellular calcium is essential in order to demonstrate a glucose-induced increase in islet cyclic AMP and insulin (198). Whatever the eventual conclusion concerning its role in the mechanism of glucose-induced insulin release, it is clear that cyclic AMP is important in the overall regulation of insulin biosynthesis and release.

Little is known about cyclic GMP in pancreatic islets, except that dibutyryl cyclic GMP has been reported recently to stimulate synthesis (200) and release (201) of insulin in islets of Langerhans, suggesting that cyclic GMP may also be involved in the regulation of insulin synthesis and release.

Cyclic Nucleotides and the Mechanism of Action of Insulin

The mechanism whereby insulin evokes its intracellular effects remains enigmatic. It has been observed that under certain circumstances insulin can decrease intracellular cyclic AMP concentration in insulin sensitive tissues (202–206). It has, however, been very difficult to relate the effects of insulin to changes in overall levels of this cyclic nucleotide (207–216). Nonetheless, physiologic concentrations of insulin can directly inhibit the basal or epinephrine-stimulated activities of adenylate cyclase in freshly prepared membranes (173, 217–221). Furthermore, a small stimulatory effect of insulin on a membrane-bound phosphodiesterase has been noted by a number of investigators (223–228). These observations provide a basis for a direct modulation of intracellular cyclic AMP levels by insulin. Another possible means by which insulin might affect cellular cyclic AMP levels, suggested by Kissebah et al. (229), is through redistribution of membrane-bound and intracellular calcium. They suggest that such redistribution of calcium might also account for the effects of insulin that do not depend on changing cyclic AMP levels.

In liver, insulin can abolish the effects of moderate levels of glucagon stimulation of glycogenolysis, but under more intensive stimulation by glucagon, even extremely high insulin levels do not prevent the glucagon induced rise in intracellular cyclic AMP levels and concomitant stimulation of glycogenolysis (202, 230, 231). It thus appears that at least in liver, the effects of insulin on cyclic AMP levels may provide a means of coordinating glucose metabolism pathways with other primary effects of insulin on glucose and amino-acid uptake. Under conditions of intensive stimulation of the major cyclic nucleotide regulated metabolic pathways (glycogenolysis, gluconeogenesis), the modulation of these pathways can be overridden by the specific stimulus.

The possibility that cyclic GMP has a role in the mechanism of action of insulin cannot be excluded. In the fat cell, physiologic concentrations of insulin cause a marked but transient rise in the level of cyclic GMP (232). Insulin also causes a rise in cyclic GMP and a fall in cyclic AMP levels in mouse fibroblasts in culture (233, 234), but these effects are observed only at much higher insulin levels than are found in vivo.

Cyclic Nucleotides and Glucagon Secretion

The role played by cyclic AMP in regulating glucagon secretion is uncertain. It has been shown that theophylline and epinephrine stimulate glucagon in pancreatic islets from several animal species (235, 238). These effects seem to be mediated by cyclic

AMP, since dibutyryl cyclic AMP also stimulates insulin release (238). More recent evidence is somewhat contradictory. Toyata et al. (239) have reported that in perfused rat pancreas, at low glucose concentrations, isoproterenol and dibutyryl cyclic AMP inhibit glucagon secretion, while norepinephrine enhances it. Theophylline, however, abolishes the stimulatory effect of norepinephrine on glucagon secretion. The effects of isoproterenol and dibutyryl cyclic AMP were glucose dependent, in that higher glucose levels abolished their inhibitory effects on glucagon secretion.

Virtually nothing is known about the behavior of cyclic GMP in the alpha cell.

Cyclic Nucleotides and the Mechanism of Action of Glucagon

It is well established that the major effect, if not the only effect, of glucagon is activation of adenylate cyclase in glucagon sensitive tissues, including liver (240), breast (241–244), pancreatic islets (180), fat (93), skeletal muscle (245), kidney (246, 247) and parathyroid glands (248). The activation of adenylate cyclase is accomplished through binding of glucagon to a specific receptor associated with the cyclase in the plasma membrane of the target cell (249). Activation of adenylate cyclase elevates cyclic AMP levels within the cell, and this, in turn, leads to increased protein phosphorylation by cyclic AMP-dependent protein kinases. The protein phosphorylations lead eventually to dramatically changed levels of activation in such enzymes as glycogen phosphorylase in liver and muscle (250) or hormone sensitive triglyceride lipase in adipose tissue (251). The mechanism of glucagon stimulation of glycogenolysis is a paradigm of cyclic-AMP–mediated regulatory mechanisms.

The Relation of Cyclic Nucleotides to the Effects of Glucagon and Insulin in Normal and Diabetic Human Subjects

Glucagon has been shown to increase both plasma and urinary cyclic AMP levels (118, 252, 263). The effect of glucagon on plasma cyclic AMP appears to be predominantly a result of increased hepatic release of the nucleotide (165, 252) while the increased urinary excretion of cyclic AMP appears to be derived by glomerular filtration of plasma, with little or no direct renal contribution (252, 254).

In a study of net splanchnic production of cyclic AMP in normal men and in juvenile-onset, ketosis-prone, insulin-dependent diabetic men by Liljenquist et al., brachial artery and hepatic vein plasma cyclic AMP levels were measured (255). In normal men, the liver was a major source of plasma cyclic AMP. In the diabetic group, no net release of cyclic AMP by the liver could be demonstrated, although arterial cyclic AMP levels were elevated. Glucagon infusion increased cyclic AMP release from the liver to the same extent in both groups indicating that there was no abnormal sensitivity of the liver to glucagon in diabetic males.

The effects of insulin and diet on urinary cyclic AMP levels in eight female subjects hospitalized for uncontrolled diabetes mellitus were studied by Tucci et al. (256). Urinary excretion of cyclic AMP was either elevated or normal, although the mean excretion of cyclic AMP for the diabetic group was not significantly different from normal. When cyclic AMP excretion was normalized in terms of creatinine (cyclic AMP/creatinine ratio) the diabetic group excreted significantly more cyclic AMP than the normal group. Insulin treatment of the diabetics reduced both cyclic AMP excretion and cyclic AMP/creatinine ratios to normal.

Hamet et al. have recently reported that insulin induced hypoglycemia in normal human subjects resulted in a fourfold increase in plasma cyclic AMP levels with only a slight rise in urinary cyclic AMP excretion (257). The increase in plasma cyclic AMP levels was thought to be secondary to beta adrenergic stimulation since the response was absent in adrenalectomized, cortisol-treated subjects, and it was abolished by propranolol administration. Broadus et al. (252) have reported that glucose loading in oral glucose tolerance tests had no effect on either plasma or urinary cyclic AMP levels, except that 3 to 4 hr after glucose administration there was a slight increase in plasma cyclic AMP when blood glucose levels had decreased below basal levels. The delayed increase in plasma cyclic AMP was thought to be secondary to an increase in glucagon release resulting from the lowered blood glucose levels. Interestingly, Das (258) has reported that in rats treated with antiinsulin serum, plasma cyclic AMP levels did not change despite an increase in blood glucose levels of more than 200 percent.

Glucagon has little or no effect on plasma cyclic GMP (115), although several investigators have reported slight increases in urinary cyclic GMP excretion (115, 118), suggesting the possibility that the glucagon-stimulated increase in cyclic GMP may be of renal origin (115). In the normal, untreated individual, virtually all of the urinary cyclic GMP appears to be derived by glomerular filtration, with little or no renal contribution (108).

Thus far, no studies have been reported of the effects of insulin on either plasma or urinary cyclic GMP levels. In a small group of uncontrolled diabetic women, urinary cyclic GMP levels were normal and were unaffected by treatment with insulin.* Oral glucose tolerance tests have been reported to have no effect on either plasma or urinary cyclic GMP levels (252).

SUMMARY

Cyclic AMP plays a vital role in the functioning of the hypothalamic–pituitary–thyroid axis, and directly or indirectly may be involved in at least some of the peripheral effects of T4 and T3. The effect of TRF upon anterior pituitary TSH secretion appears to be mediated by cyclic AMP, as are most if not all of the effects of TSH upon the thyroid. Experimental hyper- and hypothyroidism in some animal species are associated with changes in tissue adenylate cyclase, cyclic AMP concentration, and/or cyclic nucleotide phosphodiesterase activity. In hyperthyroid patients, variable increases in plasma and urinary cyclic AMP excretion have been reported along with an invariable increase in urinary cyclic AMP/creatinine ratios. In contrast, in hypothyroid patients, plasma cyclic AMP has been reported to be depressed and urinary cyclic AMP and cyclic AMP/creatinine ratios are variably diminished. In regard to cyclic GMP, much less is known of its possible role in the effects of TRF and TSH, in intrathyroidal metabolism, and in the peripheral effects of T3 and T4. Also, there have been few measurements of extracellular concentrations of this cyclic nucleotide in patients with thyroid dysfunction.

It is becoming increasingly evident that diabetes mellitus is a bihormonal disorder involving both insulin and glucagon. Cyclic AMP is intimately involved with the regulation of secretion of these hormones, and with their mechanisms of action, although the precise role of this nucleotide is not entirely clear except for the mechanism of action of glucagon. The possibility that cyclic GMP may

* Unpublished observations.

also be significant in the regulation of secretion of these hormones or in their mechanisms of action cannot be excluded. The available literature does not allow any firm conclusions about cyclic GMP in this respect.

Review of the available data in man strongly suggests that hyper- and hypothyroidism as well as uncontrolled diabetes mellitus can be associated with alterations in extracellular cyclic nucleotide levels. It appears that an increasing number of physiologic and pathologic states may in fact be characterized by such changes. The data also suggest that any change in extracellular cyclic nucleotide levels must be interpreted with caution as they may not be specific for a particular clinical condition. The changes in extracellular cyclic nucleotide levels observed in these disease states appear to be more a consequence of hormonal perturbations than of intrinsic changes in the cyclic nucleotide systems per se. It does not appear that these changes in cyclic nucleotide levels are directly responsible for any of the clinical manifestations of these particular disease processes.

ACKNOWLEDGMENTS

Our appreciation is expressed to Miss Marcia Thompson for her secretarial help, to our medical librarians, Shirley Halzel and Hadassah Stein, and to Richard Margolies for their help in locating many of the references cited in this review, and to Dr. Paul Calabresi for his overall support.

This work was supported in part by a grant from the National Institutes of Health to the National Cancer Institute, grant No. CA13943-03.

REFERENCES

1. Sutherland, E. W., Rall, T. W.: The relation of adenosine 3',5' phosphate and phosphorylase to the actions of catecholamines and other hormones. Pharmacol. Rev. 12:265-299, 1960.
2. Sutherland, E. W., Oye, I., and Butcher, R. W.: The action of epinephrine and the role of the adenyl cyclase system in hormone action. Recent Prog. Horm. Res. 21:623-646, 1965.
3. Butcher, R. W., Sutherland, E. W.: Adenosine 3',5'-phosphate in biological materials. I. Purification and properties of cyclic 3',5'-nucleotide phosphodiesterase and use of this enzyme to characterize adenosine 3',5'-phosphate in human urine. J. Biol. Chem. 237:1244-1255, 1962.
4. Ashman, D. F., Lipton, R. L., Melicow, M. M., Price, T. D.: Isolation of adenosine 3',5'-monophosphate and guanosine 3',5'-monophosphate from rat urine. Biochem. Biophys. Res Commun. 11:330-334, 1963.

5. Robison, G. A., Butcher, R. W., Sutherland, E. W.: Cyclic AMP. Annu. Rev. Biochem. 37:149-174, 1968.
6. Sutherland, E. W., Robison, G. A.: The Banting Memorial Lecture 1969: The role of cyclic AMP in the control of carbohydrate metabolism. Diabetes. 18:797-819, 1969.
7. Poirier, G., Barden, N., Labrie, F., Borgeat, P., De Lean, A. Partial purification and some properties of adenyl cyclase and receptor for TRH from anterior pituitary gland. In, Abstracts IV International Congress of Endocrinology, Washington, D. C., 1972. Excerpta Medica, Amsterdam, Congress Series No. 256, Amsterdam, 1972, p. 85.
8. Labrie, F., Barden, N., Poirier, G., De Lean, A.: Binding of thyrotropin-releasing hormone to plasma membranes of bovine anterior pituitary gland. Proc. Nat. Acad. Sci. USA. 69:283-287, 1972.
9. Dumont, J. E.: The action of thyrotropin on thyroid metabolism. Vitam. Horm. 29:287-412, 1971.
10. Field, J. B.: Thyroid-stimulating hormone and cyclic adenosine 3',5'-monophosphate in the regulation of thyroid gland function. Metabolism 24:381-393, 1975.
11. Reichlin, S.: Hypothalamic-pituitary function, In: Endocrinology, Proceedings of the Fourth International Congress of Endocrinology, Washington, D. C., 1972. R. O. Scow (ed.) Excerpta Med, pp. 1-16.
12. Pastan, I., Katzen, R.: Activation of adenyl cyclase in thyroid homogenates by thyroid-stimulating hormone. Biochem. Biophys. Res. Commun. 29:792-798, 1967.
13. Gilman, A. G., Rall, T. W.: The role of adenosine 3',5'-phosphate in mediating the effects of thyroid-stimulating hormone on carbohydrate metabolism of bovine thyroid slices. J. Biol. Chem. 243: 5872-5881, 1968.
14. Zor, U., Kaneko, T., Lowe, I. P., Bloom, G., Field, J. B.: Effect of thyroid-stimulating hormone and prostaglandins on thyroid adenyl cyclase activation and cyclic adenosine 3',5'-monophosphate. J. Biol. Chem. 244: 5189-5195, 1969.
15. Kaneko, T., Zor, U., Field, J. B.: Thyroid-stimulating hormone and prosstaglandin E_1 stimulation of cyclic 3',5' adenosine monophosphate in thyroid slices. Science 163:1062-1063, 1969.
16. Amir, S. M., Carraway, T. F., Jr., Kohn, L. D., Winand, R. J.: The binding of thyrotropin to isolated bovine thyroid plasma membranes. J. Biol. Chem. 248:4092-4100, 1973.
17. Yamashita, K., Field, J. B.: Preparation of thyroid plasma membranes containing a TSH responsive adenyl cyclase. Biochem. Biophys. Res. Commun. 40:171-178, 1970.
18. Marshall, N. J., Von Borcke, S., Malan, P. G.: Studies on inhibition of TSH stimulation of adenyl cyclase activity in thyroid plasma membrane preparations by propranolol. Endocrinoligy 96:1513-1519, 1975.
19. Burke, G.: Comparison of thyrotropin and sodium fluoride effects on thyroid adenyl cyclase. Endocrinology 86:346-352, 1970.
20. Ahn, C. S., Rosenberg, I. N.: Iodine metabolism in thyroid slices: effect of TSH, dibutyryl cyclic 3',5' AMP, NaF and prostaglandin E_1. Endocrinology 86:396-405, 1970.

21. Kariya, T., Kotani, M., Field, J. B.: Effects of sodium fluoride and other metabolic inhibitors on basal and TSH-stimulated cyclic AMP and thyroid metabolism. Metabolism 23:967-973, 1974.
22. Spaulding, S. W., Burrow, G. N.: β-adrenergic stimulation of cyclic AMP and protein kinase activity in the thyroid. Nature 254:347-349, 1975.
23. Levey, G. S., Roth, J., Pastan, I.: Effect of propranolol and phentolamine on canine and bovine responses to TSH. Endocrinology 84:1009-1015, 1969.
24. Rapoport, B., West, M. N., Ingbar, S. H.: Inhibiting effect of dietary iodine on the thyroid adenylate cyclase response to thyrotropin in the hypophysectomized rat. J. Clin. Invest. 56:516-519, 1975.
25. Adams, D. D., Purves, H. D.: Bioassay of long-acting thyroid stimulator (L.A.T.S.): the dose-response relationship. J. Clin. Endocrinol. Metab. 21:799-805, 1961.
26. McKenzie, J. M.: Studies on the thyroid activator of hyperthyroidism. J. Clin. Endocrinol. Metabolism 21:635-647, 1961.
27. Scott, T. W., Good, B. F., Ferguson, K. A.: Comparative effects of long-acting thyroid stimulator and pituitary thyrotropin on the intermediate metabolism of thyroid tissue *in vitro*. Endocrinology 79:949-954, 1966.
28. Field, J. B., Remer, A., Bloom, G., Kriss, J. P.: *In vitro* stimulation by long-acting thyroid stimulator of thyroid glucose oxidation and ^{32}P incorporation into phospholipids. J. Clin. Invest. 47:1553-1560, 1968.
29. McKenzie, J. M.: Further evidence for a thyroid activator in hyperthyroidism. J. Clin. Endocrinol. Metab. 20:380-388, 1960.
30. Shishiba, Y., Solomon, D. H., Beall, G. N.: Comparison of early effects of thyrotropin and long-acting thyroid stimulator on thyroid secretion. Endocrinology 80:957-961, 1967.
31. Brown, J., Munro, D. S.: A new *in vitro* assay for thyroid stimulating hormone. J. Physiol. 182:9P-10P, 1966.
32. Ochi, Y., DeGroot, L. J.: Stimulation of RNA and phospholipid formation by long-acting thyroid stimulator and by thyroid-stimulating hormone. Biochim. Biophys. Acta 170:198-201, 1968.
33. Kaneko, T., Zor, U., Field, J. B.: Stimulation of thyroid adenyl cyclase activity and cyclic adenosine 3',5'-monophosphate by long-acting thyroid stimulator. Metabolism 19:430-438, 1970.
34. Levey, G. S., Pastan, I.: Activation of thyroid adenyl cyclase by long-acting thyroid stimulator. Life Sci. (I) 9:67-73, 1970.
35. Yamashita, K., Field, J. B.: Effects of long-acting thyroid stimulator on thyrotropin stimulation of adenyl cyclase activity in thyroid plasma membranes. J. Clin. Invest. 51:463-472, 1972.
36. Solomon, D. H., Chopra, I. J.: Graves' disease — 1972. Mayo Clin. Proc. 47:803-811, 1972.
37. Chopra, I. J., Solomon, D. H., Johnson, D. E. and Chopra, V.: Thyroid gland in Graves' disease. Victim or culprit? Metabolism 19:760-772, 1970.
38. Orgiazzi, J., Chopra, I. J., Williams, D. E., Solomon, D. H.: Evidence for normal thyroidal adenyl cyclase, cyclic AMP-binding and protein-kinase activities in Graves' disease. J. Clin. Endocrinol. Metab. 40:248-255, 1975.
39. Onaya, T., Kotani, M., Yamada, T., Ochi, Y.: New *in vitro* tests to detect

the thyroid stimulator in sera from hyperthyroid patients by measuring colloid droplet formation and cyclic AMP in human thyroid slices. J. Clin. Endocrinol. Metab. 36:859-866, 1973.
40. Kendall-Taylor, P.: Effects of long-acting thyroid stimulator (LATS) and LATS protector on human thyroid adenyl cyclase activity. Br. Med. J. 3:72-75, 1973.
41. Field, J. B., Larsen, P. R., Yamashita, K.: In vitro demonstration of normal thyrotropin (TSH) responsiveness in thyroid tissue from patients with Graves' disease. Trans. Assoc. Amer. Physicians 86:300-309, 1973.
42. Schneider, P. B.: TSH stimulation of ^{32}P incorporation into phospholipids of thyroids from patients with Graves' disease. J. Clin. Endocrinol. Metab. 38:148-150, 1974.
43. Reichlin, S., Martin, J. B., Mitnick., et al: The hypothalamus in pituitary-thyroid regulation. Recent Progr. Horm. Res. 28:229-286, 1972.
44. Schally, A. V., Redding, T. W.: In vitro studies with thyrotropin-releasing factor. Proc. Soc. Exp. Biol. Med. 26:320-325, 1967.
45. Vale, W., Burgus, R., Guillemin, R.: Competition between thyroxine and TRF at the pituitary level in the release of TSH. Proc. Soc. Exp. Biol. Med. 125:210-213, 1967.
46. Wilbur, J. F., Utiger, R. D.: In vitro studies on mechanism of action of thyrotropin-releasing factor. Proc. Soc. Exp. Biol. Med. 127:488-490, 1968.
47. Bowers, C. Y., Schally, A. V., Reynolds, G. A., Hawley, W. D.: Interaction of L-thyroxine or L-triiodothyronine and thyrotropin-releasing factor on the release and synthesis of thyrotropin from the anterior pituitary gland of mice. Endocrinology 81:741-747, 1967.
48. Bowers, C. Y., Lee, K. L., Schally, A. V.: A study on the interaction of the thyrotropin-releasing factor and L-triiodothyronine: Effects of puromycin and cycloheximide. Endocrinology 82:75-82, 1968.
49. May, P. B., Donabedian, R. K.: Thyrotropin-releasing hormone (TRH) mediated thyroid-stimulating hormone (TSH) release from human anterior pituitary tissue in vitro. J. Clin. Endocrinol. Metab. 36:605-607, 1973.
50. Cortell, R., Rawson, R. W.: Effect of thyroxine on response of thyroid gland to thyrotropin hormones. Endocrinology 35:488-498, 1944.
51. Shellabarger, C. J., Godwin, J. T.: Effects of thyroxine or triiodothyronine on chick thyroid in presence or absence of exogenous TSH. Am. J. Physiol. 176:371-373, 1954.
52. Greer, M. A., Shull, H. F.: A quantitative study of the effect of thyrotropin upon the thyroidal secretion rate in euthyroid and thyrotoxic subjects. J. Clin. Endocrinol. Metab. 17:1030-1039, 1957.
53. Goolden, A. W. G.: Effect of thyrotropic hormone on the accumulation of radioactive iodine in thyrotoxicosis. J. Clin. Endocrinol. Metab. 19:1252-1257, 1959.
54. Takasu, N., Sato, S., Tsukui, T., Yamada, T., Furihata, R. and Makiuchi, M.: Inhibitory action of thyroid hormone on the activation of adenyl cyclase-cyclic AMP system by thyroid-stimulating hormone in human thyroid tissues from euthyroid subjects and thyrotoxic patients. J. Clin. Endocrinol. Metab. 39:772-778, 1974.
55. Melander, A., Ericson, L. E., Sundler, F.: Sympathetic regulation of thy-

roid hormone secretion. Life Sci. 14:237-246, 1974.
56. Marshall, N. J., von Borcke, S., Malan, P. G.: Studies on isoproterenol stimulation of adenyl cyclase in membrane preparations from the bovine thyroid. Endocrinology 96:1520-1524, 1975.
57. Maayan, M. L., Ingbar, S. H.: Epinephrine: effect on uptake of iodine by dispersed cells of calf thyroid gland. Science 162:124-125, 1968.
58. Melander, A., Sundler, F., Westgren, U.: Intrathyrodial amines and the synthesis of thyroid hormone. Endocrinology 93:193-200, 1973.
59. Gilman, A. G., Rall, T. W.: Factors influencing adenosine 3',5'-phosphate accumulation in bovine thyroid slices. J. Biol. Chem. 243:5867-5871, 1968.
60. Melander, A., Erickson, L. E., Ljunggren, J. -G., Norberg, K. -A., Persson, B., Sundler, F., Tibblin, S., Westgren, U.: Sympathetic innervation of the normal human thyroid. J. Clin. Endocrinol. Metab. 39:713-718, 1974.
61. Hays, M. T.: Effect of epinephrine on radioiodide uptake by the normal human thyroid. J. Clin. Endocrinol. Metab. 25:465-468, 1965.
62. Hadden, D. R., Bell, T. K., McDevitt, D. G., Sharks, R. G. Montgomery, A. D., Weaver, J. A.: Propranolol and the utilization of radioiodine by the human thyroid gland. Acta Endocrinol. 61:393-399, 1969.
63. Woolf, P. D., Lee, L. A., Schalch, D. S.: Adrenergic manipulation and thyrotropin-releasing hormone (TRH)-induced thyrotropin (TSH) release. J. Clin. Endocrinol. Metab. 35:616-618, 1972.
64. Spaulding, S. W., Burrow, G. N., Donabedian, R., Van Woert, M.: L-dopa suppression of thyrotropin-releasing hormone response in man. J. Clin. Endocrinol. Metab. 35:182-185, 1972.
65. Rapoport, B., Refetoff, S., Fang, V. S., Friesen, H. G.: Suppression of serum thyrotropin (TSH) by L-dopa in chronic hypothyroidism: Interrelationships in the regulation of TSH and prolactin secretion. J. Clin. Endocrinol. Metab. 36:256-262, 1973.
66. Ishikawa, E., Ishikawa, S., Davis, J. W., Sutherland, E. W.: Determination of guanosine 3',5'-monophosphate in tissues and of guanyl cyclase in rat intestine. J. Biol. Chem. 244:6371-6376, 1969.
67. Barmasch, M., Pisarev, M. A., Altschuler, N.: Guanyl cyclase activity in rat thyroid tissue. Acta Endocrinol. Pan. Amer. 4:19-23, 1973.
68. Yamashita, K., Field, J. B.: Elevation of cyclic guanosine 3',5'-monophosphate levels in dog thyroid slices caused by acetylcholine and sodium fluoride. J. Biol. Chem. 247:7062-7066, 1972.
69. Fallon, E. F., Agrawal, R., Furth, E., Steiner, A. L., Cowden, R.: Cyclic guanosine and adenosine 3',5'-monophosphate in canine thyroid tissue: Different sites of intracellular localization by immunofluorescence. Science 184:1089-1091, 1974.
70. Van Sande, J., Decoster, C., Dumont, J. E.: Control and role of cyclic 3',5'-guanosine monophosphate in the thyroid. Biochem. Biophys. Res. Commun. 62:168-175, 1975.
71. Macchia, V., Varrone, S.: Mechanism of TSH action. Studies with dibutyryl cyclic AMP and dibutyryl cyclic GMP. FEBS Lett. 13:342-344, 1971.
72. De Nayer, P.: Effect of cyclic 3',5'-GMP on *in vitro* protein synthesis. Biochimie 55:1507-1509, 1973.

73. Pisarev, M. A., De Groot, L. J., Wilber, J. F., Altschuler, N.: Action of cyclic guanosine monophosphate on thyroid weight and protein. Endocrinology 88:1074-1076, 1971.
74. Levey, G. S.: Catecholamine sensitivity, thyroid hormone and the heart, a reevaluation. Amer. J. Med. 50:413-420, 1971.
75. Sobel, B. E., Braunwald, E.: Cardiovascular system. In: The Thyroid, S. C. Werner, S. H. Ingbar (eds.), pp. 551, Harper and Row, 1971.
76. Spaulding, S. W., Noth, R. H.: Thyroid-catecholamine interactions. Med. Clin. North Amer. 59:1123-1131, 1975.
77. Levey, G. S.: The heart and hyperthyroidism. Use of beta-adrenergic blocking drugs. Med. Clin. North Amer. 59:1193-1201, 1975.
78. Marcus, R., Lundquist, C., Chopra, I.: *In vitro* inhibition of cyclic nucleotide phosphodiesterase (pDE) by thyroid hormones. Endocrine Society, 57th Meeting, pp. 470, 1975.
79. Leak, D., Low, M.: Effect of treatment of hypothyroidism on the circulatory response to adrenalin. Br. Heart J. 25:30-34, 1963.
80. Young, B. A., McNeill, J. H.: The effect of noradrenaline and tyramine on cardiac contractility, cyclic AMP, and phosphorylase a in normal and hyperthyroid rats. Can. J. Physiol. Pharmacol. 52:375-383, 1974.
81. Levey, G. S., Epstein, S. E.: Activation of cardiac adenyl cyclase by thyroid hormone. Biochem. Biophys. Res. Commun. 33:990-995, 1968.
82. Brodie, B. B., Davies, J. I., Hynie, S., Krishna, G., Weiss, B.: Interrelationships of catecholamines with other endocrine systems. Pharmacol. Rev. 18:273-289, 1966.
83. Levey, G. S., Epstein, S. E.: Myocardial adenyl cyclase: activation by thyroid hormones and evidence for two adenyl cyclase systems. J. Clin. Invest. 48:1663-1669, 1969.
84. Klein, I., Levey, G. S., Epstein, S. E.: Effect of reserpine on the activation of myocardial adenyl cyclase by thyroid hormone. Proc. Soc. Exp. Biol. Med. 137:366-369, 1971.
85. McNeill, J. H., Muschek, L. D., Brody, T. M.: The effect of triiodothyronine on cyclic AMP, phosphorylase, and adenyl cyclase in rat heart. Can. J. Physiol. Pharmacol. 47:913-916, 1969.
86. Frazer, A., Hess, M. E., Shanfeld, J.: The effects of thyroxine on rat heart adenosine 3',5'-monophosphate, phosphorylase b kinase and phosphorylase a activity. J. Pharmacol. Exp. Ther. 170:10-16, 1969.
87. Sobel, B. E., Dempsey, P. J., Cooper, T.: Normal myocardial adenyl cyclase activity in hyperthyroid cats. Proc. Soc. Exp. Biol. Med. 132:6-9, 1969.
88. Levey, G. S., Skelton, C. L., Epstein, S. E.: Influence of hyperthyroidism on the effects of norepinephrine on myocardial adenyl cyclase activity and contractile state. Endocrinology 85:1004-1009, 1969.
89. Levey, G. S., Skelton, C. L., Epstein, S. E.: Decreased myocardial adenyl cyclase activity in hypothyroidism. J. Clin. Invest. 48:2244-2250, 1969.
90. Buccino, R. A., Spann, J. F., Jr., Pool, P. E., Sonnenblick, E. B., Braunwald, E.: Influence of thyroid state on the intrinsic contractile properties and energy stores of the myocardium. J. Clin. Invest. 46:1669-1682, 1967.

91. Broekhuysen, J., Ghislsin, M.: Increased heart adenyl cyclase activity in the hypothyroid rat. Biochem. Pharmacol. 21:1493-1500, 1972.
92. Birnbaumer, L., Rodbell, M.: Adenyl cyclase in fat cells. J. Biol. Chem. 244:3477-3482, 1969.
93. Butcher, R. W., Baird, C. E., Sutherland, E. W.: Effects of lipolytic and antilipolytic substances on adenosine 3',5'-monophosphate levels in isolated fat cells. J. Biol. Chem. 243: 1705-1712, 1968.
94. Deykin, D., Vaughan, M.: Release of free fatty acids by adipose tissue from rats treated with triiodothyronine or propylthiouracil. J. Lipid Res. 4:200-203, 1963.
95. Krishna, G., Hynie, S., Brodie, B. B.,: Effects of thyroid hormones on adenyl cyclase in adipose tissue and on free fatty acid mobilization. Proc. Natl. Acad. Sci. USA 59:884-889, 1968.
96. Caldwell, A., Fain, J.: Triiodothyronine stimulation of cyclic adenosine 3',5'-monophosphate accumulation in fat cells. Endocrinology 89:1195-1204, 1971.
97. Mandel, L. R., Kuehl, F. A., Jr.: Lipolytic action of 3,3'5-triiodo-L-thyronine, a cyclic AMP phosphodiesterase inhibitor. Biochem. Biophys. Res. Commun. 28:13-18, 1967.
98. Armstrong, K. J., Stouffer, J. E., Van Inwegen, R. G., Thompson, W. J. and Robison, G. A.: Effects of thyroid hormone deficiency on cyclic adenosine 3',5'-monophosphate and control of lipolysis in fat cells. J. Biol. Chem. 249:4226-4231, 1974.
99. Rosenquist, U., Effendic, S., Jerob, B. and Ostman, J.: Influence of the hypothroid state on lipolysis in human adipose tissue in vitro. Acta Med. Scand. 189:381-384, 1971.
100. Rosenquist, U.: Inhibition of nonadrenaline-induced lipolysis in hypothyroid subjects by increased α-adrenergic responsiveness. An effect mediated through the reduction of cyclic AMP levels in adipose tissue. Acta Med. Scand. 192:353-359, 1972.
101. Goodman, H. M., Bray, G. A.: Role of thyroid hormones in lipolysis. Amer. J. Physiol.: 210:1053-1058, 1966.
102. Fisher, J. N., and Ball, E. G.: Studies on the metabolism of adipose tissue. XX. The effect of thyroid status upon oxygen consumption and lipolysis. Biochemistry 6:637-647, 1967.
103. Stout, B. D., Wiener, L., Cox, J. W.: Combined α and β sympathetic blockade in hyperthyroidism. Ann. Intern. Med. 70:963-970, 1969.
104. Burns, T. W., Langley, P. E., Robison, G. A.: The influence of catecholamines on human adipose tissue lipolysis: The role of adrenergic receptor sites and 3',5'-cyclic adenosine monophosphate (cAMP). Excerpta Med. International Congress (Amst.) Ser. No. 209, 45, 1970.
105. Grill, V., Rosenquist, U.: Accumulation of cyclic AMP in hypothyroidism. Decreased sensitivity to norepinephrine in rat adipocytes. Acta Endocrinol. (Kbh) 78:39-43, 1975.
106. Karlberg, B. E., Henriksson, K. G., Andersson, R. G. G.: Cyclic adenosine 3',5'-monophosphate concentration in plasma, adipose tissue and skeletal muscle in normal subjects and in patients with hyper- and hypothyroidism. J. Clin. Endocrinol. Metab. 39:96-101, 1974.

107. Sutherland, E. W., Rall, T. W., Menon, T.: Adenyl cyclase: Distribution, preparation and properties. J. Biol. Chem. 237:1220-1227, 1962.
108. Broadus, A. E., Kaminsky, N. I., Hardman, J. G., Sutherland, E. W., Liddle, G. W.: Kinetic parameters and renal clearances of plasma adenosine 3',5'-monophosphate and guanosine 3',5'-monophosphate in man. J. Clin. Invest. 49:2222-2236, 1970.
109. Singhal, R. L., Lafreniere, R., Ling, G. M.: Cerebrocortical adenyl cyclase activity following neonatal thyroid hormone deficiency. Int. J. Clin. Pharmacol. 8:1-4, 1973.
110. Schmidt, M. J., Robison, G. A.: The effect of neonatal thyroidectomy on the development of the adenosine 3',5'-monophosphate system in the rat brain. J. Neurochem. 19:937-947, 1972.
111. Schmidt, M. J.: Effects of neonatal hyperthyroidism on activity of cyclic AMP-dependent microsomal protein kinase. J. Neurochem. 22:469-471, 1974.
112. Thanassi, N. M., Newcombe, D. S.: Cyclic AMP and thyroid hormone: inhibition of epiphyseal cartilage cyclic 3',5'-nucleotide phosphodiesterase activity by L-triiodothyronine. Proc. Soc. Exp. Biol. Med. 147:710-714, 1974.
113. Chase, L. R., Melson, G. L., Aurbach, G. D.: Pseudohypoparathyroidism: Defective excretion of 3',5'-AMP in response to parathyroid hormone. J. Clin. Invest. 48:1832-1844, 1969.
114. Murad, F., Pak, C. Y.: Urinary excretion of adenosine 3',5'-monophosphate and guanosine 3',5'-monophosphate. N. Engl. J. Med. 286:1382-1387, 1972.
115. Murad, F.: Clinical studies and applications of cyclic nucleotides. In: Advances in Cyclic Nucleotide Research, P. Greengard and G. A. Robison (eds.) Raven Press, New York, 1973, pp. 355-383.
116. Hardman, J. G., Davis, J. W., Sutherland, E. W.: Measurement of guanosine 3',5'-monophosphate and other cyclic nucleotides. Variations in urinary excretion with hormonal state of the rat. J. Biol. Chem. 241:4812-4815, 1966.
117. Hardman, J. G., Davis, J. W., Sutherland, E. W.: Effects of some hormonal and other factors on the excretion of guanosine 3',5'-monophosphate and adenosine 3',5'-monophosphate in rat urine. J. Biol. Chem. 244:6354-6362, 1969.
118. Williams, R. H., Barish, J., Ensinck, J. W.: Hormone effects upon cyclic nucleotide excretion in man. Proc. Soc. Exp. Biol. Med. 139:447-454, 1972.
119. Rosen, O. M.: Urinary cyclic AMP in Graves' disease. N. Engl. J. Med. 287:670-671, 1972.
120. Estep, H., Fratkin, M., Campbell, G., Smith, E. P.: Hypercalcemia and cyclic AMP excretion in hyperthyroidism. Clin. Res. 21:490, 1973.
121. Lin, T., Kopp, L. E., Tucci, J. R.: Urinary excretion of cyclic 3',5'-adenosine monophosphate in hyperthyroidism. J. Clin. Endocrinol. Metab. 36:1033-1036, 1973.
122. Bartley, P. C., Lloyd, H. M., Willgoss, D.: Urinary cyclic AMP in diagnosis and management of hypercalcemia: studies of patients without primary hyperparathyroidism. Aust. N. Z. J. Med. 5:32-35, 1975.

123. Carter, D. J., Heath, D. A.: Urinary cyclic AMP excretion in thyroid disease. Clin. Sci. Mol. Med. 47:19-20, 1974.
124. Guttler, R. B., Shaw, J. W., Otis, C. L., Nicoloff, J. T.: Epinephrine-induced alterations in urinary cyclic AMP in hyper- and hypothyroidism. J. Clin. Endocrinol. Metab. 41:707-711, 1975.
125. Kühlback, B.: Creatine and creatinine metabolism in thyrotoxicosis and hypothyroidism. A clinical study. Acta Med. Scand. (Suppl) 331:1-65, 1957.
126. Elkeles, R. S., Lazarus, J. H., Siddle, K., Campbell, A. K.: Plasma adenosine 3',5'-cyclic monophosphate response to glucagon in thyroid disease. Clin. Sci. Mol. Med. 48:27-31, 1975.
127. Guttler, R. B., Otis, C. L., Shaw, J. W., Warren, D. W., Nicoloff, J. T.: Adenyl cyclase (AC) as a potential site for thyroid hormone action. Clin. Res. 22:340, 1974.
128. Muffelman, D. W., Mundy, J. C., Schrank, J. P., Murad, F.: The effect of exercise on cyclic AMP and cyclic GMP in human plasma and urine. Clin. Res. 22:476a, 1974.
129. Heath, D. A., Bilezikian, J. P., Fedak, S. A., Mallette, L. E., Aurbach, G. D.: The effect of posture and exercise on cyclic 3',5'-AMP in the extracellular fluid of man. Endocrine Society, 55th meeting, pp. 227, 1973.
130. Stoffer, S. S., Jiang, N.-S., Gorman, C. A., Pikler G. M.: Plasma catecholamines in hypothyroidism and hyperthyroidism. J. Clin. Endocrinol. Metab. 36:587-589, 1973.
131. Christensen, N. J.: Increased levels of plasma noradrenaline in hypothyroidism. J. Clin. Endocrinol. Metab. 35:359-363, 1972.
132. Harrison, T. S., Siegel, J. H., Wilson, W. S.: Adrenergic reactivity in hyperthyroidism. Arch. Surg. 94:396-402, 1967.
133. Levine, R. J., Oates, J. A., Vendsalu, A.: Studies on the metabolism of aromatic amines in relation to altered thyroid function in man. J. Clin. Endocrinol. Metab. 22:1242-1250, 1962.
134. Wiswell, J. G., Hurwitz, G. E., Coronko, V., et al.: Urinary catecholamines and their metabolites in hyperthyroidism and hypothyroidism. J. Clin. Endocrinol. Metab. 23:1102-1106, 1963.
135. Nishizawa, Y., Hamada, N., Fujii, S., Morii, H., Okuda, K., Wada, M.: Serum dopamine-β-hydroxylase activity in thyroid disorders. J. Clin. Endocrinol. Metab. 39:599-602, 1974.
136. Noth, R. H., Spaulding, W.: Decreased serum dopamine-beta-hydroxylase in hyperthyroidism. J. Clin. Endocrinol. Metab. 39:614-617, 1974.
137. Kaminsky, N. I., Ball, J. H., Broadus, A. E., Hardman, J. G., Sutherland, E. W., Liddle, G. W.: Hormonal effects on extracellular cyclic nucleotides in man. Trans. Assoc. Amer. Physicians 83:235-244, 1970.
138. Bouillon, R., De Moor, P.: Parathyroid function in patients with hyper- or hypothyroidism. J. Clin. Endocrinol. Metab. 38:999-1004, 1974.
139. Castro, J. H., Genuth, S. M., Klein, L.: Comparative response to parathyroid hormone in hyperthyroidism and hypothyroidism. Metabolism 24:839-848, 1975.
140. Sode, J., Georges, L. P., Santangelo, R. P., Mackin, J. F., and Canary, J. J.: Urinary cyclic adenosine monophosphate (cAMP) excretion in hyperthyroidism. Clin. Res. 22:756, 1972.

141. Georges, L. P., Sode, J., Santangelo, R. P., Mackin, J. F., Canary, J. J.: Propranolol (P) effect on urinary cyclic adenosine monophosphate (cAMP) excretion in hyperthyroidism. Endocrine Society, 55th meeting, pp. 218, 1973.
142. Guttler, R. B., Shaw, J. W., Nicoloff, J. T.: Catecholamine sensitive increase in urinary cAMP excretion in thyrotoxicosis. Endocrine Society, 55th meeting, pp. 181, 1973.
143. Eisenberg, M., Donabedian, R.: Urinary cyclic AMP and cyclic GMP in thyroid disease. Clin. Res. 22:339, 1974.
144. Unger, R. H., Madison, L. L., Muller, W. A.: Abnormal alpha cell function in diabetes. Response to insulin. Diabetes 21:301-307, 1972.
145. Muller, W. A., Faloona, G. R., Anguilar-Parada, E., Unger, R. G.: Abnormal alpha cell function in diabetes. Response to carbohydrate and protein ingestion. N. Engl. J. Med. 283:109-115, 1970.
146. Assan, R., Rosselin, G., Dolais, J.: Effets sur la glucagonemie des perfusions et ingestions d'acides amines. J. Ann. Diabet. Hotel Dieu 7:25-41, 1967 (Fr).
147. Sutherland, E. W., Robison, G. A., and Hardman, J. G.: Some thoughts on the possible role of cyclic AMP in diabetes. In: Pathogenesis of Diabetes Mellitus, E. Cerase and R. Luft (eds.) Wiley, New York, 1970, pp. 137-139.
148. Gerich, J. E., Lorenzi, M., Hane, S., Gustafson, G., Guillemin, R., Forsham, P. H.: Evidence for a physiologic role of pancreatic glucagon in human glucose homeostasis: studies with somatostatin. Metabolism 24: 175-182, 1975.
149. Gerich, J. E., Tsalikian, E., Lorenzi, M., Karam, J. H., Bier, D. M.: Plasma glucagon and alanine responses to acute insulin deficiency in man. J. Clin. Endocrinol. Metab. 40:526-529, 1975.
150. Alberti, K. G. M. M., Christensen, N. J., Iversen, J., Orskov, H.: Role of glucagon and other hormones in development of diabetic ketoacidosis. Lancet 1:1307-1311, 1975.
151. Brazeau, P., Vale, W., Burgus, R., Ling, N., Butcher, M., Riuier, J., Guillemin, R.: Hypothalamic polypeptide that inhibits the secretion of immunoreactive pituitary growth hormone. Science 179:77-79, 1973.
152. Coy, D. H., Coy, E. J., Arimura, A., Schally, A. V.: Solid phase synthesis of growth hormone-release inhibiting factor. Biochem. Biophys. Res. Commun. 54:1267-1273, 1973.
153. Polak, J. M., Pearse, A. G. E., Grimelius, L., Bloom, S. R., and Arimura, A: Growth hormone release inhibiting hormone in gastrointestinal and pancreatic D cells. Lancet 1:1220-1221, 1975.
154. Alberti, K. G., Christensen, N. J., Christensen, S. E., Hansen, A. P., Iversen, J., Lundback, K., Sayer-Hansen, K., Orskov, H.: Inhibition of insulin secretion by somatostatin. Lancet 2:524-525, 1973.
155. Mortimer, C. H., Tunbridge, W. M., Carr, D., Yeomans, L., Lind, T., Coy, D. H., Bloom, S. R., Kustin, A., Mallinson, C. N., Besser, G. M., Schally, A. V., Hall, R.: Effects of growth hormone release-inhibiting hormone on circulating glucagon, insulin, and growth hormone in normal, diabetic, acromegalic and hypopituitary patients. Lancet 1:697-701, 1974.
156. Iversen, J.: Inhibition of pancreatic glucagon release by somatostatin.

In Vitro. Scand. J. Clin. Lab. Invest. 33:125-129, 1974.
157. Gerich, J. E., Lorenzi, M., Bier, D. M., Schneider, V., Tsalikian, E., Karam, J. H., Forsham, P. H.: Prevention of human diabetic ketoacidosis by somatostatin. Evidence for an essential role of glucagon. N. Engl. J. Med. 292:985-989, 1975.
158. Howell, S. L., Taylor, K. W.: Secretion of newly synthesized insulin in vitro. Biochem. J. 102:922-927, 1967.
159. Steiner, D. F., Cunningham, D. D., Spigelman, L., Aten, B.: Insulin biosynthesis: evidence for a biosynthetic precursor. Science 157:697-700, 1967.
160. Lin, B. J., Haist, R. E.: Insulin biosynthesis: Effects of carbohydrates and related compounds. Can. J. Physiol. Pharmacol. 47:791-801, 1969.
161. Tanese, T., Lazarus, N. R., Devrins, S., Recant, L.: Synthesis and release of proinsulin and insulin by isolated rat islets of Langerhans. J. Clin. Invest. 49:1394-1404, 1970.
162. Lin, B. J., Haist, R. E.: Effect of some modifiers of insulin secretion on insulin biosynthesis. Endocrinology 92:735-742, 1973.
163. Schatz, H., Maier, V., Hinz, M., Nierle, C., Pfeiffer, E. E.: Stimulation of H-3 leucine incorporation into the proinsulin and insulin fraction of isolated pancreatic mouse islets in the presence of glucagon, theophylline and cyclic AMP. Diabetes 22:433-441, 1973.
164. Sussman, K. E., Vaughan, G. D.: Insulin release after ACTH, glucagon, and adenosine 3',5' phosphate (cyclic AMP) in the perfused isolated rat pancreas. Diabetes 16:449-454, 1967.
165. Malaisse, W. J., Malaisse-Lagae, F., and Mayhew, D.: A possible role for the adenyl cyclase system in insulin secretion. J. Clin. Invest. 46:1724-1734, 1967.
166. Levine, R. A., Oyama, S., Kagan, A., Glick, E. M.: Stimulation of insulin and growth hormone secretion by adenine nucleotides in primates. J. Lab. Clin. Med. 75:30-36, 1970.
167. Samols, E., Marri, G., Marks, V.: Promotion of insulin secretion by glucagon. Lancet 2:415-416, 1965.
168. Vecchio, D., Luyckx, A., Zahand, G. R., Renold, A. E.: Insulin release induced by glucagon in organ cultures of fetal rat pancreas. Metabolism 15:577-581, 1966.
169. Rosen, O. M., Hirsch, A. H., Goren, E. N.: Factors which influence cyclic AMP formation and degradation in an islet cell tumor of the Syrian hamster. Arch. Biochem. Biophys. 146:660-663, 1971.
170. Goldfine, I. D., Roth, J., Birnbaumer, L.: Glucagon receptors in B cells. Binding of ^{125}I glucagon and activation of adenylate cyclase. J. Biol. Chem. 247:1211-1218, 1972.
171. Howell, S. L., Montague, W.: Adenylate cyclase activity in isolated rat islets of Langerhans. Effects of agents which alter rates of insulin secretion. Biochim. Biophys. Acta 320: 44-52, 1973.
172. Lebovitz, H. E., Pooler, K.: Puromycin potentiation of corticotropin-induced insulin release. Endocrinology 80:656-662, 1967.
173. Kuo, W., Hodgins, D. S., Kuo, J. F.: Adenylate cyclase in islets of Langerhans, isolation of islets and regulation of adenylate cyclase activity

by various hormones and agents. J. Biol. Chem. 248:2705-2711, 1973.
174. Davis, B., and Lazarus, N. R.: Insulin release from pancreatic islets: properties of a membrane bound phosphokinase from cod and mouse islets. In: Proc. 8th Congr. of Internatl. Diabetes Federation., Amsterdam. Excerpta Medica, International Congress Series, No. 280:7, 1973.
175. Thompson, W. J., Johnson, D. G., Williams, R. H.: Modulation of hormonal stimulation of adenyl cyclase from isolated pancreatic islets by guanosine triphosphate. Diabetes Suppl. 1 23:297, 1973.
176. Malaisse, W. J., Malaisse-Lagae, F., Wright, P. H., Ashmore, J.: Effects of adrenergic and cholinergic agents upon insulin secretion in vitro. Endocrinology 80:975-978, 1967.
177. Vance, J. E., Buchanan, K. D., Williams, R. H.: Glucagon and insulin release, influence of drugs affecting the autonomic nervous system. Diabetes 20:78-82, 1971.
178. Lambert, A. E., Jeanrenaud, B., Renold, A. E.: Enhancement by caffeine of glucagon-induced and tolbutamide-induced insulin release from isolated feotal pancreatic tissue. Lancet 1:819-820, 1967.
179. Ashcroft, S. J. H., Bassett, J. M., Randle, P. J.: Insulin secretion mechanisms and glucose metabolism in isolated islets. Diabetes Suppl. 2 21:538-545, 1972.
180. Turtle, J. R., Kipnis, D. M.: An adrenergic receptor mechanism for the control of cyclic 3'5' adenosine monophosphate synthesis in tissues. Biochem. Biophys. Res. Commun. 28:797-802, 1967.
181. Montague, W., Cook, J. R.: The role of adenosine 3'5' cyclic monophosphate in the regulation of insulin release by isolated rat islets of Langerhans. Biochem. J. 122:115-120, 1971.
182. Cooper, R. H., Ashcroft, S. J. H., Randle, P. J.: Concentrations of adenosine 3',5'-cyclic monophosphate in mouse pancreatic islets measured by a protein binding radioassay. Biochem. J. 134:599-605, 1973.
183. Kosaka, K., Ide, T., Kuzuya, T., Niki, E., Kuzuya, N., Okinaka, S.: Insulin-like activity in pancreatic vein blood after glucose loading and epinephrine. Endocrinology 75:9-14, 1964.
184. Hertelendy, F., Machlin, A. J., Gordon, R. S., Horino, M., Kipnis, D. M.: Lipolytic activity and inhibition of insulin release by epinephrine in the pig. Proc. Soc. Exp. Biol. Med. 121:675-677, 1966.
185. Hertelendy, F., Machlin, L. J., Kipnis, D. M.: Further studies on the regulation of insulin and growth hormone secretion in the sheep. Endocrinology 84:192-199, 1969.
186. Kris, A. D., Miller, R. E., Wherry, F. E., Mason, J. W.: Inhibition of insulin secretion by infused epinephrine in rhesus monkeys. Endocrinology 78:87-97, 1966.
187. Porte, D., Jr., Graber, A. L., Kuzuya, T., Williams, R. H.: The effects of epinephrine on immunoreactive insulin levels in man. J. Clin. Invest. 45:228-236, 1966.
188. Wong, K. K., Symchowicz, S., Staub, M. S., Tabachnick, I. I. A.: The in vitro effect of catecholamines, diazoxide and theophylline on insulin release. Life Sci. 6:2285-2291, 1967.
189. Porte, D., Jr.,: A receptor mechanism for the inhibition of insulin release

by epinephrine in man. J. Clin. Invest. 46:86-94, 1967.
190. Cerasi, E., Luft, R.: Diabetes mellitus—a disease of pancreatic and extra-pancreatic origin. In: Advances in Metabolic Disorders, Vol. 7, 1974, pp. 193-212.
191. Charles, M. A., Fanska, R., Schmid, F. G., Forsham, P. H., Grodsky, G. M.: Adenosine 3',5'-monophosphate in pancreatic islets: glucose induced insulin release. Science 179:569-571, 1973.
192. Grill, V., Cerasi, E.: Activation by glucose of adenyl cyclase in pancreatic islets of the rat. FEBS Lett. 33:311-314, 1973.
193. Selawry, H., Marcks, C., Fink, G., Lavine, R., Cresto, J., Recant, L.: A mechanism of glucose-induced insulin release. Diabetes 22:295 (Abstract), 1973.
194. Zawalich, W. S., Ferrendelli, J., Matschinsky, F. M.: Glucose-induced elevation of cyclic 3',5'-AMP (CAMP) levels in pancreatic islets. Diabetes 22:331 (Abstract), 1973.
195. Grill, V., Cerasi, E.: Stimulation by D-glucose of cyclic adenosine 3'-5'-monophosphate accumulation and insulin release in isolated pancreatic islets of the rat. J. Biol. Chem. 249:4196-4201, 1974.
196. Zawalich, W. S., Karl, R. C., Ferrendelli, J., Matschinsky, F. M.: Effects of glucose, Ca++, and an ionophore on cyclic 3'5'-AMP(CAMP)and insulin release in isolated pancreatic islets. Diabetes 23:337 (Abstract), 1974.
197. Capito, K., Hedeskov, C. J.: The effect of starvation on phosphodiesterase activity and the content of adenosine 3'5' cyclic monophosphate in isolated mouse pancreatic islets. Biochem. J. 142:653-658, 1974.
198. Zawalich, W. S., Karl, R. C., Ferrendelli, J. A., Matschinsky, F. M.: Factors governing glucose-induced elevation of cyclic AMP levels in pancreatic islets. Diabetologia 11:231-235, 1975.
199. Hellman, B., Idahl, L. A., Lernmark, A., Täljedal, I. B.: The pancreatic beta cell recognition of insulin secretagogues: does cyclic AMP mediate the effect of glucose. Proc. Natl. Acad. Sci. USA 71:3405-3409, 1974.
200. Voyles, N., Gutman, R. A., Selawry, A., Fink, G., Penhos, J. C., Recant, L.: Interaction of various stimulators and inhibitors on insulin secretion in vitro. Horm. Res. 4:65-73, 1973.
201. Howell, S. L., Montague, W.: Regulation of guanylate cyclase in guinea pig islets of Langerhans. Biochem. J. 142:379-384, 1974.
202. Jefferson, L. S., Exton, J. H., Butcher, R. W., Sutherland, E. W., Park, C. R.: Role of adenosine 3',5'-monophosphate in the effects of insulin and antiinsulin serum on liver metabolism. J. Biol. Chem. 243:1031-1038, 1968.
203. Butcher, R. W., Sneyd, J. G. T., Park, C. R., Sutherland, E. W.: Effect of insulin on adenosine 3',5'-monophosphate in the rat epididymal fat pad. J. Biol. Chem. 241: 1651-1653, 1966.
204. Exton, J. H., Lewis, S. B., Ho, R. J., Robison, G. A., Park, C. R.: The role of cyclic AMP in the interaction of glucagon: insulin on the control of liver metabolism. Ann. N.Y. Acad. Sci. 185:85-100, 1971.
205. Jungas, R. L.: Role of cyclic 3',5'-AMP in the response of adipose tissue to insulin. Proc. Natl. Acad. Sci. USA 56:757-763, 1966.
206: Manganiello, V. C., Murad, F., Vaughan, M.: Effects of lipolytic and

antilipolytic agents on cyclic 3',5'-adenosine monophosphate in fat cells. J. Biol. Chem. 246:2195-2202, 1971.
207. Fain, J. N., Rosenberg, L.: Antilipolytic action of insulin on fat cells. Diabetes Suppl. 2 21:414-425, 1972.
208. Goldfine, I. D., Sherline, P.: Insulin action in isolated rat thymocytes. Independence of insulin and cyclic adenosine monophosphate. J. Biol. Chem. 247:6927-6931, 1972.
209. Fain, J.: Inhibition of adenosine cyclic 3',5'-monophosphate accumulation in fat cells by adenosine, N6-(phenylisopropyl) adenosine and related compounds. Mol. Pharmacol. 9:595-604, 1973.
210. Khoo, J. C., Steinberg, D., Thompson, B., Meyer, S. E.: Hormonal regulation of adipocyte enzymes. The effects of epinephrine and insulin on the control of lipase, phosphorylase kinase phosphorylase and glycogen synthesis. J. Biol. Chem. 248:3823-3830, 1973.
211. Kono, T., Barham, F. W.: Effects of insulin on the levels of adenosine 3',5'-monophosphate and lipolysis in isolated rat epididymal fat cells. J. Biol. Chem. 248:7417-7426, 1973.
212. Lavis, V. R., Williams, R. H.: Lipolytic effects of high concentrations of insulin on isolated fat cells. Enhancement of response to lipolytic hormones. Diabetes 22:629-636, 1973.
213. Trueheart, P. A., Herrera, M. G., Jungas, R. L.: Paradoxical effects of theophylline and its interaction with insulin on glucose metabolism in adipose tissue. Eur. J. Biochem. 38:137-145, 1973.
214. Trueheart, P. M., Herrera, M. G., Jungas, R. L.: Similar and opposite effects of dibutyryl cyclic AMP and insulin on glucose metabolism in adipose tissue. Biochim. Biophys. Acta 313:310-319, 1973.
215. Kissebah, A., Tulloch, B. R., Vydelingum, N., Hope-Gill, H. F., Clarke, P. V., Fraser, T. R.: The role of calcium in insulin action. II. Effects of insulin and procaine hydrochloride on lipolysis. Horm. Metab. Res. 6: 357-364, 1974.
216. Siddle, K., Hales, C. N.: The relationship between the concentration of adenosine 3',5'-cyclic monophosphate and the anti-lipolytic action of insulin in isolated rat fat cells. Biochem. J. 142:97-103, 1974.
217. Hepp, K. D.: Adenylate cyclase and insulin action. Effect of insulin, non-suppressible insulin-like material and diabetes on adenylate cyclase activity in mouse liver. Eur. J. Biochem. 31:266-276, 1972.
218. Hepp, K. D., Renner, R.: Insulin action on the adenyl cyclase system: antagonism to activation by lipolytic hormones. FEBS Lett. 20:191-194, 1972.
219. Illiano, G., Cuatrecasas, P.: Modulation of adenylate cyclase activity in liver and fat cell membranes by insulin. Science 175:906-908, 1972.
220. DeAsua, L. L., Surian, E. S., Flawia, M. M., Torres, H. N.: Effect of insulin on the growth pattern and adenylate cyclase activity of BHK fibroblasts. Proc. Natl. Acad. Sci. USA 70:1388-1392, 1973.
221. Flawia, M. M., Torres, H. N.: Adenylate cyclase activity in Neurospora crassa. III. Modulation by glucagon and insulin. J. Biol. Chem. 248:4517-4520, 1973.
222. Golder, M. P., Boyns, A. R.: Distribution of adrenocorticotrophic hormone-

stimulated adenylate cyclase in the adrenal cortex. J. Endocrinol. 56:471-481, 1973.
223. Loten, E. G., Sneyd, J. G. T.: An effect of insulin on adipose tissue adenosine 3',5'-cyclic monophosphate phosphodiesterase. Biochem. J. 120:187-193, 1970.
224. Manganiello, V., Vaughan, M.: An effect of insulin on cyclic adenosine 3',5'-monophosphate phosphodiesterase activity in fat cells. J. Biol. Chem. 248:7164-7170, 1973.
225. Thompson, W. J., Little, S. A., Williams, R. H.: Effect of insulin and growth hormone on rat liver cyclic nucleotide phosphodiesterase. Biochemistry 12:1889-1894, 1973.
226. Pawlson, L. G., Lovell-Smith, C. J., Manganiello, V. C., Vaughan, M.: Effects of epinephrine, adrenocorticotrophic hormone, and theophylline on adenosine 3',5'-monophosphate phosphodiesterase activity in fat cells. Proc. Natl. Acad. Sci. USA 71:1639-1642, 1974.
227. Zinman, B., Hollenberg, C. H.: Effect of insulin and lipolytic agents on rat adipocyte low Km cyclic adenosine 3',5'-monophosphate phosphodiesterase. J. Biol. Chem. 249:2182-2187, 1974.
228. Hollenberg, M. D., Cuatrecasas, P.: Insulin: interaction with membrane receptors and relationship to cyclic purine nucleotides and cell growth. Fed. Proc. 34:1556-1563, 1975.
229. Kissebah, A. H., Hope-Gill, H., Vydelingum, N., Tulloch, B. R., Clarke, P. V., Fraser, T. R.: A mode of insulin action. Lancet 1:144-146, 1975.
230. Unger, R. H.: Glucagon and insulin: glucagon ratio in diabetes and other catabolic illnesses. Diabetes 20:834-838, 1971.
231. Exton, J. H., Park, C. R.: Interaction of insulin and glucagon in the control of liver metabolism. Handbook of Physiology, Section 7 Endocrinology, Washington, American Physiological Association. 1:437-455, 1972.
232. Illiano, G., Tell, G.P.E., Siegel, J. I., Cuatrecasas, P.: Guanosine 3',5'-cyclic monophosphate and the action of insulin and acetylcholine. Proc. Natl. Acad. Sci. USA 70:2443-2447, 1973.
233. Sheppard, J. R.: Difference in the cyclic adenosine-3',5'-monophosphate levels in normal and transformed cells. Nature 236:14-16, 1972.
234. Goldberg, N. D., Haddox, M. K., Dunham, E., Lopez, C., Hadden, W.: In: Control of Proliferation in Animal Cells, B. Calrkson and R. Baserga (eds.) Cold Spring Harbor Laboratory, pp. 609, 1974.
235. Leclercq-Meyer, V. Brisson, G. R., Malaisse, W. J.: Effect of adrenaline and glucose on the release of glucagon and insulin in vitro. Nature 231:248-249, 1971.
236. Chesney, T. McC, Schofield, J. G.: Studies on the secretion of pancreatic glucagon. Diabetes 18:627-632, 1972.
237. Iversen, J.: Adrenergic receptors and secretion of glucagon and insulin from the isolated, perfused canine pancreas. J. Clin. Invest. 52:2102-2116, 1973.
238. Howell, S. L., Edwards, J. C., Montague, W.: Regulation of adenylate cyclase and cyclic AMP dependent protein kinase activities in A_2-cell rich guinea pig islets of Langerhans. Horm. Metab. Res. 6:49-52, 1974.
239. Toyota, T., Sato, S-I., Kudo, M., Abe, K., Goto, Y.: Secretory regulation

of endocrine pancreas: cyclic AMP and glucagon secretion. J. Clin. Endocrinol. Metab. 41:81-89, 1975.
240. Rall, T. W., Sutherland, E. W.: Formation of a cyclic adenine ribonucleotide by tissue particles. J. Biol. Chem. 232:1065-1076, 1958.
241. Murad, F.: Effect of glucagon on heart. N. Engl. J. Med. 279:434-435, 1968.
242. Murad, F., and Vaughan, M.: Effect of glucagon on rat heart adenyl cyclase. Biochem. Pharmacol. 18:1053-1059, 1969.
243. Levey, G. S., Epstein, S. E.: Activation of adenyl cyclase by glucagon in cat and human heart homogenates. Circ. Res. 24:151-156, 1969.
244. Levey, G. S., Prindle, K. H., Epstein, S. E.: Effects of glucagon on adenyl cyclase activity in the left and right ventricles of heart and in experimentally-produced isolated right ventricular failure. J. Mol. Cell. Cardiol. 1:403-410, 1970.
245. Murad, F., Shen, L. C., Larner, J.: Effect of glucagon on rat diaphragms. Fed. Proc. 31:889, 1972.
246. Marcus, R., Aurbach, G. D.: Bioassay of parathyroid hormone in vitro with a stable preparation of adenyl cyclase from rat kidney. Endocrinology 85:801-810, 1969.
247. Murad, F., Brewer, H. B., Vaughan, M.: Effect of thyrocalcitonin on adenosine 3',5'-cyclic phosphate formation by rat kidney and bone. Proc. Natl. Acad. Sci. USA 65:446-453, 1970.
248. Gitelman, H. J., Alderman, F. R., Dufresne, L. R.: *In vivo* regulation of parathyroid gland cyclic 3',5'-AMP content. Clin. Res. 20:35, 1972.
249. Levey, G. S.: The glucagon receptor and adenylate cyclase. Metabolism 24:277-310, 1975.
250. Robison, G. A., Butcher, R. W., Sutherland, E. W.: *Cyclic AMP*. Academic Press, New York, 1971.
251. Steinberg, D., Mayer, S. E., Khoo, J. C., Miller, E. A., Miller, R. E., Fredholm, B., Erchner, R.: Hormonal regulation of lipase, phosphorylase, and glucogen synthetase in adipose tissue. In: Advances in cyclic nucleotide research, Vol. 5, G. I. Drummond, P. Greengard and G. A. Robison (eds.) Raven Press, New York, 1975 pp. 549-568.
252. Broadus, A. E., Kaminsky, N. I., Northcutt, R. C., Hardman, J. G., Sutherland, E. W., Liddle, G. W.: Effects of glucagon on adenosine 3',5' monophosphate and guanosine 3',5' monophosphate in human plasma and urine. J. Clin. Invest. 49:2237-2245, 1970.
253. Taylor, A. L., Davis, B. B., Pawlson, L. G., Josimovich, J. B., Mintz, D. H.: Factors influencing the urinary excretion of 3',5'-adenosine monophosphate in humans. J. Clin. Endocrinol. Metab. 30:316-323, 1970.
254. Liddle, G. W., Hardman, J. G.: Cyclic adenosine monophosphate as a mediator of hormone action. N. Engl. J. Med. 285:560-566, 1971.
255. Liljenquist, J. E., Bomboy, J. D., Lewis, S. B., Sinclar-Smith, B. C., Felts, P. W., Lacy, W. W.: Effect of glucagon on net splanchnic cyclic AMP production in normal and diabetic men. J. Clin. Invest. 53:198-204, 1974.
256. Tucci, J. R., Lin, T., Kopp, L.: Urinary cyclic 3',5'-adenosine monophosphate levels in diabetes mellitus before and after treatment. J. Clin. Endocrinol. Metab. 37:832-835, 1973.

257. Hamet, P., Lowder, S. C., Hardman, J. G., Liddle, G. W.: Effect of hypoglycemia on extracellular levels of cyclic AMP in man. Metabolism 24: 1139-1144, 1975.
258. Das, I.: Plasma adenosine-3',5'-monophosphate in antiinsulin-treated rats. Experientia 30:860, 1974.

Changes of Cyclic Nucleotides in Normal and Pathological Pregnancy

WILLIAM Y. LING

Since the establishment of the second messenger role of adenosine-3',5'-monophosphate (cyclic AMP) by Sutherland's group (1), by far the greater part of the available information is compatible with the observation that certain hormone actions are mediated via a rise or fall in the intracellular levels of this nucleotide. Similarly, guanosine-3',5'-monophosphate (cyclic GMP) is now considered as another possible second messenger in the mechanism of certain hormone actions (2). Moreover, it seems most likely that both the intra- and extracellular concentration of each nucleotide is not only governed by the relative rates of its synthesis and destruction, but may be modulated by the rate at which it is extruded into the extracellular space.

The following sections outline some pertinent data concerning the physiologic actions of cyclic nucleotides in pregnancy, particularly as they relate to the preparation and maintenance of pregnancy, in the proliferation and differentiation of cells, the development of the human fetus, and the possible factors influencing the intra- and extracellular concentrations of these nucleotides in normal and abnormal human gestations.

POSSIBLE ROLE OF CYCLIC NUCLEOTIDES IN PREGNANCY

Reproduction in the human female consists of monthly ovulation, fertilization, a long gestational period, a hormonally active placenta, a growng fetus, parturition, and lactation. The interplay among the central nervous system, hypothalamus, pituitary gland, and gonads is important in maintaining the proper hormonal balance and becomes more complex as the metabolic and physiologic changes are multiplied by the added complexities of a developing fetoplacental unit.

The anterior pituitary gland secretes at least six protein hormones: adrenocorticotrophic hormone (ACTH), growth hormone (GH), thyroid-stimulating hormone (TSH), luteinizing hormone (LH), follicle-stimulating hormone (FSH), and prolactin (PRL). In a recent review, Labrie et al. (3) have documented evidence that implicates cyclic AMP as a mediator in the release of all these hormones. The first three are not directly involved in the endocrine control of gestation but are nevertheless essential in the homeostatic support of the pregnant state. There is evidence to suggest that the actions of these hormones may be mediated by cyclic AMP (4). LH, FSH, and PRL are considered to be among the numerous hormones essential in reproduction. An intricate balance of stimulation and response between the pituitary gonadotropins (LH, FSH) and ovarian steroid hormones (estrogens and progestins) is essential for the proper sequence of events that results in the preparation for the maintenance of pregnancy. There is now a large body of data indicating that the steroidogenic effect of LH on the ovary is mediated by increased intracellular levels of cyclic AMP, and although FSH has not been shown to elicit acute synthesis of steroids, it can nevertheless cause an elevation of cyclic AMP level in the ovary (5).

After fertilization, development of the zygote probably involves the participation of cyclic nucleotides (details will be discussed in the following section). Cyclic AMP may also play a role in the process of implantation. In pregnant mice, "delayed" implantation of the blastocyst can be induced by the administration of estrogens, and since cyclic AMP can duplicate this effect, it is believed that cyclic AMP acts as second messenger in this estrogen action (6). Before the development of a fully functional human placenta, the corpus luteum is a vital organ for the secretion of steroids which sustain the early implanted embryo. Several observations suggest that cyclic AMP mediates the action of LH and human chorionic gonadotropin (hCG) on steriodogenesis in the human corpus luteum (7, 8).

Other pregnancy-dependent physiologic processes may also involve the active participation of the cyclic nucleotides. For example, the development of the mammary gland during pregnancy and the transition to the lactational state at parturition depend on a complex mechanism involving numerous hormones such as estrogens, progestins, corticoids, prolactin, and insulin. Sapag-Hagar and Greenbaum (9) reported that progesterone and estradiol stimulate adenylate cyclase activity in the gland of the pregnant rat and that insulin causes a marked inhibition. In a subsequent study, the same authors showed that growth and development of the rat mammary gland is related to a coordinated change in tissue levels of both cyclic nucleotides, with the cyclic AMP level being progressively elevated during gestation and falling sharply at lactation; cyclic GMP showed the reverse pattern (10).

ROLE OF CYCLIC NUCLEOTIDES IN CELLULAR GROWTH AND DIFFERENTIATION

Once the conceptus is implanted in the uterine endometrium a period of rapid proliferation of the embryonic cells ensues. To date, almost no information is available about the function of cyclic nucleotides in the development of the human embryo. However, there is ample evidence from studies in lower organisms, on cell cultures and in certain proliferative diseases to indicate that cyclic nucleotides may play a significant role. Pastan et al. (11) documented 263 separate studies supporting the thesis that cyclic nucleotides are important biological regulators of cellular proliferation. Based on the observation that rapidly growing cells have a depressed cyclic AMP level and that elevated cyclic AMP levels inhibit accelerated cell division, Sheppard (12) has proposed a modulatory mechanism for cyclic AMP in regulating cell division in cultured animal cells. Hadden et al. (13) suggested that cyclic GMP may also act as an intracellular signal regulating cell proliferation. When lymphocytes were transformed by mitogenic concentrations of purified phytohemagglutinin, they found that the intracellular level of cyclic AMP was unchanged, while the level of cyclic GMP increased markedly. Although elevated cyclic GMP level stimulates cell division, addition of exogenous cyclic GMP and its analogs failed to promote growth in certain transformed cells (14). This seeming contradiction may be explained by the hypothesis that cellular proliferation requires a dualistic, coordinated action of both nucleotides (15). For example, Voorhees et al. (16) reported increased cyclic GMP but decreased cyclic AMP levels in the proliferating epithelium of psoriasis, and Kram et al. (17) showed that growth inhibition by cyclic AMP on cultured cells could be reversed by adding exogenous cyclic GMP. The antagonistic actions of the two nucleotides in cell proliferation have also been observed in cultured fibroblasts (18). However, inconsistency in the observation that cyclic GMP stimulates cell growth and opposes cyclic AMP action has been reported (19). The exact role of cyclic GMP in cell growth remains to be elucidated. Recently, Bloch (20) demonstrated a possible regulatory action for cytidine-3',5'-monophosphate (cyclic CMP) in the proliferation of leukemic cells, which could not be duplicated by cyclic AMP or cyclic GMP, suggesting that cell division may be regulated by yet a third cyclic nucleotide.

As the embryo proceeds to maturity, a period of cellular differentiation begins, eventually leading to the acquisition of the complete physiologic machinery characteristic of differentiated tissues. In mouse neuroblastoma cell cultures, it has been shown that dibutyryl cyclic AMP (db-cyclic AMP) can induce the formation of axons (21), and promote axon extension in the dorsal root ganglia (22). In cultures of adrenal cortical cells from fetal rats, cyclic AMP has been found to mimic certain effects induced by ACTH, namely, transformation of immature cells to the differentiated cell type with an increase in steroidogenic capacity (23). Recently it has been reported that both ACTH and db-cyclic AMP, but not dibutyryl-cyclic GMP (db-cyclic GMP) stimulate the formation of cortisol from progesterone in midterm human fetal adrenal cells in culture (24). In a study of cultured fetal rat pancreas, cyclic AMP has been implicated in the maturation of the fetal insulin-releasing mechanism (25). Several investigators have also noted ontogenic changes in the hormonally stimulated adenylate cyclase activity during the development of certain fetal tissues. Injection of glucagon, epinephrine, or cyclic AMP into rat fetuses one to 2 days before birth have led to the induction of certain hepatic enzymes (26), whereas at earlier stages only cyclic AMP could elicit similar induction (26, 27). It is likely that the fetal hepatic adenylate cyclase system would respond to stimulation by these hormones only at a particular developmental stage. In human fetal liver cultures, the addition of glucagon or db-cyclic AMP together with prednisolone has evoked increases in the tyrosine transaminase activity which is also correlated with the increasing age of the fetus (28). There is no definitive evidence for the participation of cyclic GMP in cellular differentiation but there is evidence to suggest that it stimulates protein synthesis (29). Thus, the role of cyclic GMP still remains unclear, but cyclic AMP may be an important regulator of proliferation and differentiation of embryonic cells. Based on the available information, Robison (30) has produced a very simple illustration of the changes in the relative levels of cyclic AMP that may occur during growth and development in a complex organism. Initially, the level of cyclic AMP will decrease after fertilization to permit rapid cell division, but after a time, a signal develops that prompts the cells to spend more of their energy for differentiation than proliferation. The implication of this event is that most cells will acquire a mechanism causing their cyclic AMP levels to rise. The level of cyclic AMP will rise further when the cells become fully differentiated and acquire hormone receptors that in turn stimulate the enzyme systems involved in cyclic AMP synthesis.

FACTORS AFFECTING CYCLIC NUCLEOTIDE LEVELS IN THE DEVELOPING HUMAN FETAL AND PLACENTAL TISSUES

There are as yet few reports on the factors affecting the tissue concentration of cyclic nucleotides in human fetuses. In the preceding section, we have seen that the cyclic AMP analog can mimic the stimulatory effect of ACTH and glucagon in cultures of human fetal adrenal and liver, respectively (24, 28). Under physiologic conditions, these hormones may exert their effects in a fashion similar to that in adult tissues, i.e., by elevating the cellular cyclic AMP levels. Adenylate cyclase and phosphodiesterase activities have been measured in several tissues from human fetuses 10 to 14 wk old (31). In general, the basal cyclic AMP concentration reflects the relative activities of these two enzymes, and thus, the highest cyclic AMP levels, in descending order, were found in the adrenals, heart, and brain. These levels are low compared to those found in the corresponding human adult tissues. In human fetuses at 12 to 13 wk gestation, Menon et al. (32) showed that the highest level of cyclic AMP occurred in the heart ventricle and that this tissue was most responsive to catecholamine stimulation. In addition, the liver adenylate cyclase system was responsive to glucagon and epinephrine, and this enzyme in the adrenal was specifically stimulated by ACTH. Thus, the adenylate cyclase activities appeared to correlate with the functional capacities of these fetal tissues at this stage of gestation.

The placenta is an endocrine gland that both produces several important pregnancy hormones and may serve as a target organ for some of them. The normal functioning placenta produces two major classes of hormones: protein hormones and steroids. Although very little is known about their exact influence, they are necessary for maintenance and progress of pregnancy. Studies reported by a number of investigators have implicated cyclic AMP in both the release and the action of hCG. Handwerger et al. (33) found that in cultured human placental explants, cyclic AMP, or theophylline (which increases cyclic AMP level by inhibition of cyclic nucleotide phosphodiesterase) elicited marked stimulation of hCG secretion but had no effect on the secretion of human placental lactogen (hPL). This study not only showed that cyclic AMP could modulate the secretion of hCG, but that the intracellular control mechanisms regulating the secretion of various placental hormones may be different.

The adenylate cyclase activities in normal human term placenta have been studied by several groups of investigators. Ferre and Cedard (34) reported the presence of adenylate cyclase activity in 1000xg pellets. Satoh and Ryan (35, 36) also reported this activity in the 3000xg pellets which could be stimulated by fluoride, catecholamines, prostaglandins, but not hCG. However, Menon and Jaffe (37) reported an hCG-sensitive adenylate cyclase system in a minced preparation of human term placenta. Recently, it has been shown the LH, hCG, and prostaglandin E_1 (PGE_1) could induce a marked increase in the intracellular cyclic AMP level in perfused human placenta (38). The fact that hCG can stimulate adenylate cyclase activity in the placenta suggests a role for cyclic AMP in the action of hCG. In this regard, placental perfusion studies carried out by Cedard et al. (39) demonstrated that in the first trimester, or term, human placentae, hCG, or cyclic AMP added to the perfusing medium increased the formation of estrogens from androgen and also produced a transitory but significant rise in the concentration of glucose in the perfusate.

These experiments suggest that in the trophoblast, cyclic AMP regulates the secretion of hCG. In other specific cellular sites, hCG in turn imparts its effects, via cyclic AMP, on the mechanisms which stimulate the aromatase system for estrogen synthesis and the glycogen phosphorylase to provide glucose. Thus, by regulating both the secretion and action of certain hormones, cyclic AMP may influence the physiologic functions of the human placenta, and deviations in cyclic AMP levels may reflect abnormalities in placental function.

CLINICAL STUDIES OF CYCLIC NUCLEOTIDES IN PREGNANCY

Although the cyclic nucleotides mediate intracellular events, their extracellular levels can be altered under certain physiologic or pathologic conditions and by various pharmacologic manipulations. This observation has been utilized in a number of clinical studies for the specific purpose of understanding disease states and therapy (40).

Cyclic Nucleotides in Normal Menstrual Cycle

Although the urinary levels of cyclic AMP excreted in man are fairly constant from day to day (40) they appear to vary during the menstrual cycle. Taylor and his coworkers (41) reported that when sequential 24-hr urines were collected for 1 month, a 20 to 40 percent increase in daily cyclic AMP excretion occurred during the mid menstrual cycle in three women. On the basis of parallel pregnanediol and estrogen excretions, this peak corresponded with the ovulatory phase of the cycle. This finding was confirmed by another study in which weekly 24-hr urine samples were measured in 12 women over two menstrual cycles (42). In a third report (43), only 6 of 12 women showed a peak urinary cyclic AMP excretion at mid cycle. Moreover, one subject studied over two cycles showed an increase in 24-hr cyclic AMP excretion in the first but not in the second ovulatory phase. In the same study, daily urinary excretion of cyclic GMP was higher during the second phase of the cycle in 10 of 12 subjects. The concentration of cyclic AMP in human cervical mucus was found to be highest in the periovulatory phase of the menstrual cycle (44).

Cyclic Nucleotides in the Amniotic and Maternal Fluids from Normal Human Pregnancy

The presence of cyclic AMP in human amniotic fluid was first reported in a review by Broadus et al. (45). We have measured cyclic AMP concentrations in amniotic fluid specimens taken from various stages of normal pregnancy. The lowest cyclic AMP levels were found in the fluids taken during the 4 to 14 wk of gestation. Thereafter, cyclic AMP concentration rose steadily as pregnancy progressed, reaching a maximum level at term. In addition, cyclic AMP levels in 13 samples taken from 34 to 40 wk of pregnancy showed a positive correlation with the corresponding birth weight of the newborn (46). Since the fetal compartment contributes directly to the amniotic fluid contents, and as this contribution increases in importance with the advance of gestation (47), it is likely that cyclic AMP measured in the amniotic fluid was of fetoplacental origin (the dynamics of cyclic AMP transfer during pregnancy will be discussed in a later section). Thus, our measurements of cyclic AMP may be correlated to certain aspects in fetal development, and these empirical relationships may have important clinical applications in the estimation of fetal maturity and well-being.

The urinary level of cyclic AMP in nonpregnant subjects was first reported by Butcher and Sutherland (48), and subsequently cyclic GMP was also identified in human urine (49). Serial measurements in one woman with normal pregnancy and one with idiopathic hypoparathyroidism and sustained hypocalcemia throughout gestation showed a progressive rise in the daily urinary cyclic AMP concentration from the 12th to the 36th week of gestation. This was followed by a plateau lasting until term. After the first postpartum day, urinary cyclic AMP level declined to control (male and non-pregnant female subjects) values (41).

Plasma levels of cyclic AMP and cyclic GMP in man have also been documented (45). In studies carried out in our laboratory in Miami, (Ling, W. Y., Marsh, J. M. and LeMarie, W. J.: unpub-

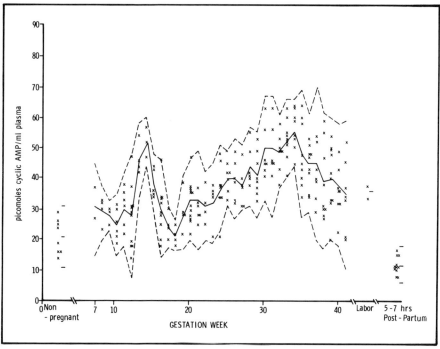

1 Cyclic AMP levels measured in plasma samples taken from women in the 7th–41st week of normal pregnancy. The solid line represents the mean of the individual values at each week of gestation and the two broken lines represent the range of 2SD above and below the mean. Plasma levels of cyclic AMP measured in menstruating women (n = 8), during labor (n = 2) and after delivery (n = 11) are also shown.

lished data) we have measured the plasma levels of cyclic AMP in nonpregnant women, in women from 7 to 41 wk of normal gestation, and in postpartum women. As shown in Fig. 1, the plasma levels showed an initial peak of short duration at the 13 to 15 weeks of pregnancy and subsequently falling rapidly to a nadir at the 18th week. The level then rises steadily, reaching a second peak value at about the 34th week of gestation. A gradual decrease in cyclic AMP levels occurred thereafter which persisted until labor. After delivery, the concentration declined within 5 to 7 hr to the nonpregnant level. In the same study, serial plasma samples were

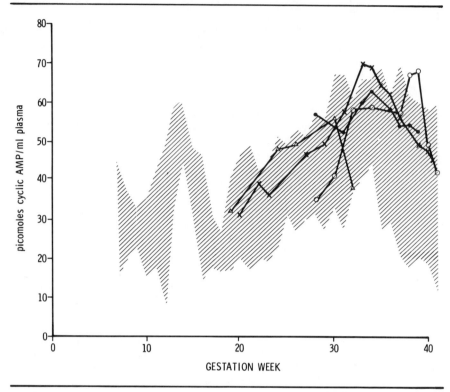

2 Plasma cyclic AMP levels measured in blood samples taken serially from four women of normal pregnancy. The shaded area represents the normal range (±2 SD from the mean) of plasma cyclic AMP levels.

collected from four normal volunteers. The cyclic AMP concentrations shown in Fig. 2 indicate that these values are within the normal range measured in the cross-sectional study.

Cyclic Nucleotides in Pregnancies Complicated by Hypertension

The classification of hypertensive pregnancies to be discussed in this section is based on the usually accepted clinical criteria described in most gynecologic textbooks (e.g., 50). In general, preeclampsia (or toxemia) is pregnancy-induced. This group of patients comprises primigravidas whose blood pressures are known to be normal before the 21st week of pregnancy, and are followed by a rise of more than 30/15 mm Hg (systolic/diastolic) above their normal values. This group is further divided into mild (the highest blood pressure not exceeding 160/110) and severe preeclampsia (blood pressure greater than 160/110 with one or a combination of the symptoms of edema, proteinuria and excessive weight gain). The chronic hypertensive group is comprised of those who have prior histories of hypertension. In this group of patients, pregnancy may have aggravated their conditions but is not considered as a precipitating cause. This group is also divided by the above criteria into mild and severe chronic hypertensives.

In our study of cyclic AMP levels in amniotic fluid from women whose pregnancies were complicated by hypertension (51), a total of 53 specimens from 45 patients in the 28 to 42 wk of gestation were analyzed. This group consisted of subjects with mild or severe preeclampsia and subjects with mild or severe chronic hypertension. All the patients were under treatment at the time of amniocentesis. A marked elevation of cyclic AMP was found in amniotic fluid from pregnancies complicated by various forms of hypertension. The cyclic AMP concentration in the combined group of chronic hypertensive patients was not statistically different from the combined preeclamptic groups. However, the level of cyclic AMP was significantly higher in the amniotic fluid from severe preeclamptic than the fluid from mild preeclamptic patients. In the combined groups of preeclamptic patients, consideration of each fluid specimen with the patient's corresponding clinical data showed a positive correlation between fluid cyclic AMP values and the single highest diastolic blood pressure recorded during pregnancy. On the other hand, there was a positive correlation between the cyclic AMP values from the chronic hypertensive groups and the diastolic blood pressure taken at the time of aminocentesis. These findings suggest that the elevation of amniotic fluid cyclic AMP concentrations in pregnancy complicated by preeclampsia may be correlated with the severity of the syndrome, whereas, in chronic hypertension, it may be related to blood pressure alone. Further study is required, however, before we will know whether the measurement of amniotic fluid cyclic AMP is useful in the management of this pregnancy complication.

The urinary excretion of cyclic AMP was studied in a group of women with toxemic pregnancy (52). These investigators found that during the period of 24 to 34 wk of gestation, the urinary excretion of cyclic AMP in toxemic patients was similar to the level at the corresponding period of uncomplicated pregnancy. However, during 35 to 40 wk of gestation, the urinary cyclic AMP excretion was decreased to half the corresponding normal values in untreated toxemic patients. Unfortunately, these authors did not correlate the onset of toxemia with the time when the urinary cyclic AMP began to fall below the normal range. Patients who received various medication, such as dihydralazine and hydrochlorothiazide, excreted normal amounts of cyclic AMP. Using urinary creatinine as an index, these authors reported that the decreased cyclic AMP excretion in toxemia was not a result of impaired glomerular filtration of this nucleotide.

We have measured the sequential plasma concentrations of cyclic AMP in five patients whose pregnancies were complicated by hypertension (Ling, W. Y., Marsh, J. M. and LeMarie, W. J.: unpublished data). The results shown in Fig. 3 indicate that the plasma levels were markedly elevated above the normal range during the gestation period of 16 to 26 wk, but became comparable to normal values for the 27 to 40 wk of gestation. It appears that in pregnancies complicated by hypertension, plasma cyclic AMP concentrations do not exhibit the biphasic pattern of normal pregnancy, but tend to remain high throughout the second and third trimester.

Of particular interest are the plasma cyclic AMP concentrations in patient C (Fig. 3), who was initially placed in the normal group but developed preeclampsia after the 31st week of her pregnancy. Her plasma cyclic AMP level was elevated in the 16 to 27 wk gestational period when she showed no clinical signs or symptoms of preeclampsia. Other indicators of impending preeclampsia have been reported. Gant et al. (53) noted that 90 percent of the young primigravidas who developed pregnancy-induced hypertension showed a significant loss in resistance of the pressor effect of angiotensin II 8 to 12 wk before the appearance of any clinical symptoms of preeclampsia. In another related study (54), it was shown that a rise in ACTH/cortisol ratio may occur in preeclamptic women before the development of edema, proteinuria or hypertension. It may be inferred from these findings that in preeclampsia, the decreased cortisol and increased ACTH levels could result in increased adrenal secretion of mineralocorticoids such as aldosterone. The association between hypertension and the renin–angio-

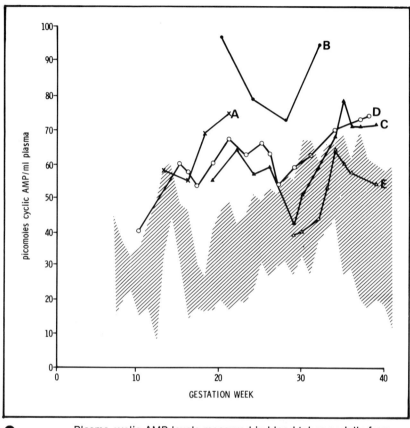

3 Plasma cyclic AMP levels measured in blood taken serially from five women whose pregnancies were complicated by hypertension. Patients A, B, D, and E were chronic hypertensives. Patient C developed clinical signs and symptoms of preeclampsia after the 31st week of pregnancy. Note the elevated cyclic AMP levels before the onset of detectable clinical symptoms. The shaded area represents the normal range (±2 SD from the mean) of plasma cyclic AMP levels.

tensin–aldosterone system has long been recognized, and cyclic AMP has been shown to mediate the secretion of these hormones (1). Therefore, it is possible to see the three observations above as part of the manifestation for preeclampsia.

We have also carried out preliminary studies (Ling, W., Le-

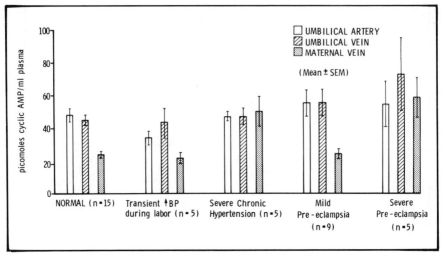

4 Plasma cyclic AMP levels measured in the matched samples of umbilical arterial, umbilical venous and maternal venous plasma from women of normal or hypertensive pregnancies.

Marie, W., Marsh, J., and Holsinger, K.: unpublished data) of the levels of cyclic AMP in matched samples of umbilical arterial, venous and maternal plasma in normal pregnancy and pregnancies complicated by hypertension. The results (Fig. 4) showed an elevated maternal plasma cyclic AMP level in the severe forms of hypertensive pregnancies. In cord plasma, cyclic AMP levels were not significantly different among all five groups of patients.

Could errors in cyclic AMP metabolism play a role in hypertensive pregnancies? The answer, at best, is that it is speculatively conceivable. The evidence that might justify this answer is as follows: It is generally accepted that one of the possible etiologic factors in toxemic pregnancy involves the regulatory mechanism for the uteroplacental blood flow. In a review on toxemia, Speroff (55) postulated a role for the prostaglandins in the vasoregulation of the uteroplacental circulation. The ability of PGE_1 to stimulate adenylate cyclase systems has been documented (56). One of the known effects of PGE_1 is that it is hypotensive and has been shown to produce a relaxation in the umbilical and placental blood vessels in vitro (57). Moreover, it was found that in aortic strips, β-adrenergic stimulation produced an increase in cyclic AMP content as well as muscle relaxation (58). There are two studies that support the probability that the mechanism regulating uteroplacental circulation by prostaglandins involves the participation of cyclic AMP. The first showed that, among all the prostaglandins tested, PGE_1 is the most potent stimulator of adenylate cyclase activity in human term placenta (36). The second showed that the rate of PGE_1 metabolism in placental tissue from toxemic preganacies was depressed in direct relation to the severity of the disease (59). Therefore, one would expect a positive correlation between cyclic AMP levels in the placenta and the severity of the syndrome; we have found such a correlation in the amniotic fluid from preeclamptic patients. Thus, from the standpoint of cyclic AMP action, the following sequence of events may be postulated in toxemic pregnancy: PGE_1 elaborated in the placenta in response to angiotensin II stimulation exerts its hypotensive effect by stimulating the placental adenylate cyclase system. The increased cyclic AMP led to events which cause vasodilation and thus maintains a reduced resistance to the blood flowing through the uteroplacental bed.

The results reviewed thus far deal mainly with the possible role of cyclic AMP in toxemic pregnancy. Although the function of cyclic GMP remains somewhat obscure, increasing evidence suggests that cyclic GMP at times functions dualistically with cyclic AMP to regulate certain physiologic processes, and alterations in the relative levels of these two nucleotides may figure in the etiology of certain diseases. It has been observed that cyclic GMP can modulate smooth muscle contraction, and cyclic AMP influences relaxation (60). Based on this finding, Amer studied the action of cyclic nucleotides in aortas from spontaneously hypertensive and stress hypertensive rats. He found that the intracellular cyclic AMP level was lowered by a diminished adenylate cyclase response to hormonal stimulations and a much elevated phosphodiesterase activity (61), and that there was also a marked elevation in the intracellular levels of cyclic GMP (62). Thus, in hypertensive aortas, the cyclic GMP/cyclic AMP ratio is much increased over the normotensive one.

In events more closely related to human pregnancy, Karim (63) reported that of all the agents tested (e.g., isoproterenol), only PGE_1 could effect relaxation of the human umbilical artery. It was also found that only PGE_1 was capable of inducing an accumulation of cyclic AMP in this tissue. Moreover, the arterial contraction caused by various agents was always accompanied by a rise in cyclic GMP levels without alteration of the cyclic AMP content.

Clinical studies of cyclic nucleotides in other pathologic pregnancies are not available in the literature to mid-1975.

POSSIBLE SOURCES AND TRANSFER OF CYCLIC NUCLEOTIDES IN PREGNANCY FLUIDS

Broadus et al. (64) in studies of normal young adult males found that under basal conditions, the urinary cyclic AMP was derived from both plasma and kidneys. One-half to two-thirds of the total level of cyclic AMP in urine resulted from glomerular filtration and the remainder was contributed primarily by proximal nephrons. Furthermore, other studies have shown that the contribution of either source may be altered by administering hormones such as parathyroid hormone or glucagon (65, 66). Under basal conditions, virtually all the cyclic GMP in human urine appears to be derived from glomerular filtration (67). Urinary excretion of cyclic GMP is also influenced by hormonal changes: in the rats it is reduced drastically by hypophysectomy, but a mixture of six anterior pituitary hormones can reverse this effect; adrenalectomy also lowers cyclic GMP excretion and is restored to normal by hydrocortisone (67).

The sources of basal levels of plasma cyclic nucleotide in man are unknown. Wehmann et al. (68) found that, in anesthetized dogs, the lungs, and intestines are probably sites of net addition to the circulating levels of both nucleotides. Although the liver removes the nucleotides from the plasma, the kidney is by far the major site of this removal. The authors concluded that basal plasma levels of each nucleotide are the result of small net contributions from many tissues or bidirectional fluxes between tissues and plasma, or both.

During normal human pregnancy, the fetoplacental unit may contribute to the elevated levels of cyclic AMP in maternal urine and plasma. This is inferred from the study of Taylor et al. (41) showing that the elevated urinary cyclic AMP level returned to the nonpregnant level 1 day after delivery, and it is confirmed by our finding that plasma cyclic AMP levels in normal pregnancy (Fig. 1) fall to nonpregnant values 5 to 7 hr after delivery. Further evidence can be deduced from the results in Fig. 4, which shows that in normal pregnancy the cyclic AMP levels measured in the umbilical cord plasma at delivery were significantly higher than the level in the corresponding maternal plasma. In pregnancies complicated by severe chronic hypertension or preeclampsia, however, the maternal plasma cyclic AMP level is also elevated, but the factor responsible for this is unknown.

To learn more about the significance of these increasing cyclic AMP levels in urine, amniotic fluid, and plasma during the course of normal pregnancy, we investigated the possible sources and the fate of this nucleotide in the amniotic fluid of rhesus monkeys in late pregnancy (69). The results indicated that an extremely small fraction of the injected cyclic AMP was available for transport across the placenta and into the amniotic fluid, but once in the amniotic fluid, it was not readily metabolized or transported out. Recently, we completed a study on the dynamics of cyclic AMP exchange among the mother, amniotic fluid, and the fetus in 10 monkeys (70). In separate animals, a single dose of ^{3}H-labeled cyclic AMP was injected into the circulation of the mother or fetus, or into the amniotic fluid. The pattern of transfer after each injection was followed by simultaneous measurements of ^{3}H-labeled cyclic AMP in samples taken from each compartment at pre-determined time intervals. A schematic diagram summarizing the results of this study is shown in Fig. 5. Similar to the findings in young men (64) and dogs (71), ^{3}H-cyclic AMP injected into either the maternal or fetal circulation was rapidly and exponentially cleared. A very small fraction of the injected dose was bidirectionally exchanged between the maternal and fetal circulation. Both compartments could contribute cyclic AMP to the amniotic fluid, independently or in concert. One hour after injection, 12 percent of the injected dose was recovered in the maternal urine. This is in agreement with the results reported by Broadus et al. (64) indicating that in man, approximately 85 percent of the elimination of the cyclic nucleotide was due to extrarenal clearance. In the dogs, less than 20 percent of the plasma clearance could be accounted for by urinary excretion (71). In the present study, only 4.5 percent of the injected dose was recovered in the fetal urine after the labeled cyclic AMP was injected into the fetal circulation. The bulk of the radioactivity injected into the maternal or fetal circulation was unaccounted for. Presumably, in short-term experiments, it was irreversibly taken up by the tissues. When ^{3}H-cyclic AMP

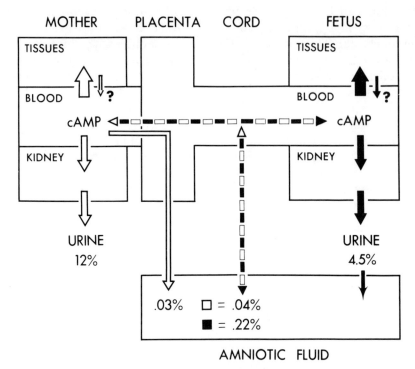

5 Schematic diagram showing the transport of cyclic AMP among the mother, fetus and the amniotic fluid of rhesus monkeys. The open arrows indicate the route when ³H-cyclic AMP was injected into the maternal circulation, the closed arrows show the route of transport when ³H-cyclic AMP was injected into the fetal circulation.

was injected into the amniotic sac, more than 65 percent of the injected dose remained unchanged in this compartment after 1 hr, and only a minimal transfer into the maternal and fetal circulations occurred.

SUMMARY AND SPECULATION

Cyclic nucleotides may play a central role in the physiologic process of pregnancy. A great deal of evidence has implicated cyclic AMP as the mediator in the secretions and actions of several hormones that are vital to the preparation and maintenance of normal gestation. Considerable interest has been focused on the possibility that the participation of both cyclic AMP and cyclic GMP may be important in the growth and development of complex organisms.

In normal human pregnancy, the cyclic AMP level in urine and amniotic fluid appears to increase as pregnancy progresses. However, the maternal plasma cyclic AMP concentration shows a biphasic pattern with peak values at the 14th and 34th week of gestation. In pregnancies complicated by hypertensive disorders, the levels of this nucleotide are dramatically altered in all three pregnancy fluids. Although cyclic AMP is readily transferred across the placenta, and both the maternal and fetal compartments can contribute cyclic AMP to the amniotic fluid, the factors responsible for altered fluid levels in hypertensive pregnancies are unknown. Evidence suggests that the fetoplacental unit is a source of the increasing cyclic AMP levels in the pregnancy fluids of normal gestation and that the patterns are altered during hypertensive pregnancies by factors within the maternal organism.

Thus, the continued study of cyclic nucleotides in human pregnancy may have at least two important implications: cyclic nucleotide measurement in pregnancy fluids could be potentially useful as an aid to antenatal diagnosis, and it may improve our understanding of the control of the gestation process. The overall result may lead to better methods in the management and therapy of pregnancy complications.

ACKNOWLEDGMENT

I am indebted to Dr. John M. Marsh and Dr. William J. LeMaire who have initiated my interest in the research of cyclic nucleotides in human pregnancy, and whose collaboration has led to the realization of most of my own findings reported in this review.

I would like to express my appreciation to Dr. William Moger for critically reviewing the manuscript, to Elizabeth Shapter for her editorial help in its preparation, and to Charlotte MacDonald for her care in the typing of this manuscript.

REFERENCES

1. Robison, G. A., Butcher, R. W., and Sutherland, E. W.: Cyclic AMP, Academic Press, New York, 1971.

2. Goldberg, N. D., O'Dea, R. F., and Haddox, M. K.: Cyclic GMP. Adv. Cyclic Nucleotide Res. 3:155-223, 1973.
3. Labrie, F., Borgeat, P., LeMay, A., Lemaire, S., Braden, N., Drouin, J., Lemaire, I., Jolicoeur, P., and Belanger, A.: Role of cyclic AMP in the action of hypothalamic regulatory hormones. Adv. Cyclic Nucleotides Res. 5:787-801, 1975.
4. Major, P. W. and Kilpatrick, R.: Cyclic AMP and hormone action. J. Endocrinol. 52:593-630, 1972.
5. Marsh, J. M.: The role of cyclic AMP in gonadal function. Adv. Cyclic Nucleotide Res. 6:137-199, 1975.
6. Holmes, P. V., and Bergstrom, S.: Induction of blastocyst implantation in mice by cyclic AMP. J. Reprod. Fert. 43:329-332, 1975.
7. Hermier, C., Santos, A. A., Wisnewsky, C., Netter, A., and Justisz, M.: Role de l'AMPc et d'une proteine regulatrice dans l'action, in vitro, de la gonadotropine choriale humaine (HCG) sur le corps jaune humaine. Comp. Rend. Acad. Sci. Paris 275:1415-1418, 1972.
8. Marsh, J. M., and LeMaire, W. J.: Cyclic AMP accumulation and steroidogenesis in the human corpus luteum: Effect of gonadotropins and prostaglandins. J. Clin. Endocr. 38:99-106, 1974.
9. Sapag-Hagar, M., and Greenbaum, A. L.: Adenosine-3',5'-monophosphate and hormone interrelationships in the mammary gland of the rat during pregnancy and lactation. Eur. J. Biochem. 47:303-312, 1974.
10. Sapag-Hagar, M., and Greenbaum, A. L.: The role of cyclic nucleotides in the development and function of rat mammary tissue. FEBS Letters 46: 180-183, 1974.
11. Pastan, I. H., Johnson, G. S., and Anderson, W. B.: Role of cyclic nucleotides in growth control. Ann. Rev. Biochem. 44:491-522, 1975.
12. Sheppard, J. R.: The role of cyclic AMP in the control of cell division. In: Cyclic AMP, Cell Growth and the Immune Response, Braun, W., Lichtenstein, L. M., and Parker, C. W. (eds.) pp. 290-301, Springer-Verlag, New York, 1974.
13. Hadden, J. W., Hadden, E. M., Haddox, M. K., and Goldberg, N. D.: Gaunosine-3',5'-cyclic monophosphate: A possible intracellular mediator of mitogenic influences in lymphocytes. Proc. Natl. Acad. Sci. (U.S.A.) 69: 3024-3027, 1972.
14. Carchman, R. A., Johnson, G. S., Pastan, I., and Scolnick, E. M.: Studies on the levels of cyclic AMP in cells transformed by wild-type and temperature-sensitive kirsten sarcoma virus. Cell 1:59-64, 1974.
15. Goldberg, N. D., Haddox, M. K., Dunham, E., Lopez, C., and Hadden, J. W.: The Ying Yang hypothesis of biological control: Opposing influences of cyclic GMP and cyclic AMP in the regulation of cell proliferation and other biological processes. In: Control of Proliferation in Animal Cells, Clarkson, B., and Baserga, R. (eds.) pp. 609-625, Cold Spring Harbor Laboratory, 1974.
16. Voorhees, J. J., and Duel, E. A.: Imbalanced cyclic AMP-cyclic GMP levels in psoriasis. Adv. Cyclic Nucleotide Res. 5:735-758, 1975.
17. Kram, R., Mamont, P., and Tomkins, G. M.: Pleiotypic control by adenosine-3',5'-cyclic monophosphate: A model for growth control in animal cells. Proc. Natl. Acad. Sci. (U.S.A.) 70:1432-1436, 1973.

18. Seifert, W., and Rudland, P. S.: Possible involvement of cyclic GMP in growth control of cultured mouse cells. Nature 248:138-140, 1974.
19. Miller, Z., Lovelace, E., Gallo, M., and Pastan, I.: Cyclic guanosine monophosphate and cellular growth. Science 190:1213-1215, 1975.
20. Bloch, A.: Isolation of cytidine-3′,5′-monophosphate from mammalian tissues and body fluids and its effect on leukemia L-1210 cell growth in culture. Adv. Cyclic Nucleotide Res. 5:331-338, 1975.
21. Prasad, K. N., and Hsie, A.: Morphologic differentiation of mouse neuroblastoma cells induced in vitro by dibutyryl adenosine-3′,5′-cyclic monophosphate. Nature 233:141-142, 1971.
22. Roisen, F. J., Murphy, R. A., Pichichero, M. E., and Braden, W. G.: Cyclic adenosine monophosphate stimulation of axonal elongation. Science 175: 73-74, 1972.
23. Milner, A. J.: Cyclic AMP and the differentiation of adrenal cortical cells grown in tissue culture. J. Endocrinol. 55:405-413, 1972.
24. Kan, K., Caspin, S., and Solomon, S.: Stimulation of cortisol formation in human fetal adrenal cells in culture. Abstract from the Endocrine Society 57th Annual Meeting Program. p. 57, New York, 1975.
25. Lambert, A. E., Kanazawa, Y., Burr, I. M., Orci, L., and Renold, A. E.: On the role of cyclic AMP in insulin release: I. Overall effects in cultured fetal rat pancreas. Ann. N. Y. Acad. Sci. 185:232-244, 1971.
26. Greengard, O.: The hormonal regulation of enzymes in prenatal and postnatal rat liver. Biochem. J. 115:19-24, 1969.
27. Wicks, W. D.: Induction of hepatic enzymes by adenosine-3′,5′-monophosphate in organ culture. J. Biol. Chem. 244:3941-3950, 1969.
28. Kirby, L., and Hahn, P.: Enzyme response to prednisolone and dibutyryl adenosine-3′,5′-monophosphate in human fetal liver. Pediat. Res. 8:37-41, 1974.
29. Varrone, S., DiLauro, R., and Macchia, V.: Stimulation of polypeptide synthesis by cyclic 3′,5′-guanosine monophosphate. Arch. Biochem. Biophys. 157:334-338, 1973.
30. Robison, G. A.: The biological role of cyclic AMP: An updated overview. In: Prostaglandins and Cyclic AMP: Biological Actions and Clinical Applications, Kahn, R. H., and Lands, W. E. M. (eds.) pp. 229-247, Academic Press, New York, 1973.
31. Weiss, B., and Strada, S. J.: Adenosine-3′,5′-monophosphate during fetal and postnatal development. In: Fetal Pharmacology, Boreus, L. O. (ed.) pp. 205-235, Raven Press, New York, 1973.
32. Menon, K. M. J., Giese, S., and Jaffe, R. B.: Hormone- and fluroide-sensitive adenylate cyclases in human fetal tissues. Biochem. Biophys. Acta 304:203-209, 1973.
33. Handwerger, S., Barret, J., Tyrey, L., and Schomberg, D.: Differential effect of cyclic adenosine monophosphate on the secretion of human placental lactogen and human chorionic gonadotropin. J. Clin. Endocr. 36: 1268-1270, 1973.
34. Ferre, F., and Cedard, L.: Etude du mechanisme de la regulation hormonale dans le placenta humain I. Mise en evidence d'une activite adenyl cyclasique dans les homogenats de placenta a terme. Biochem. Biophys. Acta 237:316:319, 1971.

35. Satoh, K., and Ryan, K. J.: Adenyl cyclase in human placenta. Biochim. Biophys. Acta 244:618-624, 1971.
36. Satoh, K., and Ryan, K. J.: Prostaglandins and their effects on human placental adenyl cyclase. J. Clin. Invest. 51:456-458, 1972.
37. Menon, K. M. J., and Jaffe, R. B.: Chorionic gonadotropin-sensitive adenyl cyclase. J. Clin. Endocr. 36:1104-1109, 1973.
38. Levilliers, J., Alsat, E., Landat, Ph., and Cedard, L.: Hormone-stimulated cAMP production in human placenta perfused *in vitro*. FEBS Letters 47: 146-148, 1974.
39. Cedard, L., Alsat, E., Urtasun, M. J., and Varaugot, J.: Studies on the mode of action of LH and hCG on estrogenic biosynthesis and glycogenolysis by human placenta perfused *in vitro*. Steroids 16:361-375, 1970.
40. Murad, F.: Clinical studies and applications of cyclic nucleotides. Adv. Cyclic Nucleotide Res. 3:355-383, 1973.
41. Taylor, A. L., Davis, B. B., Pawlson, L. G., Josimovich, J. B., and Mintz, D.: Factors influencing the urinary excretion of 3',5'-adenosine monophosphate in humans. J. Clin. Endocr. 30:316-324, 1970.
42. Hamadah, K., Holmes, H., Stokes, M. L., Hartman, G. C., and Parke, D. V.: Variation in urinary adenosine-3',5'-monophosphate content during the human menstrual cycle. Biochem. Soc. Trans. 2:461-462, 1974.
43. Lebeau, M., Dumont, J. E., and Goldstein, J.: Urinary cyclic AMP and cyclic GMP excretion during the menstrual cycle. Horm. Metab. Res. 7:190-194, 1975.
44. Beck, K. J., Schonhofer, P., and Schlebusch, H.: Studies on cyclic adenosine-3',5'-monophosphate concentrations in human cervical mucus. Acta Endocr. (Suppl. 173) 43:43, 1973.
45. Broadus, A. E., Hardman, J. G., Kaminsky, N. I., Ball, J. H., Sutherland, E. W., and Liddle, G. W.: Extracellular cyclic nucleotides. Ann. N.Y. Acad. Sci. 185:50-66, 1971.
46. Ling, W., LeMaire, W., d'Adesky, C., Cleveland, W. W., and Marsh, J. M.: Adenosine-3',5'-monophosphate in amniotic fluid from normal pregnancy. J. Clin. Endocr. 37:56-62, 1973.
47. Ostergard, D. R.: The physiology and clinical importance of amniotic fluid: A review. Obst. Gynec. Survey 25:297-319, 1970.
48. Butcher, R. W., and Sutherland, E. W.: Adenosine-3',5'-phosphate in biological materials. I. Purification and properties of cyclic 3',5'-nucleotide phosphodiesterase and use of this enzyme to characterize adenosine-3',5'-phosphate in human urine. J. Biol. Chem. 237:1244-1250, 1962.
49. Price, T. D., Ashman, D. F., and Melicow, M. M.: Organophosphates in urine, including adenosine-3',5'-monophosphate and guanosine-3',5'-monophosphate. Biochim. Biophys. Acta 138:452-465, 1967.
50. Chesley, L. C.: Hypertensive disorders in pregnancy. In: Williams Obstetrics 14th ed., Hellman, L. M., and Pritchard, J. A. (eds.) pp. 685-689, Appleton-Century-Crofts, New York, 1971.
51. Ling, W. Y., Marsh, J. M., Spellacy, W. N., Thresher, A. J., and LeMarie, W. J.: Adenosine-3',5'-monophosphate in amniotic fluid from pregnancy complicated by hypertension. J. Clin. Endocr. 39:479-486, 1974.
52. Raij, K., Harkonen, M., Castren, O., Saarikoski, S., and Adlercreutz, H.:

Urinary cyclic AMP excretion in toxemia of pregnancy. Scand. J. Clin. Lab. Invest. 32:193-197, 1973.
53. Gant, N. F., Daley, G. L., Chand, S., Whalley, P. J., and MacDonald, P. C.: A study of angiotensin II pressor response throughout primigravid pregnancy. J. Clin. Invest. 52:2682-2689, 1973.
54. Mukherjee, K., and Swyer, G. I. M.: Plasma cortisol and adrenocorticotrophic hormone in normal men and non-pregnant women, normal pregnant women and women with pre-eclampsia. J. Obstet. Gynaecol. Br. Commonw. 79:504-512, 1972.
55. Speroff, L.: An essay: Prostaglandins and toxemia of pregnancy. Prostaglandins 3:721-728, 1973.
56. Kuehl, F. A. Jr., Humes, J. L., Cirillo, V. J., and Ham, E. A.: Cyclic AMP and prostaglandins in hormone action. Adv. Cyclic Nucleotide Res. 1:493-502, 1972.
57. Hillier, K., and Karim, S. M. M.: Effects of prostaglandins E_1 E_2 $F_{1}\alpha$ $F_{2}\alpha$ on isolated human umbilical and placental blood vessels. J. Obstet. Gynaecol. Br. Commonw. 75:667-673, 1968.
58. Triner, L., Vulliemoz, Y., Verosky, M., Habif, D. V., and Nahas, G. G.: Adenyl cyclase-phosphodiesterase system in arterial smooth muscle. Life Sci. 11 (Pt. 1): 817-824, 1972.
59. Alam, N. A., Clary, P., and Russel, P. T.: Depressed placental prostaglandin E_1 metabolism in toxemia of pregnancy. Prostaglandins 4:363-370, 1973.
60. Schultze, G., Hardman, J., and Sutherland, E. W.: Cyclic nucleotides in smooth muscle function. In: Asthma: Physiology, Immunology, and Treatment, Austen, K. F., and Lichenstein, L. M. (eds.) pp. 123, Academic Press, New York, 1973.
61. Amer, M. S.: Cyclic adenosine monophosphate and hypertension in rats. Science 179:807-809, 1973.
62. Amer, M. S.: Possible involvement of cyclic nucleotide system in hypertension. Meeting Abstract from Clinical Applications of Cyclic Nucleotides, Philadelphia, 1974.
63. Karim, S. M.: The identification of prostaglandins in human umbilical cord. Br. J. Pharmacol. 29:230-237, 1967.
64. Broadus, A. E., Kaminsky, N. I., Hardman, J. G., Sutherland, E. W., and Liddle, G. W.: Kinetic parameters and renal clearances of plasma adenosine-3',5'-monophosphate and guanosine-3',5'-monophosphate in man. J. Clin. Invest. 49:2222-2236, 1970.
65. Kaminsky, N. I., Broadus, A. E., Hardman, J. G., Jones, D. J. Jr., Ball, J. H., Sutherland, E. W., and Liddle, G. W.: Effects of parathyroid hormone on plasma and urinary adenosine-3',5'-monophosphate. J. Clin. Invest. 49:2387-2395, 1970.
66. Broadus, A. E., Kaminsky, N. I., Northcutt, R. C., Hardman, J. G., Sutherland, E. W., and Liddle, G. W.: Effects of glucagon on adenosine-3',5'-monophosphate and guanosine-3',5'-monophosphate in human plasma and urine. J. Clin. Invest. 49:2237-2245, 1970.
67. Hardman, J. G., Davis, J. W., and Sutherland, E. W.: Effects of some hormonal and other factors on the excretion of guanosine-3',5'-monophos-

phate and adenosine-3',5'-monophosphate in rat urine. J. Biol. Chem. 244: 6354-6362, 1969.
68. Wehmann, R. E., Blonde, L., and Steiner, A. L.: Sources of cyclic nucleotides in plasma. J. Clin. Invest. 53:173-179, 1974.
69. Ling, W. Y., LeMarie, W. J., Jones, G. L., Marsh, J. M., and Little, W. A.: Dynamics of cyclic AMP exchange in the amniotic fluid of the pregnant rhesus monkey. Gynecol. Invest. 6:33, 1975.
70. Ling, W. Y., LeMarie, W. J., Jones, G. L., Marsh, J. M., and Little, W. A.: Dynamics of adenosine-3',5'-monophosphate transfer among mother, fetus and amniotic fluid in the rhesus monkey. Am. J. Obstet. Gynecol. (in press).
71. Blonde, L., Wehmann, R. E., and Steiner, A. L.: Plasma clearance of ^{3}H-labeled cyclic AMP and ^{3}H-labeled cyclic GMP in the dog. J. Clin. Invest. 53:163-172, 1974.

4

Role of Cyclic Nucleotides in the Etiology and Therapy of Polyuric Disorders

MARSHAL P. FICHMAN

INTRODUCTION

The ability to elaborate a concentrated urine is fundamental to the maintenance of water balance and normal serum osmolality in mammalian species including man. The capacity to concentrate the urine is a function of the release of vaospressin (ADH) from the hypothalamic supraoptic nuclei and the peripheral action of ADH at the collecting duct of the kidney which facilitates the back diffusion of water from the tubular urine to the hypertonic medullary interstitial fluid (1, 2) (Fig. 1).

The release of ADH is stimulated by a rise in serum osmolality which leads to movement of water out of the osmoreceptor cells of the supraoptic nuclei producing a contraction of intracellular volume (3). There are also a variety of nonosmotic stimuli for the release of ADH including an isotonic contraction of the extracellular or intravascular volume stimulating intrathoracic volume

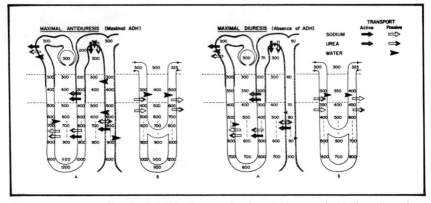

1 Renal concentrating mechanism during maximal diuresis and maximal antidiuresis. The numbers represent the osmolalities of tubular urine and intestinal fluid (A), and in the vasa rectae (B). (Reprinted from Kleeman, C. R. and Fichman, M. P.: New Eng. J. Med. 24:1303, 1967.)

receptors located near the carotid body and the left atrium (4, 5), higher cortical stimuli resulting from pain and fright, general anesthetics, narcotics (6–13), nicotine (14), barbiturates, acetyl choline (15), prostaglandins (16), and beta-adrenergic amines such as isoproterenol (17), and possibly the renin–angiotensin system (18, 19). The actions of catecholamines, prostaglandins, and nicotine are probably mediated by stimulation of the baro-receptors in the chest (16, 17) but may also involve a direct stimulation of the hypothalamus. The release of ADH is inhibited by a rise in intracellular volume secondary to extracellular fluid hypoosmolality (3), expansion of the extracellular or intravascular volume (5), and by the pharmacologic action of ethanol (20) and diphenylhydantoin (21), and alpha-adrenergic catecholamines such as norepinephrine (22, 23). There is no direct evidence that cyclic nucleotides are involved in the synthesis or release of ADH; however, adenosine 3′,5′-monophosphate (cyclic AMP) (24–29) and prostaglandins (30) may be involved in the releasing action on pituitary hormones of several hypothalamic releasing factors. It is also conceivable, therefore, that the action of catecholamines (31) and prostaglandins (32) on ADH release might be cyclic AMP mediated in the central nervous system, since these compounds interact with cyclic AMP in many organ systems throughout the body.

ADH is released in the form of neurovesicles, and attached to a carrier protein, neurophysine, and transported by a process of axonal streaming along the supraoptic-hypophyseal tract to the posterior pituitary (33). From there, the hormone is released into the circulation, and then exerts its peripheral action in the renal collecting duct in allowing the back diffusion of water into the hypertonic interstitium.

The capacity to concentrate the urine (3, 4) is dependent on the creation of hypertonic medullary interstitial fluid, which is the consequence of the active transport of chloride and sodium out of the relatively water impermeable segment of the ascending limb of the loop of Henle facilitated by the operation of the renal countercurrent multipler system (35–37) (Fig. 1). The effectiveness of ADH in producing maximal diffusion of water out of the collecting duct is modified not only by the magnitude of the gradient between the interstitium and the tubular lumen, but the delivery rate of filtrate from the proximal tubule to the distal nephron. With increased delivery rates, which occurs with an osmotic diuresis, such as with mannitol, there is limited time for maximal back diffusion of water out of the collecting duct because of an overwhelming of the capacity to extract water by ADH (34, 37). In this situation there is a tendency to excrete a large volume of isoosmotic or mildly hypotonic urine. With marked hydropenia, on the other hand, for example with volume depletion due to diuretics, there is avid proximal tubular reabsorption of salt and water with marked limitation of delivery of filtrate to the collecting duct. There may then be minimal back diffusion of water even in the absence of ADH, and in the instance of reduced delivery rates, this may be sufficient to concentrate the urine to isotonicity (1, 34, 36–39).

ROLE OF CYCLIC AMP IN THE ACTION OF ADH

ADH was one of the first peptide hormones for which the generation of cyclic AMP could be demonstrated to function as a molecular mediator of hormone action at its peripheral effector site (40). In the isolated toad bladder system, ADH applied to the serosal solution stimulates the transport of water (41, 42) in the presence of an osmotic gradient, sodium (43) (which can be measured by

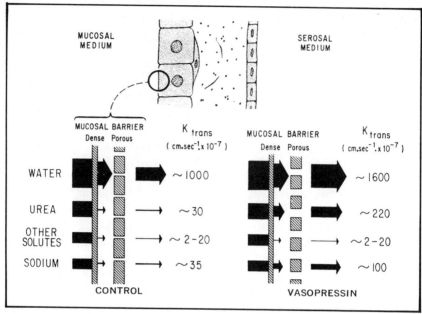

2 Action of vasopressin on a double diffusion barrier in the amphibian toad bladder. (Reprinted from Lichenstein, N. S., Leaf, A.: J. Clin. Invest. 44:1328, 1965.)

alterations in short circuit current [44]) and small molecules (45) such as urea (42, 45, 46) from the apical mucosal to the basal serosal side. The mechanism of the increase in bulk flow of water by ADH in part fits a pore hypothesis (47, 48) of enlargement or increase in number of intra- or intercellular pores rather than the stimulation of a simple increase in diffusion of water alone.* On the other hand, the effects of ADH on sodium and urea as well as water transport may be most consistent with the presence of a double diffusion barrier (42, 49, 50) (Fig. 2), the inner serosal barrier being a porous one in which the pores enlarge in response to ADH allowing greater bulk water flow. The outer mucosal barrier is a dense diffusion barrier that responds to ADH by increasing the transport of sodium and urea. The application of amphoteracin B

* More recent data, however, has demonstrated that ADH may increase water diffusion across the toad bladder as much as 20-fold, and that the effects of ADH on H_2O and urea transport may be dissociated (449). These latter findings suggest an important role of increase in water diffusion by ADH.

to the mucosal side has been demonstrated to destroy the barrier to sodium and urea transport but preserve the porous barrier, which retains its dependence on ADH for water transport (49, 50). Likewise the application of ADH to the serosal side of the isolated rabbit tubule has been shown to increase movement of water and sodium but not urea from the mucosal to the serosal side of the tubular cell (51). ADH does however increase urea transport in the rat papilla (52). The action of ADH on water, sodium, and urea transport in the toad bladder (53–56) and on water and sodium in the isolated rabbit tubule (51) is duplicated by the addition of cyclic AMP to the serosal media and also by the addition of methylanthines such as theophylline (53–56), the most potent known inhibitor of cyclic AMP phosphodiesterase (57). Addition of ADH or theophylline to the toad bladder is associated with increased levels of cyclic AMP in tissues analyses (58). Similarly, in the isolated hamster kidney, ADH increases levels of adenylate cyclase primarily in the papilla and renal medulla (Fig. 3) but not the renal cortex (59), and also increases the generation of cyclic AMP in the renal medulla but not the renal cortex of the rat (60, 61).* Infusion of ADH in dosages sufficient to produce an antidiuretic response has been shown by some investigators to increase urinary cyclic AMP (expressed as μM/gm creatinine) by 50 to 100 percent over levels during water diuresis in patients with hypothalamic diabetes insipidus or sustained water loaded normal subjects (62–65), but no rise in urinary cyclic AMP with ADH has been detected by others (66, 67). A modest increase in urinary cyclic AMP with ADH, when detected, is in striking contrast to the marked rise in urinary cyclic AMP with parathyroid hormone (62, 66, 68–70) due to its effect on the renal cortex (61, 71–75). Thus the action of ADH on water transport is duplicated by cyclic AMP and theophylline and associated with increased renal medullary tissue levels of adenylate cyclase and cyclic AMP, but relatively minor increases in urinary cyclic AMP.†

* A marked increase in adenylate cyclase activity has been observed with ADH when applied to both cortical and medullary portions of the collecting duct, but not the distal tubule, of single nephron segments of the rabbit kidney (450).

† Minor variations in urinary cyclic AMP may reflect poorly more marked increases in renal medullary production of cyclic AMP. Utilization of cyclic AMP clearances (446), along with accurate determination of glomerular filtration rate, would be a much more precise method of determining nephrogenous production of cyclic AMP as opposed to urinary cyclic AMP which may reflect filtered cyclic AMP.

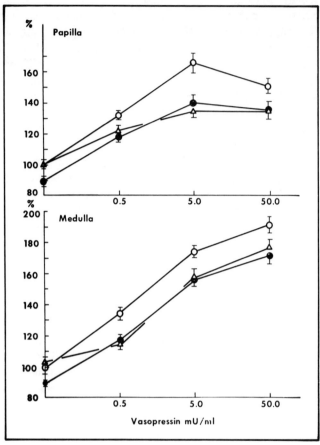

3 Inhibition of vasopressin induced generation of adenylate cyclase (open circles) by calcium (closed circles) and PGE_1 (open triangles) in the hamster renal medulla and papilla. (Reprinted from Marumo, F. and Edelman, I. S.: J. Clin. Invest. 50:1616, 1971.)

The mechanism by which cyclic AMP carries out the effect of ADH on water transport, may involve the stimulation of the production of a protein kinase (76) inducing a protein phosphorylation. Indeed, cyclic AMP has increased the activation of a phosphorylase in the toad bladder (77) and a cyclic AMP dependent kinase has been found in frog bladder epithelial cells (78).

IONIC MODIFIERS OF THE ACTION OF ADH

A variety of factors may alter the sensitivity of the toad bladder to the action of ADH and cyclic AMP (56, 78–80). The effects of pH of the serosal media in the toad bladder model differ with reference to the action of ADH and cyclic AMP with pH of 8.0 being optimum for ADH and a pH of 7 optimum for the action of cyclic AMP on water transport (80–84). The acidity of urine may also influence the renal excretion of cyclic AMP, which may be secreted as an organic acid (447).

Addition of calcium to the serosal media inhibits the stimulation of ADH production of adenylate cyclase in the medulla or papilla but not the cortex of the hamster (Fig. 3), as well as the action of PTH on the renal cortex (59). Calcium also has an inhibitory action on ADH or theophylline-induced water (85–86) and urea transport but not on sodium transport (86, 87), but is not inhibitory to the action of cyclic AMP on water, sodium, or urea transport in the toad bladder (86). Thus the inhibitory action of calcium on ADH induced water transport appears to be due to inhibition of the generation of cyclic AMP by ADH.

An increase in magnesium concentration of the serosal media has a similar effect to increased calcium of inhibiting the effect of ADH on water and urea transport but not on sodium transport (85, 87, 88). The dissociation of calcium and magnesium inhibitory action on water transport from their lack of inhibition of ADH induced sodium transport suggests that there may be two adenylate cyclase receptor sites in the renal medullary cells (or toad bladder epithelium): one for sodium transport and one for water and urea transport (56, 86) (Fig. 4). The possibility that a different nucleotide might be involved in ADH induced Na^+ transport has been considered. Guanosine 3′,5′-monophosphate (cyclic GMP) has also been shown in one study to stimulate Na^+ transport, but not water transport, across the toad bladder (89). There is no other evidence implicating this nucleotide in the action of ADH however.

In contrast to calcium, removal of potassium from the serosal media blocks the action of ADH on sodium transport, rather than water transport (85, 90). With prolonged removal of potassium, however, for greater than 60 min from the serosal media, there is an inhibitory effect on ADH and cyclic-AMP–induced water transport as well (91). Thus, potassium depletion is also inhibitory to

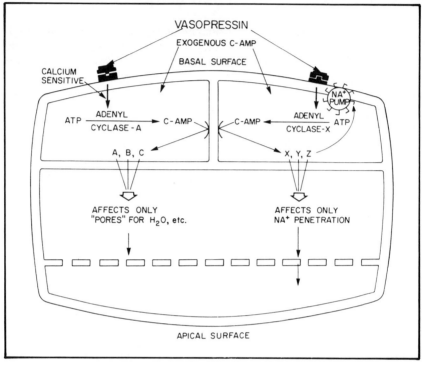

4 Model of dual adenyl cyclase receptor sites for vasopressin in the toad bladder, one for Na+ transport, and the other for H₂O and urea transport. (Reprinted from Orloff, J. and Handler, J.: Am. J. Med. 42:765, 1967.)

the action of cyclic AMP, suggesting an antagonistic action distal to cyclic AMP generation on water transport. The effects of high calcium and low potassium in the serosal media may be additive in their inhibitory effects on ADH-induced water transport with calcium producing an action proximal to the generation of cyclic AMP and potassium depletion acting distal to cyclic AMP generation (80). In potassium-deprived rats there is also marked inhibition of ADH-stimulated cyclic AMP production in renal medullary slices accompanied by both adenylate cyclase inhibition and phosphodiesterase stimulation (92).

Another ion which can modify the generation of cyclic AMP (93) and inhibit the antidiuretic action of ADH is lithium

(94). Lithium stimulates thirst in the rat independent of its effects on the renal concentrating mechanism (95, 96), which may involve inhibition of generation of cyclic AMP in the central nervous system or be due to stimulation of renin release by lithium (97, 98), which might activate the thirst mechanism (99, 100). Lithium also produces depletion of the neurosecretory material in the hypothalamus of the rat (101, 102), so it might, therefore, interfere with the release of antidiuretic hormone, resulting in a partially ADH responsive diabetes insipidus (103, 104). Lithium, however, much more commonly produces an ADH-resistant concentration defect, or "nephrogenic" diabetes insipidus (103, 104). The effect of lithium-inhibiting sodium transport from the ascending limb of the loop of Henle to the interstitial fluid of the renal medulla could decrease interstitial fluid hyperosmolality and reduce the gradient between the collecting duct urine and the interstitium, interfering with urinary concentration by that mechanism. Lithium does inhibit renal Na^+ transport in man (105), in rats (102), and across toad bladders (103, 106, 107). However, studies in dogs (108) and rats (104) treated with lithium failed to show an effect of lithium on the corticomedullary sodium gradient; also there is no impairment of free water clearance in man (103) or rats (109) treated with lithium. On the other hand, lithium probably blocks ascending limb chloride transport in the rat (110), which is inferred from its inhibition of the generation of negative free water clearance (TcH_2O). The primary action of lithium, however, in relation to water metabolism is its inhibitory effect on the generation of cyclic AMP and water transport by ADH, as demonstrated in toad bladders (103, 111, 112) and rats (102, 104, 112, 113). Lithium has no effect on cyclic AMP or dibutyryl cyclic-AMP–induced water transport in most studies (103, 111). However, in one study (112) 1 mM serosal lithium inhibited the water flow response to 1.1 mM cyclic AMP, and in lithium-induced polyuric rats, dibutyryl cyclic AMP was only partially effective (104) in reversing the polyuria in doses that were effective in reversing the polyuria of hereditary diabetes insipidus (Brattleboro) rats (39). Lithium also blocks the antidiuretic response to isoproterenol and inhibits cyclic AMP generation by isoproterenol in the rat renal medulla (114). Lithium has been shown to inhibit ADH-induced generation of cyclic AMP in dogs (115), rats (102) and rabbit (116–118) renal homogenates, but does not affect basal levels of cyclic AMP. Likewise, lithium impairs ADH stimulation of, but not basal levels of, adenylate cyclase

(119). Lithium had no effect on phosphodiesterase levels in renal homogenates (119) and produced no inhibitory effect on dibutyryl cyclic AMP induced water flow (103, 111) or theophylline induced sodium transport (106) in toad bladders. There was a slight inhibitory effect of lithium on cyclic AMP induction of protein kinase in human renal medullary tissue (119). These data suggest that while lithium's primary effect on water metabolism is inhibition of ADH generation of cyclic AMP, it may also have some inhibitory effect distal to cyclic AMP.

Fluoride ion has been shown to be a potent stimulator of adenylate cyclase in broken cell preparation (120) including renal cell adenylate cyclase (71, 121, 122). Indeed, fluoride is the most potent known stimulator of adenylate cyclase in broken cell preparations exceeding the action of parathyroid hormone by sixfold (71), but generally has shown no effect on intact cells (31, 123, 124).* Magnesium and manganese ions (125, 126) can be stimulators to the generation of adenylate cyclase by fluoride in broke cell preparation, while copper, zinc, and cobalt are inhibitors (125, 126).

HORMONAL MODIFIERS OF THE ACTION OF ADH

Prostaglandins

Prostaglandins have been shown to interact with numerous hormonal-induced cyclic-AMP–mediated reactions in a variety of tissues, sometimes stimulating and other times inhibiting cyclic-AMP–mediated reactions (32, 127). PGE_1 has been shown to inhibit the antidiuretic action of ADH, but not cyclic AMP, in toad bladders (128, 129) and isolated rabbit collecting tubules (130). These observations have led to the postulate that renal medullary prostaglandins may have a physiologic role as antagonists or modulators of ADH-induced antidiuresis by interacting with the generation of cyclic AMP (131). PGE_1 has been shown to inhibit the production of adenylate cyclase by ADH in the hamster renal medullas and papillas, but not in the renal cortex (59) (Fig. 3), and the production of cyclic AMP by ADH in rat renal medullas (60), and in theophylline-treated toad bladders (132). Two studies, however, failed to show an inhibitory effect on adenylate cyclase

---

* One recent report, however, describes an increase with fluoride in intracellular cyclic AMP in the epithelial cells of the toad bladder, but an antagonism to the effect of vasopressin on water permeability by fluoride (443).

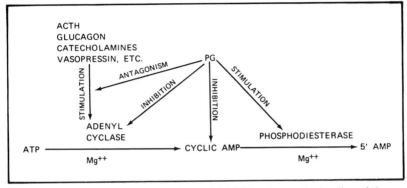

5 Possible mechanisms of inhibition by prostaglandins of the generation of cyclic AMP in adipose tissue and toad bladder. (Reprinted from Higgins, C. B. and Braunwald, E.: Am. J. Med. 53:97, 1972.)

production by ADH in untreated toad bladders (133, 134). Prostaglandins could interfere with the action of ADH by interfering with the attachment of ADH to the renal tubular adenylate cyclase receptor site by inhibiting adenylate cyclase generation, inhibiting the action of cyclic AMP, or stimulating phosphodiesterase (135) (Fig. 5). No inhibitory effect of prostaglandins on phosphodiesterase has been demonstrated and, indeed, PGE_1 inhibits theophylline-induced water flow in toad bladders (128, 129). Paradoxically, PGE_1 $10^{-7}M$ has a slight stimulatory effect on water transport in untreated isolated rabbit tubules while inhibiting the effect of ADH on water transport (130). PGE_1 $10^{-6}M$ also increased cyclic AMP in the outer renal rat medullas (60). In toad bladders, while PGE_1 $10^{-7}M$ to $10^{-8}M$-inhibited ADH or theophylline-induced water transport, with $10^{-5}M$ or $10^{-6}M$ PGE_1, water transport and cyclic AMP production was enhanced in the presence of aminophylline (136). The explanation for these discrepancies may lie in the ability of prostaglandins to stimulate an adenylate cyclase system not concerned with water transport, for example, concerned with sodium transport. At high concentrations of PGE_1, there may be a spill over of cyclic AMP into the water transport compartment, producing a facilitory effect. However, the basic action of prostaglandin E at lower concentrations is inhibition of cyclic AMP generation and water transport by ADH, perhaps by competing with ADH for its adenylate cyclase receptor site. PGA_1, PGA_2, PGE_1, and PGE_2 have been shown to have a natriuretic action when administered intravenously to animals (137–139) or man (140–144), either because of renal cortical vasodilatation (137, 138, 144, 145) or inhibition of proximal tubular reabsorption of sodium by a direct tubular effect that might be related to activation of cAMP (136, 143) or inhibition of Na^+-K^+–dependent ATPase (146). The resultant increase in delivery of filtrate from the proximal tubule to the collecting duct could modify the response to ADH by increasing free water clearance (137, 138, 141–143) independent of any action on cyclic AMP generation by ADH. Infusion of PGA_1 (less than 0.5 μgm/kgm/min) in man failed to show any inhibitory action of PGA_1 on the antidiuretic effect of endogenous or exogenous vasopressin (143). The failure to observe an inhibitory action of PGA_1 on ADH-induced antidiuresis may have been related to the potent natriuretic effect of prostaglandin A_1 increasing distal delivery and also the observation that pharmacologic doses of PGE_1 increase ADH release (16), probably by stimulating baro-receptors secondary to its vasodepressor effect (140–144).

Attempts to demonstrate a modulatory physiologic role of medullary prostaglandins to the action of ADH in man has involved the use of inhibitors of prostaglandin synthesis, in particular the nonsteroidal antiinflammatory drugs, such as indomethacin (147, 148). In steroid replaced hypophysectomized dogs both indomethacin and meclofenamate increase significantly the antidiuretic response to 100 mU of iv ADH (149). The antidiuretic response to 40 mU of ADH in man, administered by iv bolus, has also been amplified by treatment with indomethacin (150, 151). In one study, the increase in antidiuretic response to ADH in diabetes insipidus or sustained water-loaded normal subjects by indomethacin was associated with a reduction in urine sodium, but no greater increase in urinary cyclic AMP generation than with ADH alone (151). These data suggest that the inhibition by indomethacin of the antagonistic effect of PGE or PGA on ADH was related to blocking the inhibitory effect of prostaglandins on tubular reabsorption of sodium rather than on ADH-induced cyclic AMP generation, although minor changes in urinary cyclic AMP may not accurately reflect greater alterations in intracellular cyclic AMP in the renal medullary cell. In this same study, ADH was shown to produce a two- to fourfold rise in blood PGA measured by radioimmunoassay (152), which suggests that ADH increased the synthesis of prostaglandins, the modulator of its antidiuretic effect. Generation of prostaglandins by ADH might explain the phenomenon observed in toad bladders of a decrease in hydroosmotic responsiveness with repetitive applications of ADH, but not cyclic AMP, to the serosal media, suggesting that ADH induced the synthesis of an inhibitor of its effect on water transport (81, 153). A rise in PGE_2 has also been shown in vitro in rabbit renal medullary slices incubated with 100 mU/ml of ADH (154). It must be mentioned, however, that indomethacin is also a potent inhibitor of phosphodiesterase (155), so that its potentiation of ADH and cyclic AMP could be due to inhibition of phosphodiesterase rather than prostaglandin generation. However, indomethacin not only potentiates the effects of ADH and dibutyryl cyclic AMP on water transport, but also the effects of high doses of theophylline, which itself is a phosphodiesterase inhibitor (155). This latter action indicates that indomethacin is exerting an additional effect, most likely inhibition of PGE_2 reversal of ADH-induced cAMP generation, besides its action as a phosphodiesterase inhibitor.*

Catecholamines

The beta and alpha adrenergic nervous system has important effects on renal handling of water and sodium (156, 157) and the physiological and pharmacological effects of catecholamines in general involve the activation of cyclic nucleotides in various tissues (31). Infusion of norepinephrine inhibits the antidiuretic response to ADH infusion in diabetes insipidus rats (158) and in man (159). This inhibitory response is abolished by alpha adrenergic blockade with phentolamine (160). On the other hand, infusion of isoproterenol produces an antidiuretic effect in diabetes insipidus rats and this response is abolished by beta adrenergic blockade with propranolol (161). Isoproterenol and epinephrine have also been shown to increase cortical and medullary cyclic AMP in dog kidneys (162, 163), and this effect is blocked by beta adrenergic blockade by propranolol. However, alpha adrenergic stimulation with norepinephrine inhibits the generation of cyclic AMP by ADH (163). Carrying this concept further has been the demonstration that alpha

* However, the possible role of the potent prostaglandin endoperoxides (PGG_2 and PGH_2), precursors of PGE and PGF, as modulators of ADH, has not been determined and the degree of inhibition of PGG_2 and PGH_2 synthesis by indomethacin remains uncertain (444).

adrenergic agents inhibit adenylate cyclase and that beta adrenergic agents stimulate adenylate cyclase (164). Studies in isolated tubule preparations indicate a stimulatory effect of isoproterenol on cyclic AMP in the dog, but not in the rat tubule (163). Norepinephrine inhibited the generation of cyclic AMP by ADH in rat tubules, and this effect was abolished by phentolamine (163). The possibility that adrenergic catecholamines might stimulate PGE production as an intermediary of their effects on cyclic-AMP–generated water excretion or sodium excretion must also be considered (165–169).

On the other hand there is substantial evidence that the effects of catecholamines on water excretion may be due to actions other than interaction with cyclic-AMP generation by ADH in the collecting duct cell (156, 157). The antidiuretic response to isoproterenol in dogs (170) was abolished by hypophysectomy and while an antidiuretic response occurred following isoproterenol in water-loaded normal subjects, there was no antidiuretic response in patients with hypothalamic diabetes insipidus (23) (Fig. 6). Norepinephrine administration on the other hand, to ADH-infused dogs (171) and water deprived normal subjects (23) inhibited the antidiuretic effect, but had no effect on the antidiuretic response to ADH infusion in diabetes insipidus patients (23). These data are compatible with the primary effect of beta adrenergic stimuli increasing and alpha adrenergic stimuli decreasing ADH release from the hypothalamus, rather than modifying its peripheral action on the collecting duct. The alteration and release of ADH might be due to a baroreceptor stimulus by beta and inhibition by alpha adrenerigc agents or due to a direct hypothalamic effect of catecholamines on ADH release perhaps involving cyclic nucleotides. Denervation of the carotid baro-receptors in dogs, however, abolished the observed effects of alpha and beta adrenergic amines on water excretion (17, 172), whereas direct carotid infusion of catecholamines had no effect on water excretion (17, 172). These data were consistent with a baro-receptor effect of catecholamines, with ADH release probably responding thereby to changes in intraarterial pressure.

Also, isoproterenol, when infused in the renal artery of the dog, increased free water clearance, consistent with an inhibitory effect on proximal tubular sodium reabsorption increasing distal delivery to the diluting segment (170, 173). It appears likely that proximal tubular rejection of sodium may involve the activation of cyclic AMP in the renal cortex since not only isoproterenol but acetazolamide (174), parathyroid hormone (175), glucagon (176,

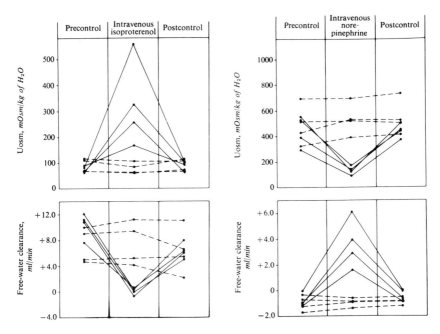

6 Absence in diabetes insipidus (broken lines) of isoproterenol antidiuresis (left) and norepinephrine diuresis (right) observed in normals (solid lines). (Reprinted from Berl, T., Harbottle, J. A., and Schrier, R. W.: Kidney Int. 6:250-251, 1974.)

177), thyrocalcitonin (178, 179), vasopressin (180, 181), prostaglandins (137–144), dibutyryl cyclic AMP (182), extracellular fluid volume expansion (183), and a natriuretic factor in uremic serum (184) have been associated with a natriuretic response and with the generation of cyclic AMP. While isoproterenol, which is natriuretic, increases renal cortical cyclic AMP in the rat (162), norepinephrine decreases cyclic AMP in the dog renal cortex (163) and appears to increase proximal tubular sodium reabsorption (185).

Infusion of cyclic GMP (186) has a similar effect to norepinephrine (171) in decreasing distal delivery consistent with increasing proximal tubular sodium reabsorption. This information brings up the attractive "ying-yang" hypothesis that the renal tubular reabsorption of sodium may be facilitated by a beta adrenergic-PGE-cyclic AMP mediated system and counterbalanced by an opposing effect, i.e., sodium reabsorption by an alpha adrenergic-PGF-cyclic GMP system.

In addition, beta adrenergic agents produce an increase in cardiac output (187), which is cyclic AMP mediated (188), and peripheral vasodilatation which may indirectly increase renal excretion of sodium by increasing renal blood flow (187). Alpha adrenergic agents, on the other hand, produce systemic and renal vascular vasoconstriction (187) which would be expected to decrease renal sodium excretion (185).

Clearly, the effects of alpha and beta adrenergic catecholamines on water excretion involve a complex interrelationship of their effects on ADH release, centrally or by influencing baro-receptor tone, cardiac output, systemic blood pressure, renal perfusion pressure, renal blood flow, glomerular filtration rate, proximal tubular sorium reabsorption and distal delivery of urine, and cyclic AMP generation by ADH in the collecting duct.

Adrenal Steroids

Aldosterone stimulates active sodium transport across the toad bladder epithelium as measured by short circuit current (189, 190). In addition, aldosterone and other adrenal steroids, including cortisol, appear to enhance the effect of vasopressin and cyclic AMP on the stimulation of both water and sodium transport in toad bladders (191–194). This permissive effect of adrenal steroids associated with an increase in intracellular cyclic AMP (195, 196) may be due to an inhibitory effect of both aldosterone (195, 196) and glucocorticoid (197) on phosphodiesterase.* On the other hand, glucocorticoids appear to have a permissive effect in animals and man on facilitation of free water excretion and urinary dilution (198) in addition to their effect maintaining normal glomerular filtration and renal blood flow (199–204). In the absence of glucocorticoids, as in hypopituitarism or adrenal insufficiency, their is impaired urinary dilution (203, 205, 206) and the ability to excrete a water load (207) and the development of hyponatremia (208). This permissive action is probably a renal effect, possibly related to stimulating sodium chloride transport out of the ascending limb

* In addition both mineralcorticoids and glucocorticoids appear to restore impaired binding of ADH to its receptor site in the renal medulla of the adrenalectomized rat to normal, while glucocorticoids (dexamethasone) facilitated the impaired adenylate cyclase-receptor coupling efficiency (445).

of the loop of Henle into the interstitium (206, 209, 210), which produces serial dilution of urine as it approaches the distal tubule, and also facilitates urinary concentration. The contention that the permissive effect of glucocorticoids is due to inhibition of the release of ADH is supported by bioassay data showing elevated levels of ADH in adrenal insufficiency (211, 212), radioimmunoassay data which shows that dexamethasone may suppress plasma ADH levels (213), and other data showing an elevated threshold for ADH release following administration of cortisol (214). The thesis that the effect of glucocorticoids facilitating water excretion depends only on suppression of ADH release is not tenable, however, in view of contrary bioassay data which show low or undetectable levels of ADH in hypopituitarism (215, 216), and the ameliorating effect of adrenalectomy on the polyuria of congenital diabetes insipidus rats (Brattleboro strain) which are incapable of synthesizing ADH (217), and the resumption of polyuria when cortisol is given. Similarly, withdrawal of cortisol therapy from patients with combined hypopituitarism and diabetes insipidus also produces a lessening of polyuria which is resumed when cortisal is readministered (208).

Thyroid Hormone

Thyroid hormone has complex effects on renal handling of Na^+ and water (218, 219). Thyroid hormone appears to have a primary action on activation of the Na^+-K^+–dependent ATPase system within the cell (220), and thyroid deficient rats appear to have reduced renal tissue levels of this enzyme (221) which parallels a reduced tubular Na^+ transport capacity. In addition, however, utilizing the isolated toad bladder, an effect of thyroxine on ADH-induced cyclic AMP–mediated water transport is also evident. L-thyroxine, $10^{-7}M$, in the serosal media has a stimulatory effect on both sodium and water transport (222). The effect on water transport is synergistic to that of ADH with a 40 percent greater effect than for ADH and thyroxine separately, while the effects on sodium transport are additive. Thyroid hormone also enhances lipolysis induced by lipolytic hormones which activate cAMP (223). The facilitory effect of thyroid on cAMP-mediated processes may be due to the action of triiodothyronine as a phosphodiesterase inhibitor (224) or as a stimulator of adenylate cyclase (225–227). The stimulatory effect of thyroid on Na^+-K^+–dependent ATPase could also be a cAMP mediated reaction (228).

Clinically however, deficiency of thyroid hormone is associated with impairment in water excretion in man (229–231) and the rat (232) and the development of hyponatremia (233–236). Hypothyroidism is associated with a decrease in cardiac output, possibly secondary to a decrease in thyroid stimulated cyclic AMP systems within the heart (237, 238), and a decrease in glomerular filtration rate and renal blood flow (229, 230, 238, 239). In general, clinical disorders of decreased cardiac output, including congestive heart failure, are characterized by avid proximal tubular reabsorption of sodium.

Consequent to avid tubular sodium reabsorption would be a decrease in delivery of urine to the distal cortical diluting segment where the generation of dilute urine takes place. It appears likely that impaired urinary dilution occurs on this basis in hypothyroidism rather than due to a specific action on ADH, since the use of osmotic diuretics, such as mannitol or proximal tubular acting diuretics like acetezolamide, correct the impaired water excretion in hypothyroidism by increasing distal delivery (229, 238). In addition, increased levels of circulating ADH measured by radioimmunoassay (240) have been reported in the thyroidectomized hypothyroid sheep. The elevated ADH levels may not represent an inappropriate ADH syndrome due to hypothyroidism (234–236), but an expected release of ADH in response to stimulation of volume receptors (4, 5) by the decreased cardiac output occurring in hypothyroidism.

Hyperthyroidism, on the other hand, is associated with increased activation of the cyclic AMP generating system by epinephrine, which is partly related to its beta adrenergic action, since its augmenting effect on blood and urinary cAMP is blocked, in part, by propranolol (241). Basal urinary levels of cAMP are also slightly greater than normal in hyperthyroidism, although some of the rise was attributable to a decrease in urinary creatinine excretion, which would increase the urinary cAMP/creatinine ratio (241, 242). The effect of thyroid hormone excess on water excretion would be a combination of the effects of increased cardiac output and increased glomerular filtration rate and renal plasma flow observed (243–245), and alterations in tubular sodium reabsorption, as well as any possible stimulatory effects on ADH generation of cyclic AMP (222–227). Hyperthyroidism is associated with polyuria (246–249) due to the increased cardiac output, glomerular filtration rate, renal blood flow, and possibly beta adrenergic catecholamine facilitation (241) leading to inhibition of proximal tubular reabsorption of Na^+, increasing delivery to the distal diluting segment. With fluid restriction or vasopressin administration, there is only a mild concentrating defect demonstrated (250), however, which may be due to osmotic diuresis, increased delivery from the proximal tubule (251) or washout of the renal medullary Na^+ content by increased renal blood flow (252). More severe concentrating defects accompany hyperthyroidism only when there is associated hypercalcemia or hypercalciuria (252, 253).

PHARMACOLOGIC MODIFIERS OF THE ACTION OF ADH

Pharmacologic agents may modify the ADH cyclic AMP generating system either by stimulating or inhibiting the release of ADH, or by stimulating or inhibiting cyclic AMP generation by altering adenylate cyclase or altering its destruction by phosphodiesterase or by acting distal to cyclic AMP generation, by inhibiting the production of a protein kinase (Table 1). Drugs may also affect the concentrating mechanism by producing changes in renal hemodynamics and glomerular filtration rate, altering delivery of urine to the distal diluting site and collecting duct where urinary concentration takes place, or by altering the transport of sodium chloride out of the ascending limb, which is needed to establish a hypertonic interstitial fluid which is necessary to provide a gradient for the back diffusion of water.

MODIFIERS OF THE ACTION OF ADH ON CYCLIC AMP GENERATION

Metabolic Inhibitors

Agents that have been shown in vitro to antagonize the effects of ADH on water transport in amphibious membranes, in addition to the aforementioned calcium, lithium, and PGE_1 included ouabain (91); metabolic inhibitors such as dinitrophenol, iodoacetate, fluoroacetic azide, which block both ADH and cyclic-AMP–induced water transport (254, 255); and thiols such as N-ethylmaleimide

Table 1. Agents Modifying the Release or Peripheral Action of ADH

Agent	Interferes with H$_2$O load excretion	Effect inhibited by ethanol	Effect on ADH release by bioassay or radioimmunoassay or indirect evidence	Effect on peripheral action of ADH	Effect on adenylate cyclase	Effect on phosphodiesterase
I A. PRIMARILY STIMULATES ADH RELEASE						
Isoproterenol	+	+	+	+	+	
Prostaglandin E			+	-	-	
Halothane						
Ether						
Cyclopropane	+	+	+			
Thiopental						
Nitrous oxide						
Morphine	+	+	+			
Barbiturates	+	+	+			
Nicotine	+	+	+			
Acetyl choline	+		+			
Carbamazepine	+	+	+			
Clofibrate	+	+	+			
Vincristine	+		+	-		
Cyclophosphamide	+	+		-		
I B. PRIMARILY FACILITATES PERIPHERAL ACTION OF ADH						
Chlorpropamide	+	+	+	+	+	-
Tolbutamide (large doses)				+	+	-
Phenformin				+		
Acetaminophen			+	+		-
Thiazides	+		+			-
Diazoxide	+					-
Indomethacin				+		-
Oxytocin (large doses)	+					
Thyroid				+		-(T$_3$)
Aldosterone						-
Theophylline						-
I C. PRIMARILY INHIBITS ADH RELEASE						
Ethanol			-			
Diphenylhydantoin			-	-		
Norepinephrine			-	? -	-	
Phenothiazines			? -			
Atropine			? -			
Cortisol			-	+		-

Effect distal to cAMP	Effect on proximal tubular reabsorption of Na^+	Effect on medullary hypertonicity	Therapeutic effect in primary diabetes insipidus	Therapeutic effect in nephrogenic diabetes insipidus	Produces hyponatremia	Produces polyuria	Effective therapy in SIADH
-		-					
-							
					+		
					+		
					+		
			+	-	+		
			+	-			
-					+		
					+		
+			+	-	+		
			-	-	+		
			+	?			
			+	-			
	+		+	+	+		
			+	+	+		
			+	+	+		
+	-						
		-					+
		-					
						+	
						+	+
	+					+	
		+					

Table 1. Agents Modifying the Release or Peripheral Action of ADH *Continued*

Agent	Interferes with H_2O load excretion	Effect inhibited by ethanol	Effect on ADH release by bioassay or radioimmunoassay or indirect evidence	Effect on peripheral action of ADH	Effect on adenylate cyclase	Effect on phosphodiesterase
I D. PRIMARILY INHIBITS PERIPHERAL ACTION OF ADH						
↑ Ca^{++}					-	-
↑ Mg^{++}					-	-
↓ K^+			? -		-	-
Lithium			? -		-	-
PGE			+		-	-
Phenacetin					-	-
Propoxyphene					-	-
Demethylchlortetracycline					-	
Methoxyfluorane					-	+(Flouride)
Ethacrynic acid			+		-	-
Furosemide			+		-	-
Oabain					-	-
Colchicine					-	
Amphotericin B					-	
Dinitrophenol						
Iodoacetic acid					-	
Fluoroacetic acid						
N-ethylmaleimide						
Cysteine					-	
Thioglycolate						
Glutathione						

(256), cysteine (257), and thioglycolate and glutathione (256), which antagonize ADH- and theophylline-induced water transport, but not cyclic-AMP–induced water transport.

Ethacrynic acid ($10^{-4}M$) inhibited the stimulatory effect of ADH on sodium transport (i.e., short circuit current) and water transport in toad bladders, but not the stimulatory effect of theophylline and cyclic AMP (258). Furosemide, on the other hand, inhibited the stimulatory effect of ADH on water transport, but not on short circuit current in toad bladders (259). The inhibitory effects of furosemide and ethacrynic acid on ADH-induced water

Effect distal to cAMP	Effect on proximal tubular reabsorption of Na$^+$	Effect on medullary hypertonicity	Therapeutic effect in primary diabetes insipidus	Therapeutic effect in nephrogenic diabetes insipidus	Produces hyponatremia	Produces polyuria	Effective therapy in SIADH
		−				+	
−					+	+	
−		? −				+	+
						+	
						+	
−					+	+	+
−						+	
	+	−	+	+	+	+	+
	−	−	+	+	+	+	+
−							
−							
−							

transport, as well as their natriuretic effect, may be, in part, explained by the increased production of PGE in the renal medulla, which may mediate the natriuretic and anti-ADH effects of these diuretics (260). The effects of furosemide as an antihypertensive and natriuretic agent and as a renal cortical vasodilator may be blocked by indomethacin (261). Alternatively, ethacrynic acid, furosemide, or prostaglandins may interfere with the binding of ADH to its receptor site on the collecting duct tubule epithelial cell surface.

On the other hand, thiazide diuretics, which are useful drugs in the treatment of primary and nephrogenic diabetes insipidus, have been shown in vitro to act as weak phosphodiesterase inhibitors (262, 263). This action could partially account for the hypercalcemic effects of thiazide diuretics by potentiating the action of PTH-generated cyclic AMP on kidney or bone (264) but is probably not important in terms of the antidiuretic action of thiazide diuretics. The antidiuretic effect of thiazides (265) and other diuretics appears to be related to sodium depletion secondary to natriuresis, resulting in increased proximal tubular reabsorption of sodium, leading to limitation of delivery of urine to the collecting duct, where even in the absence of ADH a minimal degree of back diffusion of water takes place sufficient to raise urine osmolality to isotonic levels, when delivery rate is low to the collecting duct (266, 267). Increased proximal tubular sodium reabsorption with decreased delivery of urine to the collecting duct may also be the basis for the antidiuretic effects of diazoxide (268, 269), a sodium-retaining, vasodepressor-thiazide–type drug, which is alos a phosphodiesterase inhibitor (262, 263, 270), and of indomethacin, in primary (151 and nephrogenic diabetes insipidus (269, 271).

Sulfonylurea Drugs

Sulfonylurea drugs, specifically chlorpropamide and tolbutamide, have an antidiuretic action in patients with primary diabetes insipidus (271–287). This was discovered serendipitously in a diabetes insipidus patient who took chlorpropamide, mistakenly thinking that he had diabetes mellitus. Nonetheless these was a reduction in his polyuria (273). Chlorpropamide is ineffective in nephrogenic diabetes insipidus (273, 277, 287), and in diabetes insipidus rats (288, 289) incapable of synthesizing ADH (39) and in water-loaded dogs (290). It has no direct stimulatory effect on water transport in toad bladders (291, 292). It potentiates the action of ADH on water transport in toad bladders (292), DI rats (288) and water-loaded dogs (290), and in patients with diabetes insipidus (293, 294). All these data suggest that the antidiuretic action of chlorpropamide is due to potentiation of the peripheral action of ADH on water transport, and requires at least a minimal amount of ADH present to exert its effect (292). In addition, chlorpropamide may stimulate ADH release analagous to its action thereby stimulating insulin release, since their is impaired water diuresis during oral water loading in patients receiving chlorpro-

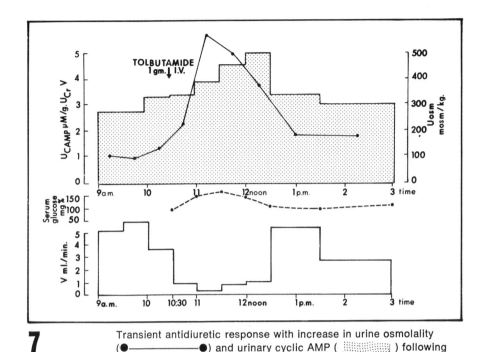

7 Transient antidiuretic response with increase in urine osmolality (●————●) and urinary cyclic AMP (▨) following intravenous tolbutamide and glucose loading in a patient with primary diabetes insipidus. (Reprinted from Fichman, M. and Brooker, G.: J. Clin. Endocrinol. Metab. 35:42, 1972.)

pamide (284, 293), which may be inhibited by ethanol (295). Occasionally, hyponatremia develops in patients with diabetes mellitus, or even diabetes insipidus, who are receiving chlorpropamide (296–301). Radioimmunoassay for ADH in the urine (302) showed an increase with chlorpropamide (293), whereas, in another laboratory, measurements of plasma arginine vasopressin measured by radioimmunoassay (303) did not show an increase with chlorpropamide (304). The increased sensitivity to ADH antidiuresis noted with chlorpropamide (293, 294) or tolbutamide in diabetes insipidus is associated with an increase in urinary cyclic AMP of the same order of magnitude as occurs with ADH (62) (Fig. 7). These findings could be attributed to sulfonylurea drugs increasing the binding of ADH to its adenylate cyclase receptor site, to a stimulatory effect on adenylate cyclase, or an inhibitory effect on phosphod-

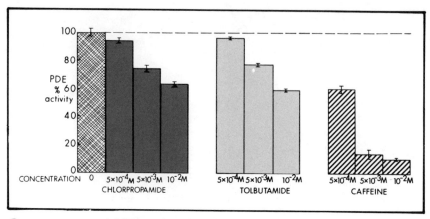

8 Inhibition of cyclic AMP phosphodiesterase in rat renal homogenates by chlorpropamide and tolbutamide in comparison to caffeine.

diesterase, or an effect distal to cyclic AMP (i.e., induction of a protein kinase, etc.). There is evidence supporting each of these possibilities. Chlorpropamide ($10^{-4}M$) enhances the water transporting effect of submaximal (291, 292) but not saturating doses of ADH in toad bladders, and actually increases the inhibitory effect of PGE_1 on ADH induced water transport (305). These data suggest that chlorpropamide facilitates binding of ADH or PGE_1 to their competed for receptor site. In addition, a stimulatory effect on chlorpropamide and tolbutamide ($10^{-5}M$) has been demonstrated on adenylate cyclase in rabbit myocardia associated with an increase in myocardial contractility (306). Chlorpropamide and tolbutamide ($10^{-4}M$) have also been shown to inhibit phosphodiesterase in rat renal medullary slices (307) (Fig. 8) and pancreatic tissue (308). Higher concentrations of chlorpropamide ($10^{-3}M$) also have a direct stimulatory effect on water transport in toad bladders in the absence of ADH or with saturating concentrations of ADH (305, 309, 310), which is consistent with a phosphodiesterase inhibitory effect. An effect of these drugs distal to cyclic AMP is also suggested by the stimulatory effect on cyclic-AMP–induced water transport when chlorpropamide is added to the mucosal media (309) but an inhibitory effect when added to the serosal media (291). These data support the hypothesis that chlorpropamide decreases cell membrane permeability to cyclic AMP

(310), which would lead to increased levels of intracellular cyclic AMP.

Tolbutamide is the only other sulfonylurea drug that shares the antidiuretic action of chlorpropamide (296, 311, 312), and its effect is observed only with very large doses orally or transiently following a bolus iv injection (62) (Fig. 7). The potentiating action of sulfonylurea drugs on cyclic-AMP–mediated reactions may explain not only their antidiuretic effect but their insulin-releasing action which is potentiated by caffeine (313), and the increased incidence of cardiovascular deaths cited in the UGDP study (314) with tolbutamide, which might be chronically facilitating the stimulatory action of adrenergic catecholamines on the myocardium (308).

Tolazamide, acteohexamide, and glyburide actually increase water excretion in water loaded normal subjects (315) or patients with primary diabetes insipidus (316). This effect is not due to ADH antagonism in man or diabetes insipidus rats (315, 317) but presumably is related to inhibition of proximal tubular reabsorption of sodium increasing delivery of urine distally to the diluting segment. This effect on inhibiting tubular reabsorption of sodium might again be cyclic-AMP mediated as postulated previously (183).

Phenformin, a biguanide, which is used in the treatment of diabetes mellitus, also appears to potentiate the action of ADH and, therefore, has an antidiuretic effect in mild primary diabetes insipidus (318, 319), but no antidiuretic action in water loaded normal subjects (272). An antidiuretic effect has also been reported in nephrogenic diabetes insipidus (320).

Analgesics

Phenylacetamide analgesics have a mild antidiuretic action in patients with primary diabetes insipidus (321). Acetaminophen increases water transport across the toad bladder in the absence of ADH, with submaximal and saturating levels of ADH (322, 323) and following addition of cyclic AMP (305), which is consistent with an inhibitory effect on phosphodiesterase, which has been demonstrated in vitro (305) in high concentrations. As with chlorpropamide, acetaminophen increases the PGE_1 inhibitory effect on ADH-induced water transport, which suggests also a facilitory effect on the adenylate cyclase receptor site for binding by PGE_1 or ADH by acetaminophen (305).

Phenacetin (322) and propoxyphene (324), on the other hand,

have inhibitory effects on ADH but not cyclic-AMP–induced water transport across toad bladders, which may partly explain the polyuria associated with phenacetin nephropathy (325) or with propoxyphene overdosage (326). In this same study, salicylates were without effect on water transport in toad bladders (322).

The antidiuretic actions of indomethacin as discussed previously may be related to inhibition of prostaglandin, synthesis, decreasing inhibition by prostaglandin of ADH generation of cyclic-AMP–mediated antidiuresis, or related to inhibition of prostaglandin-mediated proximal tubular rejection of sodium decreasing distal delivery of urine to the collecting duct (151, 271).

DRUGS ACTING DISTAL TO CYCLIC AMP

Drugs that may affect the action of ADH on water transport distal to cyclic AMP generation, which have already been mentioned, include, possibly, lithium, which may antagonize cyclic-AMP–induced water transport (119), and chlorpropamide, which, under certain conditions, may inhibit cyclic-AMP–mediated water transport as well (291). In addition, tetracyclines in vitro appear to antagonize the action of ADH and cyclic AMP on water transport, perhaps by interfering with the synthesis of a protein kinase (327). This in vitro effect has a clinical counterpart in the ADH-resistant polyuric syndrome, which may occur in patients treated with demethylchlortetracycline and is dose related (327–330).

As previously postulated, amphotericin B destroys the dense outer diffusion barrier on the mucosal surface of toad bladder epithelialcells, which leads to increased sodium and urea transport but does not alter the porous barrier which is concerned with water transport (49).

Methoxyfluorane anesthesia can produce diverse effects on renal function, including ADH-resistant polyuria (331) and acute renal failure (332). Renal biopsy specimens show extensive oxylate deposition in patients with renal failure on this basis. Renal functional impairment appears to correlate with the blood levels of fluoride and oxylate, which are metabolites of the anesthetic. In animals, inorganic fluoride can actually produce polyuria (331). Since fluoride is a potent stimulus of adenylate cyclase in broken-cell preparations, its inhibitory effect on ADH induced antidiuresis is therefore probably due to an inhibitory action on steps distal to the generation of cyclic AMP analogous to its inhibitory effects on synthesis of other proteins, for example, liver phosphorylase phosphatase (333).

Colchicine inhibits the action of ADH on water transport in toad bladders (334) by interacting with the assemblage of cytoplasmic microtubules which are probably involved in the action of ADH on water transport. Both colchicine and vinblastine also produce microtubular disruption and resistance to ADH without affecting phosphodiesterase or adenylate cyclase, in the rat renal medulla, which suggests an action distal to cyclic AMP (335).

DRUGS AFFECTING WATER EXCRETION PRIMARILY BY MODIFYING THE RELEASE OF ADH

Stimulators of ADH release

As cited previously, beta adrenergic amines such as isoproterenol (17) and prostaglandin-E_1 (16) stimulate ADH release probably by an action at the baro-receptor level, since these effects are abolished by hypophysectomy and cervical denervation (Table 1). Other agents that produce an antidiuretic response, which might involve an interplay of direct hypothalamic stimulation of ADH release and stimulation of baro-receptors or volume receptors in response to decreased effective blood volume, and changes in renal hemodynamics producing decreased glomerular filtration rate and renal plasma flow, include general anesthetic agents, morphine and narcotics, nicotine, and barbiturates (6–13). Ether, cyclopropane, thiopentol, and nitrous oxide are all associated with sodium and water retention in man, and the latter three have been shown to stimulate short circuit current, i.e., sodium transport across toad bladders (336). Halothane, on the other hand, inhibits sodium transport across the toad bladder and etheyl ether produces a diaphasic response, initially stimulatory and then inhibitory (336).*
The antidiuretic response of halothane is blocked by ethanol (6) which is consistent with a primary effect on ADH release. The antidiuretic response to morphine in dogs is abolished by hypophysectomy (8) and may occur with local installation of morphine into the supraoptic nucleus of dogs (8).

Carbamazepine, an anticonvulsant drug, used in this country as treatment for trigeminal neuralgia has a potent antidiuretic action

* Methohexital, methoxyflurane and halothane have also been shown to inhibit vasopressin induced water transport across toad bladders (448).

(337–339) which is similar to that of chlorpropamide, and indeed additive to chlorpropamide in the treatment of diabetes insipidus (340, 341), and has produced hyponatremia (342). Carbamazepine increases plasma levels of ADH (343, 344) in patients with partial hypothalamic diabetes insipidus, is ineffective in the diabetes insipidus rat (345, 346), does not stimulate water transport across toad bladders (346), does not augment the antidiuretic effect of submaximal doses of ADH in animals (345–347), is ineffective in nephrogenic diabetes insipidus (345, 348), and has it antidiuretic effect inhibited by ethanol, an inhibitor of ADH release (349).

Clofibrate likewise is effective in the treatment of mild cases of primary diabetes insipidus (350–352), increases urinary ADH levels measured by radioimmunoassay (353), inhibits the excretion of a water load in normal subjects (353) with its antidiuretic effect blocked by ethanol (353), is ineffective in the diabetes insipidus rat (353) and in nephrogenic diabetes insipidus (349), and does not potentiate the antidiuretic action of ADH in diabetes insipidus rats (353) or man (354). Clofibrate has no effect on basal or ADH-induced water transport in the toad bladder (79). Thus both carbamazepine and clofibrate appear to act as stimuli for ADH release rather than augmenting its peripheral effect on the collecting duct.

An antidiuretic effect and hyponatremia have been reported in association with both vincristine (355–361) and cyclophosphamide (362, 363), drugs used to treat neoplastic diseases, especially lymphomas and leukemia. Because vincristine is a neurotoxin, its ADH-like effect may be considered a neurologic cause of the syndrome of inappropriate ADH secretion (364, 365), and rat neurohypophyes incubated with vincristine have shown microscopic changes, including precipitation of microtubules (366). Elevated plasma levels of ADH have been reported (367) with vincristine-induced hyponatremia. Intravenous cyclophosphamide may produce an antidiuretic action 4 to 12 hr after administration, with the effects lasting 24 hr, and correlated with the excretion of alkalating metabolites of the drug (363). This effect may be inhibited by ethanol (362). Conversely, in rats, cyclophosphamide appears to inhibit the antidiuretic response to endogenous or exogenous ADH (368). It would seem more likely, therefore, that, in man, an antidiuretic effect due to cyclophosphamide would result from increased ADH release.

Inhibitors of ADH release

The list of pharmacologic agents known to inhibit ADH release includes water, ethanol (20, 295), intravenous diphenolhydantoin

(21), alpha adrenergic agents, such as epinephrine and norepinephrine, probably by inhibiting baro-receptor stimulation of ADH release (22, 23), and possibly chlorpromazine (8).

Parenteral diphenylhydantoin will improve the urinary dilutional response to acute water loading in patients with inappropriate ADH syndrome of nontumorous etiology (22), and inhibit the antidiuretic response to water deprivation, but not exogenous vasopressin (22) in normal subjects. The release of ADH measured in the plasma by radioimmunoassay is inhibited by diphenylhydantoin (personal communication, R. Skowsky), which is analogous to the inhibitory effect of diphenyldantoin on insulin release (369). In addition, diphenylhydantoin inhibits the effect of ADH on water transport in the toad bladder (22).

In man, chlorpromazine may increase urinary dilution and free water clearance without altering glomerular filtration rate or osmolar clearance. Bioassay of ADH in the urine of the rat showed decreased levels following chlorpromazine (8).

RENAL RESPONSE TO CYCLIC AMP

The renal response to intravenously administered cyclic AMP or dibutyryl cyclic AMP, in general, parallels the in vitro renal tissue effects of cyclic AMP, including mediation of the natriuretic and phosphaturic effects of parathyroid hormone, the antidiuretic effects of ADH, and the possible cyclic-AMP–mediated effects on proximal tubular sodium transport. Intravenous cyclic AMP in patients with primary diabetes insipidus produced a response similar to ADH, with a reduction in urine volume, an increase in urine osmolality, and a decrease in free water clearance (370). Dibutyryl cyclic AMP, on the other hand, produced an increase in the clearance of phosphate and an increase in free water clearance, which suggested that it was inhibiting proximal tubular reabsorption of sodium (possibly PTH mediated), as well as phosphate, thereby increasing delivery of urine to the distal diluting sites (370). When injected directly into the renal artery of the dog, both cyclic AMP and dibutyryl cyclic AMP increased urine volume and free water clearance proportionately without changing total urinary sodium (182). These effects again suggest inhibition of proximal tubular reabsorption of sodium increasing delivery of filtrate to the distal tubule, where compensatory increased distal tubular reabsorption of sodium takes place. This effect on increasing distal delivery would presumably override a facilatory effect on ADH-induced antidiuresis. Micropuncture studies confirm an inhibitory effect on proximal tubular sodium and phosphate reabsorption by dibutyryl cyclic AMP (175). Cyclic GMP, on the other hand, injected into the renal artery of the dog increased Na^+ reabsorption (186).

POLYURIC SYNDROMES

Polyuria may result from increased fluid intake, a renal concentrating defect secondary to deficient ADH release, or congenital or acquired failure of the kidney to respond to ADH, or as a result of the action of osmotic or other diuretic agents (Table 2).

Primary Polydipsia

Primary polydipsia may be of "psychogenic" origin or may possibly be due to an organic lesion of the thirst center. There is suggestive evidence that lithium may be stimulatory to thirst (95, 96) and that chlorpropamide might stimulate thirst in patients with hypodipsia (371). The role of cyclic AMP in their pharmacologic action is conjectural, since lithium is generally inhibitory to cyclic AMP, whereas chlorpropamide appears to stimulate cyclic AMP by either stimulating adenylate cyclase or inhibiting phosphodiesterase.

Hypothalamic Diabetes Insipidus

Primary hypothalamic diabetes insipidus results from destructive lesions of the hypothalamus including head trauma, posthypophysectomy, skull fractures, craniopharyngioma, metastatic tumors (especially breast cancer), histiocytosis X, tuberculosis, sarcoid, lues, sickle cell disease, aneurysms, encephalitis, and, in 50 percent of cases, idiopathic etiology (1, 2). Polyuric syndromes due to diabetes insipidus respond to the use of ADH and other antidiuretic drugs.

Nephrogenic Diabetes Insipidus

Congenital Nephrogenic Diabetes Insipidus

"Nephrogenic" diabetes insipidus may be congenital or acquired. Congenital nephrogenic diabetes insipidus is a familial disorder in which the kidney is incapable of responding to vasopressin

(372). The failure of the kidney to respond to the antidiuretic action of ADH suggested a parallelisum with the disorder, pseudohypoparathyroidism, in which the kidney not only fails to develop a normal phosphaturic response to parathyroid hormone, but also fails to augment urinary cyclic AMP in response to parathyroid hormone (66). Consequently, the responsiveness of urinary cyclic AMP to infusions of ADH (10 mU/min $\times$ 4 hr) was determined in sustained water-loaded normal subjects, in patients with hypothalamic diabetes insipidus, and in two patients with nephrogenic diabetes insipidus (62) (Fig. 9). The antidiuretic response to ADH in normal subjects and primary diabetes insipidus was accompanied by a mean of a 70 percent increase in urinary cyclic AMP (μM/gm creatinine) over the values during water diuresis. By contrast, in the two patients with nephrogenic diabetes insipidus, there was neither a rise in urinary osmolality or any increase in urinary cyclic AMP following ADH infusion. Furthermore, while there was a 60 percent rise in urinary cyclic AMP accompanying the antidiuretic response to 1 gm of tolbutamide intravenously in a glucose-loaded primary diabetes insipidus patient (Fig. 7), there was neither a rise in urinary osmolality or urinary cyclic AMP in the patient with nephrogenic diabetes insipidus. It was also of interest that a bolus iv injection of 200 Un of parathyroid hormone produced only a 10-fold and 30-fold increase in urinary cyclic AMP in the following hour in the two patients with nephrogenic diabetes insipidus compared to a 71- to 121-fold increase in urinary cyclic AMP in six normal subjects following parathyroid hormone. These findings were consistent with the hypothesis that nephrogenic diabetes insipidus resulted from an inability of the renal medullary cells to respond to ADH by generating cyclic AMP, but also suggested that there might be a partial defect in the ability of the renal cortical cells to respond to PTH, though not impaired to the degree of the failure to increase urinary cyclic AMP by PTH in pseudohypoparathyroidism (66).

The inability of urinary cyclic AMP to be augmented by ADH in nephrogenic diabetes insipidus was confirmed by Bell and co-workers (373). They also observed the failure of urinary osmolality to increase or free water clearance decrease following infusions of cyclic AMP in nephrogenic diabetes insipidus, whereas an antidiuretic response was observed in primary diabetes insipidus. These studies suggest that the basic abnormality in nephrogenic diabetes insipidus may result from an inability of ADH to attach to the

Table 2. Polyuric Disorders

	Urine Volume	Urine Osmolality	Response to Vasopressin	Role of Impaired Cyclic AMP	
				Generation	Action
I. Primary Polydipsia	↑↑	↓↓	+		
II. Primary Diabetes Insipidus	↑↑	↓↓	+	+	
A. Head trauma, skull fractures					
B. Pituitary surgery					
C. Craniopharyngioma, cerebral tumors					
D. Metastatic tumors, e.g., breast					
E. Granulomas:					
1. Sarcoid					
2. Tuberculosis					
3. Lues					
F. Histiocytosis X					
G. Sickle cell disease					
H. Encephalitis					
I. Meningitis					
J. Vascular, aneurysms					
K. Idiopathic - 50%					
III. "Nephrogenic" Diabetes Insipidus	↑↑	↓↓	-		
A. Congenital nephrogenic DI				+	+
B. Hypercalcemia				+	+
C. Hypokalemia				+	?+
D. Lithium intoxication				?+	+
E. Phenacetin nephropathy				+	
F. Demethylchlortetracycline					?+
G. Propoxyphene intoxication					
H. Methoxyfluorane					

I. Uric acid nephropathy
J. Medullary cystic disease
K. Post-relief of urinary tract obstruction
L. Post-successful renal transplant
M. Diuretic phase of acute tubular necrosis
N. Hypergammaglobulinemic disorders:
   1. Multiple myeloma
   2. Sarcoid
   3. Sjogren's syndrome
   4. Amyloidosis
O. Sickle cell anemia

IV. Osmotic Diuresis
   A. Mannitol
   B. Urea
      1. Chronic renal failure
      2. Diuretic phase of acute tubular necrosis
      3. Relief of urinary tract obstruction
      4. Successful renal transplant
      5. High protein tube feeding
      6. Intravenous hyperalimentation
   C. Glucose
      1. Diabetes mellitus out of control
      2. Non-ketotic hyperglycemic hyperosmolar coma
   D. Sodium
      1. Diuretics
      2. Excessive administration of Na^+ containing intravenous fluids

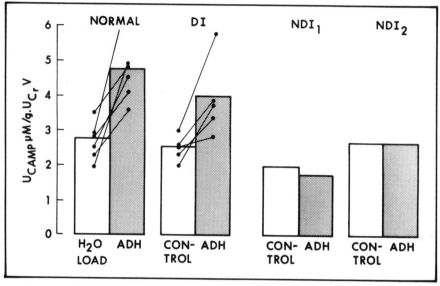

9 Failure to augment urinary cyclic AMP with ADH infusion (10 mu/min × 4 hr) in two patients with nephrogenic diabetes insipidus, in comparison to increase in urinary cyclic AMP with ADH in primary diabetes insipidus and H_2O-loaded normal subjects. Height of bars represents mean values. (Reprinted from Fichman, M. and Brooker, G.: J. Clin. Endocrinol. Metab. 35:39, 1972.)

adenylate cyclase receptor site, an inability to stimulate adenylate cyclase, a deficiency of adenylate cyclase, or the presence of some substance in the renal medulla that inhibited adenylate cyclase or stimulated phosphodiesterase. In addition, the possibility exists that there may be a defect distal to cyclic AMP. In hereditary "nephrogenic" diabetes insipidus rats, there is also impaired stimulation of renal medullary adenylate cyclase by ADH (374).

In view of the antagonistic effect of prostaglandins (PGE_1) on the generation of adenylate cyclase and cyclic AMP in vitro and the inhibition of the antidiuretic response to ADH in toad bladders and rabbit tubule, it is attractive to postulate that excessive renal medullary prostaglandins might be etiologic in the pathogenesis of nephrogenic diabetes insipidus. The potentiation of the antidiuretic response to diuretics in nephrogenic diabetes insipidus by indomethacin (271), a potent inhibitor of prostaglandin synthesis

(147), suports this hypothesis. However, neither an increase in urine osmolality nor an increase in urinary cyclic AMP was observed in response to ADH infusions in indomethacin-treated patients with nephrogenic diabetes insipidus. It is possible that the failure to increase urinary cyclic AMP did not reflect increases in renal medullary cell intracellular cyclic AMP with ADH during indomethacin treatment. On the other hand, the therapeutic response to indomethacin might be due to blocking prostaglandin- (PGE- or PGA-) induced tubular rejection of sodium leading to increased proximal tubular reabsorption of sodium with decreased delivery of urine to the collecting ducts. This effect on the tubular reabsorption of sodium of prostaglandins and other humoral agents, such as beta adrenergic catecholamines, parathyroid hormone, ADH, thyrocalcitonin, and glucagon, are possibly also mediated by cyclic AMP. Indomethacin itself, of course, is a phosphodiesterase inhibitor (155), so that it could increase sensitivity of the generation of intracellular cyclic AMP by ADH by a mechanism independent of its effect as an inhibitor of prostaglandin synthesis.

Acquired Nephrogenic Diabetes Insipidus

The known causes of acquired ADH resistant nephrogenic diabetes insipidus, include hypokalemia; hypercalcemia and hypercalcuia; medullary cystic disease; hypergammglobulinemic states, including multiple myeloma, amyloidosis, Sjogren's syndrome, analgesic nephropathy, sickle cell anemia, lithium intoxication, methoxyfluorane anesthesia, demethylchlortetracycline, propoxyphene intoxication, and the diuresis that occurs following relief of urinary tract obstruction; the diuretic phase of acute tubular necrosis; and immediately following a successful renal transplant (1, 2, 375, 376).

Hypercalcemic Nephropathy

Hypercalcemic nephropathy may result from any of the known causese of hypercalcemia, including primary hyperparathyroidism, ectopic hyperparathyroidism due to nonparathyroid neoplasms (377), tumors metastatic to bone such as breast cancer (378), prostaglandin-secreting tumors (379), multiple myeloma (380), lymphoma (381), leukemia (382), sarcoidosis (383), vitamin D intoxication (384), milk alkalae syndrome (385), hyperthyroidism (386), and tertiary hyperparathyroidism, most commonly seen in patients with chronic renal failure following a successful renal transplant

(387). The renal manifestations of hyerpcalcemia include nephrolithiasis and nephrocalcinosis, hypertension, renal failure and, most characteristically, ADH-resistant polyuria or acquired nephrogenic diabetes insipidus (388). The concentrating defect is reversible if the hypercalcemia is corrected. The most proable explanation for the concentrating defect of hypercalcemia, is the inhibitory effect of calcium on the generation of adenylate cyclase and cyclic AMP and the antidiuretic action of ADH in the collecting duct, as demonstrated in the analogous preparations of toad bladders, rabbit tubules and renal medulla papillas in rats (59, 85, 86, 389). Calcium may also inhibit the action of Na^+-K^+–dependent ATPase (390, 391) which may be involved in the transport of sodium and chloride out of the water-impermeable segment of the ascending limb of the loop of Henle into the interstitial fluid, which is necessary for the creation of hypertonic interstitial fluid, which provides the gradient for the back diffusion of water out of the collecting duct (392). Thus hypercalcemia may inhibit both the creation of hypertonic interstitial fluid and the action of ADH on the collecting duct allowing the back diffusion of water into the hypertonic interstitium.

Hypokalemic Nepropathy

Hypokalemic nephropathy may result from any of the causes of prolonged potassium depletion (393) including (1) gastrointestinal losses of potassium due to vomiting, nasogastric suction, diarrheal states (394), chronic laxative ingestion (395), fistulae, ureterosigmoidostomy (396), pancreatic adenomas (397), villious adenomas (398), cholera (399), and the carcinoid syndrome; (2) primary hyperaldosteronism (400) due to an adrenal adenoma or nodular hyperplasia (401). The polyuria associated with primary hyperaldosteronism and other mineralocorticoid excess syndrome is due, in part, to hypokalemic nephropathy, but it is also related to the volume expansion due to sodium retention leading to inhibition of proximal tubular reabsorption of sodium, increasing delivery of urine to the diluting sites of the nephron facilitating free water clearance (198); (3) secondary aldosteronism due to renal artery stenosis (402), malignant hypertension (403), renin secreting tumors (404), edematous states, especially cirrhosis, and ascites (405), and oral contraceptives (406); (4) other mineralocorticoid excess syndromes due to 17-hydroxylase deficiency (407): Cushing's syndrome due to extraadrenal ACTH-producing tumors (408); adrenal hyperplasia, adenoma, or carcinoma; and exogenous steroids, chronic licorice ingestion (409), and Bartter's syndrome (410). (5) Other renal K wasting states, most commonly secondary to diuretic therapy: metabolic alkalosis, proximal and distal renal tubular acidosis (411), amphotericin B, carbenacillin (412), magnesium depletion (413), myleomonocytic leukemia with lysozymuria (414), Liddle's syndrome (415), diuretic phase of acute tubular necrosis or following successful renal transplantation or following relief of urinary tract obstruction, osmotic diuresis, and renal tumors (416).

With prolonged potassium depletion there are characteristic vacuolar changes in the renal tubules (417) and a renal concentration defect in which the maximum ability to concentrate the urine is the elaboration usually of isotonic or mildly hypotonic urine (418, 419). The concentrating defect is reversible within a few weeks of potassium repletion. The interference of potassium depletion with normal renal concentrating ability may be due to impedance of the normal delivery of sodium chloride to the ascending limb of the loop of Henle, or interference with the transport of sodium chloride to the interstitial fluid, and thereby inhibition of the creation of a hypertonic medullary interstitial fluid, which normally provides the gradient for the back diffusion of water out of the collecting duct under the influence of ADH (395, 418–421). Eliminating potassium from the serosal media of toad bladders, however, interferes with the action of ADH on sodium transport, which may represent a parallel inhibitory process (85, 90). On the other hand, prolonged elimination of potassium will also inhibit the effect of ADH and cyclic AMP on water transport (91). Therefore, the effect of potassium depletion in producing polyuria may not only be due to interference with the establishment of hypertonic interstitial fluid but also be related to an interference with a back diffusion of water out of the collecting duct by ADH by inhibiting the mechanisms distal to cyclic AMP. In vitro studies of potassium depletion in the rat have demonstrated an inhibition of adenylate cyclase and a stimulation of phosphodiesterase (92).

Lithium Intoxication

Lithium carbonate is a drug used in the treatment of manic disorders. With the attainment of toxic blood levels, polyuria characteristically ensues, produced by several possible mechanisms but probably primarily by inhibition of ADH production of cyclic AMP (94). The possible mechanism by which lithium might produce polyuria as previously cited include (1) stimulation of thirst by a central action (95, 96), (2) inhibition of the release of ADH (101–104), (3) blockage of sodium chloride transport from the ascending limb to the interstitial fluid (103, 105–107), (4) inhibition of the generation of cyclic AMP and adenylate cyclase by ADH in the collecting duct (102–104, 111–113, 119), (5) blackade of beta adrenergic catecholamine stimulation of cyclic AMP in the renal tubule (114), (6) inhibition of the action of cyclic AMP on water transport by inhibiting the generation of a protein kinase (119). The antagonism of lithium to the antidiuretic action of ADH has been the basis of the therapeutic use of this drug in the syndrome of inappropriate antidiuretic hormone secretion (422).

Sickle Cell Anemia

Sickle cell anemia and sickle cell trait may produce a polyuria rarely as a cause of primary diabetes insipidus secondary to infarctions in the area of the hypothalamus, but most commonly as a cause of a transient ADH resistant concentrating defect. It is theorized that sickling is facilitated by the hypertonic fluid in the renal medulla (423) and that sludging of the sickle cells leads to impaired blood supply and interference with oxidative metabolism in the ascending limb which provides energy for the active transport of chloride and sodium into the interstitial fluid, leading to a decrease in medullary interstitial fluid hypertonicity (376). An in vitro assay of cyclic AMP in the renal medulla of a sickle cell patient failed to reveal any deficiency of cyclic AMP (424).

Drug-induced Polyuric Syndromes

Phenacetin nephritis may be associated with polyuria and renal acidification defects, as well as the development of papillary necrosis and chronic renal failure (325). Phenacetin inhibits the action of ADH but not cyclic AMP on water transport in toad bladders (322), which may account, in part, for the polyuria that occurs in the clinical syndrome. A fatal case of propoxyphene overdosage also had massive polyuria (326), and this drug also inhibits ADH-induced water transport in toad bladders (324).

A polyuric syndrome has been described in association with

the use of demethylchlortetracycline (327–330). The tetracyclines, in general, are antianabolic agents which intrefere with protein synthesis. Tetracyclines may interfere with both ADH- and cyclic AMP induced water transport, perhaps by interferring with the generation of cyclic AMP and with the synthesis of a protein kinase distal to cyclic AMP (425). The antagonism of ADH antidiuresis by demethylchlortetracycline has been utilized in the successful treatment of two patients with the chronic syndrome of inappropriate ADH secretion (426, 427).

Methoxyfluorane anesthesia may produce not only acute renal failure associated with deposition of oxylate crystals in the kidney, but also a polyuric syndrome, which appears to be correlated with high levels of inorganic fluoride (331). Since fluoride is a potent stimulator of adenylate cyclase, its inhibitory effect on the antidiuretic action of ADH is probably distal to cyclic AMP generation (443) and probably involves inhibition of the synthesis of a protein kinase (333).*

Other Causes of Acquired Nephrogenic Diabetes Insipidus

The other causes of ADH resistant polyuria (428) listed in Table 2 including hypergammaglobulinemic states, amyloidosis, Sjogren's syndrome (429), uric acid nephropathy, and medullary cystic disease (376) are probably related to structural changes in the renal medullary cells concerned with the urinary concentration process. The diuresis following relief of urinary tract obstruction is, in part, the result of an osmotic diuresis, proximal tubular injury with increased delivery of sodium chloride and is also partly due to ADH resistance which may be on a structural basis (430).

Osmotic Diuresis

The polyuria which accompanies osmotic diuresis from mannitol, glucose, or urea is characterized by the excretion of moderately large volumes of urine that tend to be isotonic or mildly hypotonic. With the introduction by glomerular filtration into the proximal tubule of nondiffusable, e.g., mannitol, or partially diffusable, e.g., glucose or urea molecules, there is osmotically induced

---

* Methoxyfluorane also has been shown to inhibit ADH induced water transport across the toad bladder (448).

back diffusion of water and increased delivery of urine which overwhelms the capacity of the ascending limb to provide sodium chloride transport into the interstitium, and the capacity of the collecting duct to allow back diffusion of water under the influence of ADH. The final urine is, therefore, in general, like an ultrafilitrate of plasma, neither concentrated nor diluted (431). Mannitol, in addition, increases renal medullary blood flow which tends to wash out the hypertonicity of the renal medulla and thereby also interferes with urinary concentration (431).

Polyuria Due to Natriuretic Effect of Diuretics

Other diuretic agents, e.g., thiazides, mercurials, ethacrynic acid, furosemide, and acetazolamide produce polyuria by acting primarily to increase sodium chloride or bicarbonate excretion. Some of these effects could be related to a postulated stimulatory effect on cyclic AMP tubular rejection of sodium. Indeed, acetazolamide has been shown to act as a stimulator of cyclic AMP, and dibutyryl cyclic AMP can duplicate its effect as an inhibitor of carbonic anhydrase (179). Furosemide (432) and ethacrynic acid (433) act at the water-impermeable segment of the ascending limb of the loop of Henle, and therefore interfere both with serial dilution of the urine as it approaches the cortical segment of the ascending limb and interfere with concentration of the urine by blocking the establishment of hypertonic interstitial fluid. These drugs may act by stimulating the production of renal medullary prostaglandins (PGA_2 or PGE_2) which may then produce natriuresis by inhibiting tubular sodium transport perhaps by means of stimulating cyclic AMP induced rejection of sodium or by producing renal cortical vasodilatation (261). In addition, both ethacrynic acid (258) and furosemide (259) inhibit the effect of ADH but not cyclic AMP or theophylline on water transport in the toad bladder, so these drugs may have a direct effect on water as well as sodium excretion.

The "diuretic" effect of theophylline, the prime inhibitor of the destruction of ADH generated cyclic AMP by phosphodiesterase (57), can best be explained by its renal action stimulating sodium excretion (434) independent of its ADH facilitory action on water transport which would be expected. It appears likely, in view of the natriuretic effects of dibutyryl cyclic AMP (182), prostaglandin (137–144), isoproterenol (170, 173), parathyroid hormone (175), thyrocalcitonin (178, 179), ADH (180, 181), glucagon (176, 177), acetazolamide (174), and theophylline (434), all of which are cyclic AMP facilitory, that tubular rejection of sodium may be a proximal tubular and possible ascending limb cyclic AMP-mediated process leading to a sodium diuresis, whereas ADH-mediated concentration of the urine is a cyclic AMP stimulated action concerned primarily with water retention as opposed to sodium.

ANTIDIURETIC DRUGS

The theoretical basis for the antidiuretic action of drugs that may be useful in polyuric states producing their antidiuretic effect by stimulating ADH release or by augmenting its peripheral action has been described earlier in this article, and is summarized in Table 1. Some of the drugs that might be useful therapeutically in the treatment of primary diabetes insipidus could conversely produce excessive water retention in patients without polyuric disorders leading to hyponatremia.

The primary antidiuretic drug is, of course, vasopressin, which activates renal medullary cyclic AMP as an intermediary of its antidiuretic effect. Chlorpropamide and large oral doses or intravenous bolus administration of tolbutamide have an ADH-like action associated with a rise in urinary cyclic AMP (62). Evidence has been previously cited supporting the following possible mechanisms of the antidiuretic action of these sulfonylurea drugs, although a facilitory action on ADH generation of cyclic AMP is best supported by the evidence available. (1) Increase the release of ADH (293); (2) augment the peripheral action of ADH (288–294), (a) by facilitating binding of ADH to its adenylate cyclase receptor site (291, 292, 295), (b) stimulating adenylate cyclase (306), (c) inhibiting phosphodiesterase (307), (d) decrease membrane permeability to cyclic AMP (310), (e) act distal to cyclic AMP (309). Of all the antidiuretic drugs other than ADH, chlorpropamide is the most clinically effective agent in the treatment of diabetes insipidus (272–287). This drug is especially efficacious in combination with thiazide diuretics. Since the antidiuretic action of these two types of drugs is based on different mechanisms, their antidiuretic effects are additive or synergistic in the treatment of diabetes insipidus. Furthermore, thiazides, by elevating blood sugar, may lessen the likelihood of hypoglycemia occurring with the use of chlorpropamide, which is potentially a serious problem especially

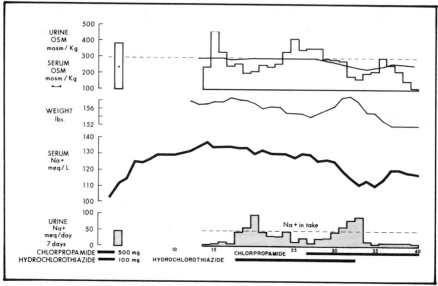

10 Chlorpropamide and hydrochlorothiazide induced hyponatremia. (Reprinted from Fichman, M., Vorherr, H., Kleeman, C. R., and Telfer, N.: Ann. Intern. Med. 75:860, 1971.)

in patients who have associated hypopituitarism (283). By the same token, while chlorpropamide may occasionally produce an inappropriate ADH-like hyponatremic state in patients with diabetes mellitus (297–301), the likelihood of hyponatremia is much greater with a combination of thiazides and chlorpropamide than with either drug alone (301) (Fig. 10). The likelihood of chlorpropamide contributing to the development of hyponatremia may be due to its augmentation of ADH release or as a possible stimulator of thirst (371), in addition to its ADH potentiating effect.

Phenformin also appears to augment the peripheral action of ADH by a mechanism which has not been elucidated (318, 319). Its antidiuretic effect is modest and, therefore, of limited usefulness in the treatment of diabetes insipidus and, at the same time, provides minimal risk toward the development of hyponatremia in patients with diabetes mellitus unless they are also taking chlorpropamide and/or thiazides.

Acetaminophen has a mild antidiuretic effect (311) and appears to potentiate the action of ADH both by inhibiting phosphodiesterase and facilitating binding of ADH to the adenylate cyclase receptor site (305). Again, this drug is minimally effective in the treatment of diabetes insipidus and not likely to produce hyponatremia because of its modest antidiuretic effect.

Carbamazepine is a highly potent antidiuretic drug of comparable potency but more rapid onset to chlorpropamide and, indeed, has an additive effect to the antidiuretic action of chlorpropamide in diabetes insipidus (340, 341). Its high level of toxicity in a variety of organ systems has precluded its general usage in the United States where FDA regulations limits its application to the treatment of trigeminal neuralgia. It appears to stimulate ADH release rather than enhance its peripheral effect (343–349). A case of hyponatremia has been reported with its use (342) as well as following a related drug, amitriptyline (435), which is used as an antidepressant.

Clofibrate is also a modestly effective antidiuretic drug that acts by increasing ADH release rather than by inhibiting its peripheral action (350–354). Hyponatremia would be an unlikely but possible accompanying problem in its use as a cholesterol lowering agent in view of its modest antidiuretic effect. Carbamazepine and clofibrate may be useful alternative drugs to chlorpropamide for use in the therapy of diabetes insipidus associated with hypopituitarism to avoid the increased likelihood of hypoglycemia occurring with chlorpropamide.

Both vincristine (355–361) and intravenous cyclophosphamide (362, 363) may produce an inappropriate ADH-like syndrome with hyponatremia. In the case of vincristine, which is highly neurotoxic, the greater likelihood is that it augments "inappropriate" release of ADH (367). Since these are highly toxic drugs and used primarily to treat malignancies such as lymphoma, Hodgkin's, myeloma, and so forth, they have had no application to the treatment of diabetes insipidus.

Oxytocin has only a small fraction of the antidiuretic potency of ADH, and there is no evidence that it activates cyclic AMP. However, when administered as a continuous drip to induce labor, at a rate greater than 40 μ/min with simultaneous administration of large volumes of intravenous fluids, an antidiuretic effect and severe hyponatremia may result from oxytocin (436).

Since chlorpropamide, pheformin, acetaminophen, carbamazepine, and clofibrate act either by stimulating the release of ADH

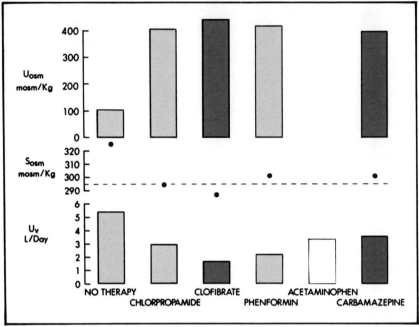

11 Twenty-four hr urine volume and serum and urinary osmolalities following 8-hr water deprivation on different oral antidiuretic drugs in a patient with mild idiopathic diabetes insipidus. The dotted line represents the upper limit of elevation of serum osmolality (295 mosm/kg) in normal subjects following 8 hr of water deprivation.

or augmenting its peripheral action, they may improve the antidiuretic response to water deprivation (Fig. 11) and also produce a reduction in total 24-hr daily urine volume in patients with mild-to-moderate diabetes insipidus much in the manner that ADH would, when released endogenously (Figs. 11 and 12).

These drugs are relatively ineffective in the treatment of severe diabetes insipidus and have no effect in the diabetes insipidus rat where ADH is totally absent and, like ADH, are totally ineffective in the treatment of nephrogenic diabetes insipidus. Thiazide and other diuretics, on the other hand, produce an antidiuretic response in both primary diabetes insipidus and nephrogenic diabetes insipidus (265), probably by producing sodium depletion rather than interacting with the action of ADH. The resultant sodium depletion

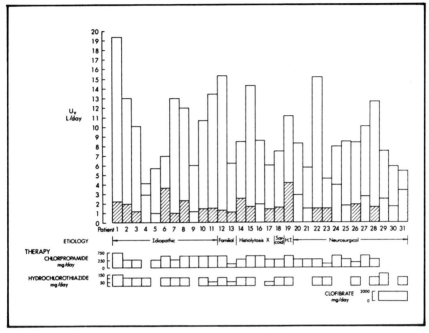

12 Reduction in 24-hr urine volume (total height of bars) with oral agents (shaded bar height) in 31 patients with diabetes insipidus of varied etiology.

leads to more avid proximal tubular reabsorption of sodium, limiting delivery of urine to the distal nephron and collecting duct, where in the absence of ADH there is a limited degree of back diffusion of water sufficient to produce a mild degree of urinary concentration (266). With sodium chloride loading there is an escape from the antidiuretic effects of diuretics and with salt restriction a continuation of the antidiuretic response temporarily even after the discontinuance of the diuretics (267). Thiazides, in addition, interfere with urinary dilution by their specific locus of action inhibiting sodium transport out of the cortical diluting segment, where the greatest urinary dilution usually takes place (431).

The occasional development of severe hyponatremia in association with thiazide diuretics is probably the result of a combination of factors (215). (1) Impaired urinary dilution due to the action of thiazides on the cortical diluting segment not allowing adequate free water excretion to prevent hyponatremia in patients with primary polydipsia (437). (2) Sodium depletion stimulating ADH might be contributory to the development of the hyponatremia with more powerful diuretics such as ethacrynic acid and furosemide, but not an important factor in patients on thiazide diuretics. Hyponatremic patients receiving thiazides show no evidence of extracellular fluid volume contraction on physical examination, have normal glomerular filtration rates, show only trivial sodium loss on balance studies, and have exchangeable sodiums using ^{22}Na which are not significantly less than predicted (215). (3) Profound potassium depletion due to diuretics may allow the movement of sodium intracellularly to replace the missing osmol, potassium (215, 438). Potassium depletion may also account for stimulation or sensitizing volume receptor to release ADH excessively in response to rather minimal degrees of sodium depletion. This may be related to an effect of potassium depletion decreasing cardiac output (439). (4) The increased levels of ADH in diuretic induced hyponatremia (215), as well as the antiduiretic effect of thiazides in diabetes insipidus (440), may be secondary to cyclic-AMP–mediated (441) stimulation of the renin–angiotensin system (442), which may, in turn, increase ADH release (18). (5) Combination of the antidiuretic effect of thiazides and chlorpropamide in patients with diabetes mellitus who are receiving both drugs (Fig. 10).

The antidiuretic action of diazoxide is also probably related to its enhancement of proximal tubular sodium reabsorption (268, 269) limiting delivery of sodium distally, which may account for its antidiuretic effect in both diabetes insipidus (151) and nephrogenic diabetes insipidus (271), although this drug also is a phosphodiesterase inhibitor (159, 263, 270).

Indomethacin is the only drug other than diuretics (and diazoxide) that has an antidiuretic effect in nephrogenic diabetes insipidus. Its antidiuretic action is minimal by itself, but additive to the effect of thiazides (271). In two siblings with nephrogenic diabetes insipidus, 24-hr urine volume on no therapy (16.8 and 15 liters) decreased modestly with indomethacin alone (7.9 and 7.7 liters) or diuretics alone (8.4 and 9.2 liters) but markedly with the combination of indomethacin and diuretics (4.2 and 5.5 liters). The antidiuretic effect of thiazides plus indomethacin was associated with a rise in urine osmolality to isotonicity (288 mosmols/kg) in one patient following 8 hr of water deprivation. There was a marked reduction with indomethacin in urine sodium and circulating PGA levels measured by radioimmunoassay, but no enhancement of the antidiuretic response or cyclic AMP generating effect of infusions of exogenous ADH. These data suggest that the antidiuretic effect of indomethacin in nephrogenic diabetes insipidus (271) (also observed in primary diabetes insipidus [51]) was due to inhibition of prostaglandin mediated proximal tubular rejection of sodium further limiting delivery of urine to the collecting duct.

SUMMARY

The action of vasopressin on water transport in isolated toad bladders, rabbit tubules and mammalian renal medullae involves the activation of adenylate cyclase and the production of cyclic AMP. Inhibitors of the generation of cyclic AMP by ADH include potassium depletion, hypercalcemia, lithium, prostaglandin E_1, and possibly norepinephrine, all of which may result thereby in inhibiting water transport and producing polyuria. Stimuli of adenylate cyclase, including beta adrenergic catecholamines and sulfonylurea drugs, may thereby produce an antidiuretic response. Inhibition of the degradation of cyclic AMP by phosphodiesterase inhibitors, including sulfonylurea drugs and acetaminophen, may partly explain their antidiuretic effect. Catecholamines and prostaglandins may serve as intrarenal modulators of the action of ADH by interfering with the generation of cyclic AMP or by altering delivery of Na^+ from the proximal tubule. The effect of ADH on sodium transport may involve a different receptor site than its effect on water, or be involved with the action of ADH on a double diffusion barrier, one porous and the other dense. Polyuric disorders result from the decreased release of ADH or interference with peripheral action of ADH on the collecting duct. Congenital nephrogenic diabetes insipidus is associated with a failure to augment urinary cyclic AMP in response to ADH and possibly also a failure of the medullary cells to respond to cyclic AMP. Polyuria due to impaired action of ADH-generating cyclic AMP may result from hypercalcemia, lithium intoxication, propoxyphene intoxication, hypokalemia, and pathological disorders that disrupt the tubular mechanisms for renal concentration. Augmentation of the release or generation of cyclic AMP by ADH or inhibition of its degradation by phosphodiesterase may be the basis for the antidiuretic effects in the treatment of diabetes insipidus or the potential for producing water intoxication and hyponatremia of sulfonylurea drugs, carbamazapine, acetaminophen, clofibrate, or beta-adrengeric amines. The pharmacologic modification of the generation of cyclic AMP by ADH may result in polyuria or antidiuresis, which may be both of etiologic and therapeutic significance in relationship to diseases of water metabolism.

REFERENCES

1. Kleeman, C. R.: Water metabolism. In: Clinical Disorders of Fluid and Electrolyte Metabolism, M. M. Maxwell and C. R. Kleeman (eds.) pp. 215-295, McGraw-Hill, New York, 1972.
2. Kleeman, C. R. and Fichman, M. P.: Clinical physiology of water metabolism. New Eng. J. Med. 277:1300-1307, 1967.
3. Verney, E. B.: Croonian Lecture: Antidiuretic hormone and factors which determine its release. Proc. Roy. Soc. London S.B. 135:27-106, 1947.
4. Share, L.: Acute reduction in extracellular fluid volume and concentration of antidiuretic hormone in blood. Endocrinol. 69:925-933, 1961.
5. Gauer, O. H., Henry, J. P., and Sieker, H. O.: Cardiac receptors and fluid volume control. Prog. Cardiovasc. Dis. 4:1-26, 1961.
6. Deutsch, S., Goldberg, M., and Dripps, R. D.: Postoperative hyponatremia with the inappropriate release of ADH. Anaesthesiology 27:250-256, 1966.
7. Moran, W. H., Miltenburger, A., Shuayb, A., and Zimmerman, B.: The relationship of antidiuretic hormone secretion to surgical stress. Surgery 56:99-108, 1964.
8. Papper, S. and Papper, E. B.: The effects of pre-anesthetic, anesthetic and postoperative drugs on renal function. Clin. Pharm. Therapeut. 5: 204-215, 1964.
9. Scott, J. C., Welch, J. S., and Berman, I. B.: Water intoxication and sodium depletion in surgical patients. Obstet. Gynec. 26:168-175, 1965.
10. Hayes, M. A. and Goldbenberg, I. S.: Renal effects of anesthesia and operation mediated by endocrines. Anesthesiology 24:487-499, 1963.
11. Wright, H. K. and Gann, D. S.: A defect in urinary concentrating ability during postoperative antidiuresis. Surg. Gynec. and Obstet. 121:47-50, 1965.
12. Zimmerman, B.: Pituitary and adrenal function in relation to surgery. Surgical Clinics of No. America 45:299-315, 1965.
13. Mielke, J. E. and Kirklin, J. W.: Renal function during and after surgery. Med. Clin. N. A. 50:978-983, 1966.
14. Cadnapaphornchai, P., Boykin, J. L., Berl, T., McDonald, K. M., and Schrier, R. W.: Mechanism of effect of nicotine on renal water excretion. Am. J. Physiol. 227:1216-1220, 1974.
15. Grossman, S. P.: Direct adrenergic and cholinergic stimulation of hypothalamic mechanisms. Am. J. Physiol. 202:572-576, 1962.
16. Berl, T. and Schrier, R. W.: Mechanism of effect of prostaglandin E_1 on

renal water excretion. J. Clin. Invest. 52:463-471, 1973.
17. Berl, T., Cadnapaphornchai, P., Harbottle, J. A., and Schrier, R. W.: Mechanism of stimulation of vasopressin release during beta-adrenergic stimulation with isoproterenol. J. Clin. Invest. 53:857-867, 1974.
18. Bonjour, J. P. and Malvin, R. L.: Stimulation of ADH release by the renin-angiotensin system. Am. J. Physiol. 218:1555-1559, 1970.
19. Malvin, R. L.: Possible role of the renin-angiotensin system in the regulation of antidiuretic hormone secretion. Fed. Proc. Fed. Am. Soc. Exp. Biol. 30:1383-1386, 1971.
20. Kleeman, C. R. and Cutler, R.: The neurohypophysis. Ann. Rev. Physiol. 25:385-432, 1963.
21. Fichman, M. P., Kleeman, C. R., and Bethune, J. E.: Inhibition of antidiuretic hormone by diphenylhydantoin. Arch. Neurol. 22:45-53, 1970.
22. Schrier, R. W. and Berl, T.: Mechanism of alpha-adrenergic stimulation with norepinephrine on renal water excretion. J. Clin. Invest. 52:502-511, 1973.
23. Berl, T., Harbottle, J. A., and Schrier, R. W.: Effect of alpha- and beta-adrenergic stimulation on renal water excretion in man. Kidney Int. 6: 247-253, 1974.
24. Wilber, J. F., Peake, G. T., and Utiger, R. D.: Thyrotropin release in vitro: Stimulation by cyclic 3',5'-adenosine monophosphate. Endocrinology 84:758-760, 1969.
25. Gagliardino, J. J. and Martin, J. M.: Stimulation of growth hormone secretion in monkeys by adrenalin, pitressin and adenosine-3',5'-cyclic monophosphoric acid. Acta Endocrinol. 59:390-396, 1968.
26. Fleishcer, N., Donald, R. A., and Butcher, R. W.: Involvement of adenosine 3',5'-monophosphate in release of ACTH. Am. J. Physiol. 217:1287-1291, 1969.
27. Jutisz, M. and de la Llosa, M. P.: Requirement of Ca^{++} and Mg^{++} ions for the in vitro release of follicle-stimulating hormone from rat pituitary glands and in its subsequent biosynthesis. Endocrinology 86:761-768, 1970.
28. Ratner, A.: Stimulation of luteinizing hormone release in vitro by dibutyryl-cyclic-AMP and theophylline. Life Sci. (I) 9:1221-1226, 1970.
29. Levine, R. A., Oyama, S., Kagan, A., and Glick, S. M.: Stimulation of insulin and growth hormone secretion by adenine nucleotides in primates. J. Lab. Clin. Med. 75:30-36, 1970.
30. Zor, U., Kaneko, T., Schneider, H. P. G., McCann, S. M., and Field, J. B.: Further studies of stimulation of anterior pituitary cyclic adenosine 3',5'-monophosphate formation by hypothalamic extract prostaglandins. J. Biol. Chem. 245:2883-2888, 1970.
31. Robison, G. A., Butcher, R. W., and Sutherland, E. W.: Cyclic AMP, Academic Press, N.Y., pp. 146-225, 1971.
32. Kahn, R. H. and Lands, W. E. M.: Prostaglandins and cyclic AMP. Biological action and clinical application. Academic Press, N.Y., pp. 1-306, 1973.
33. Sachs, H.: Biosynthesis and release of vasopressin. Am. J. Med. 42:687-700, 1967.

34. Berliner, R. W. and Bennett, C. M.: Concentration of urine in the mammalian kidney. Am. J. Med. 42:777-789, 1967.
35. Wirz, H., Hargitay, B., and Kuhn, W.: Lokalisation des konzentrierungs prozesses in der niere durch direkte kryoskopie. Helvet. Physiol. et Pharmacol. Acta 9:196-207, 1951.
36. Gottschalk, C. W. and Mylle, M.: Micropuncture study of mammalian urinary concentrating mechanism: Evidence for countercurrent hypothesis. Am. J. Physiol. 196:927-936, 1954.
37. Gottschalk, C. W.: Osmotic concentration and dilution of the urine. Am. J. Med. 36:670-685, 1964.
38. Berliner, R. W. and Davidson, D. G.: Production of hypertonic urine in the absence of pituitary antidiuretic hormone. J. Clin. Invest. 36:1416-1427, 1951.
39. Valtin, H.: Hereditary hypothalamic diabetes insipidus in rats (Brattleboro strain): A useful experimental model. Am. J. Med. 42:814-827, 1967.
40. Sutherland, E. W. and Rall, T. W.: The relation of adenosine 3'-5' phosphate and phosphorylase to the actions of catecholamines and other hormones. Pharmacol. Rev. 12:265-299, 1960.
41. Hays, R. M. and Leaf, A.: Studies on the movement of water through the isolated toad bladder and its modification by vasopressin. J. Gen. Physiol. 45:905-919, 1962.
42. Leaf, A.: Membrane effects of antidiuretic hormone. Am. J. Med. 42:745-756, 1967.
43. Leaf, A., Anderson, J., and Page, L. B.: Active sodium transport by the isolated toad bladder. J. Gen. Physiol. 41:657-668, 1958.
44. Leaf, A. and Dempsey, E. F.: Some effects of mammalian neurohypopyseal hormones on metabolism and active transport of sodium by the isolated toad bladder. J. Biol. Chem. 235:2160-2163, 1960.
45. Maffly, R. H., Hays, R. M., Lamdin, E., and Leaf, A.: The effects of neurophypophyseal hormones on the permeability of the toad bladder to urea. J. Clin. Invest. 39:630-641, 1960.
46. Leaf, A. and Hays, R. M.: Permeability of the isolated toad bladder to solutes and its modification by vasopressin. J. Gen. Physiol. 45:921-932, 1962.
47. Koefoed-Johnsen, V. and Ussing, H. H.: The contributions of diffusion and flow to the passage of D_2O through living membranes. Effect of neurohypophyseal hormone on isolated anuran skin. Acta Physiol. Scand. 28:60-76, 1953.
48. Leaf, A.: Some actions of neurohypophyseal hormones on living membranes. J. Gen. Physiol. 43:175-189, 1960.
49. Lichtenstein, N. S. and Leaf, A.: Effect of amphotericin B on the permeability of the toad bladder. J. Clin. Invest. 44:1328-1342, 1965.
50. Leaf, A.: Transport properties of water. Ann. N.Y. Acad. Sci. 125:559-571, 1965.
51. Grantham, J. J. and Burg, M. B.: Effect of vasopressin and cyclic AMP on permeability of isolated collecting tubules. Am. J. Physiol. 211:255-259, 1966.

52. Gardner, K. D. and Maffly, R. H.: An in vitro demonstration of increased collecting tubular permeability to urea in the presence of vasopressin. J. Clin. Invest. 43:1968-1975, 1964.
53. Orloff, J. and Handler, J. S.: Vasopressin-like effects of adenosine 3'-5' phosphate (cyclic 3'-5' AMP) and theophylline in the toad bladder. Biochem. Biophys. Res. Comm. 5:63-66, 1961.
54. Orloff, J. and Handler, J. S.: The similarity of effects of vasopressin, adenosine 3'-5' phosphate (cyclic AMP) and theophylline in the toad bladder. J. Clin. Invest. 41:702-709, 1962.
55. Orloff, J. and Handler, J. S.: The cellular mode of action of antidiuretic hormone. Am. J. Med. 36:686-697, 1964.
56. Orloff, J. and Handler, J. S.: The role of adenosine 3'-5' phosphate in the action of antidiuretic hormone. Am. J. Med. 42:757-768, 1967.
57. Butcher, R. W. and Sutherland, E. W.: Adenosine 3'-5' phosphate in biological materials. 1. Purification and properties of cyclic 3'-5' nucleotide phosphodiesterase and use of this enzyme to characterize adenosine 3'-5' monophosphate in human urine. J. Biol. Chem. 237:1244-1250, 1962.
58. Handler, J. S., Butcher, R. W., Sutherland, E. W., and Orloff, J.: The effect of vasopressin and of theophylline on the concentration of adenosine 3'-5'-phosphate in the urinary bladder of the toad. J. Biol. Chem. 240:4524-4526, 1965.
59. Marumo, F. and Edelman, I. S.: Effects of Ca^{++} and prostaglandin E_1 on vasopressin activation of renal adenyl cyclase. J. Clin. Invest. 50: 1613-1620, 1971.
60. Beck, N. P., Kaneko, T., Zor, U., Field, J. B., and Davis, B.: Effect of vasopressin and prostaglandin E_1 on the adenyl cyclase-cyclic 3'-5' adenosine monophosphate system of the renal medulla of the rat. J. Clin. Invest. 50:2461-2465, 1971.
61. Chase, L. R. and Aurbach, G. D.: Renal adenyl cyclase: Anatomically separate sites for parathyroid hormone and vasopressin. Science 159: 545-547, 1968.
62. Fichman, M. P. and Brooker, G.: Deficient renal cyclic adenosine 3'-5' monophosphate production in nephrogenic diabetes insipidus. J. Clin. Endocrinol. Metab. 35:35-47, 1972.
63. Takahashi, K., Kaminura, M., Shinko, T., and Tsuji, S.: Effects of vasopressin and water load on urinary adenosine 3'-5' cyclic monophosphate. Lancet 2:967, 1966.
64. Taylor, A. L., Davis, B. B., Paulson, G., Jasimovich J. B., and Mintz, D. M.: Factors influencing the urinary excretion of 3'-5' monophosphate in humans. J. Clin. Endocrinol. Metab. 30:316-324, 1970.
65. Davis, B., Zor, V., Kaneko, T., Mintz, D. M., and Field, J. B.: Effects of parathyroid extract (PTE), arginine vasopressin (AVP) and prostaglandin E_1 (PGE_1) on urinary and renal tissue cyclic 3'-5' adenosine monophosphate (cAMP). Clin. Res. 17:458, 1969.
66. Chase, L. R., Melson, G. L., and Aurbach, G. D.: Pseudohypoparathyroidism: Defective excretion of 3'-5' AMP in response to parathyroid hormone. J. Clin. Invest. 48:1832-1844, 1969.

67. Kaminsky, N. I., Ball, J. H., Broadus, A. E., Hardman, J. G., Sutherland, E. W., and Liddle, G. W.: Hormonol effects on extracellular cyclic nucleotides in man. Trans. Am. Assoc. Phy. 83:235-244, 1970.
68. Chase, L. R. and Aurbach, G. D.: Parathyroid function and the renal excretion of 3' 5'-adenylic acid. Proc. Nat. Acad. Sci. U.S. 58:518-525, 1967.
69. Kaminsky, N. I., Broadus, A. E., Hardman, J. G., Ginn, H. E., Sutherland, E. W., and Liddle, G. W.: Effects of glucagon and parathyroid hormone on plasma and urinary 3',5'-adenosine monophosphate in man. J. Clin. Invest. 48:42a, 1969.
70. Kaminsky, N. I., Broadus, A. E., Hardman, J. G., Jones. D. J., Jr., Ball, J. H., Sutherland, E. W., and Liddle, G. W.: Effects of parathyroid hormone on plasma and urinary adenosine 3',5'-monophosphate in man. J. Clin. Invest. 49:2387-2395, 1970.
71. Melson, G. L., Chase, L. R., and Aurbach, G. D.: Parathyroid hormone sensitive adenyl cyclase in isolated renal tubules. Endocrinology 86:511-518, 1970.
72. Nagata, N. and Rasmussen, H.: Parathyroid hormone 3',5' AMP, Ca^{++}, and renal gluconeogenesis. Proc. Nat. Acad. Sci. U.S. 65:368-374, 1970.
73. Beck, M. P., Field, J. B., and Davis, B.: Effect of prostaglandin E_1 (PGE_1) chlorpropamide (CPM) and vasopressin (VP) on cyclic 3',5'-adenosine monophosphate (cAMP) in renal medulla of rats. Clin. Res. 18: 494, 1970.
74. Murad, F. and Pak, C. Y. C.: Urinary excretion of adenosine 3'-5' monophosphate and guanosine 3'-'5 monophosphate. New Eng. J. Med. 286: 1382-1387, 1972.
75. Kurokawa, K. and Massry, S. G.: Evidence for two separate adenyl cyclase systems responding independently to parathyroid hormone and beta-adrenergic agents in the renal cortex of the rat. Proc. Soc. Exp. Biol. Med. 143:123-126, 1973.
76. Kuo, J. F. and Greengard, P.: An adenosine 3',5'-monophosphate dependent protein kinase from Escherichia coli. J. Biol. Chem. 244:3417-3419, 1969.
77. Handler, J. S. and Orloff, J.: Activation of phosphorylase in toad bladder mammalian kidney by antidiuretic hormone. Am. J. Physiol. 205:298-302, 1963.
78. Jard, S. and Bastide, F.: A cyclic AMP-dependent protein kinase from frog bladder epithelial cells. Biochem. Biophys. Res. Comm. 39:559-566, 1970.
79. Hays, R. M. and Levine, S. D.: Vasopressin. Kidney Int. 6:307-322, 1974.
80. Schwartz, I. L. and Walter, R.: Factors influencing the reactivity of the toad bladder to the hydroosmotic action of vasopressin. Am. J. Med. 42: 769-776, 1967.
81. Edelman, I. S., Peterson, M. J., and Gulyassy, P. F.: Kinetic analysis of the antidiuretic action of vasopressin and adenosine 3'-5' monophosphate. J. Clin. Invest. 43:2185-2194, 1964.
82. Gulyassy, P. F. and Edelman, I. S.: The pH dependence of the antidi-

uretic action of vasopressin in vitro. In: Proceedings of the 2nd International Congress of Nephrology, Prague, August 24-28, 1963. p. 605, Amsterdam, 1964, Excerpta Medica Foundation.
83. Gulyassy, P. F. and Edelman, I. S.: Hydrogen ion dependence of the antidiuretic action of vasopressin, oxytocin, and deaminooxytocin. Biochem. Biophys. Acta 102:185-197, 1965.
84. Bentley, P. J.: The effects of neurophypophyseal extracts on water transfer across the wall of the isolated urinary bladder of the toad. Bufo marinus. J. Endocrinol. 17:201-209, 1958.
85. Bentley, P. J.: The effects of ionic changes on water transfer across the isolated urinary bladder of the toad. Bufo marinus.
86. Peterson, M. J. and Edelman, I. S.: Calcium inhibition of the action of vasopressin on the urinary bladder of the toad bladder. J. Clin. Invest. 43:583-594, 1964.
87. Bentley, P. J.: The effects of vasopressin on the short circuit current across the wall of the isolated bladder of the toad, Bufo marinus. J. Endocrinol. 21:161-170, 1960.
88. Argy, W. P., Handler, J. S., and Orloff, J.: Ca^{++} and Mg^{++} effects on toad bladder response to cyclic AMP, theophylline and ADH analogues. Am. J. Physiol. 213:803-808, 1967.
89. Bourgoignie, J., Guggenheim, S., Kipnis, D. M., and Klahr, S.: Cyclic guanosine monophosphate: Effects on short-circuit current and water permeability. Science 165:1362-1363, 1969.
90. Lamdin, E., Schwarz, L., Livingston, L., and Schwartz, I. L.: Effects of Na^+, K^+, and Ca^{++} on the reactivity of the toad bladder to vasopressin. In: Abstracts of the 18th Annual Meeting of the Society of General Physiologists, Woods Hole, Mass., 1963, Marine Biology Lab.
91. Finn, A. L., Handler, J. S., and Orloff, J.: Relationship between toad bladder potassium content and permeability response to vasopressin. Am. J. Physiol. 210:1279-1284, 1968.
92. Beck, N., Reed, S. W., and Davis, B. B.: Urinary concentrating defect and vasopressin cyclic AMP in potassium depletion. VI. International Congress of Nephrology, Abstracts of Free Communication 55, Florence, Italy, June 8-12, 1975.
93. Singer, I. and Rotenberg, D.: Mechanism of lithium action. New Eng. J. Med. 289:254-260, 1973.
94. Cox, M. and Singer, I.: Lithium and water metabolism. Am. J. Med. 59: 153-157, 1975.
95. Smith, D. F. and Balagura, S.: Sodium appetite in rats given lithium. Life Sci. 11:1021-1029, 1972.
96. Smith, D. F., Balagura, S., and Lubrau, M.: "Antidotal thirst": A response to intoxication. Science 167:297-298, 1970.
97. Gutman, Y., Benzakein, F., and Livneh, P.: Polydipsia induced by isoprenaline and by lithium: Relation to kidneys and renin. Eur. J. Pharmacol. 16:380-384, 1971.
98. Gutman, Y., Tamir, N., and Benzakein, F.: Effect of lithium on plasma renin activity. Eur. J. Pharmacol. 24:347-351, 1973.

99. Epstein, A. N., Fitzsimmons, J. T., and Simons, J. B.: Drinking caused by the intracranial injection of angiotensin in the rat. J. Physiol. 200: 98P-100P, 1964.
100. Buggy, J. and Fisher, A. E.: Evidence for a dual central role for angiotensin in water and sodium intake. Nature 250:733-735, 1974.
101. Ellman, G. L. and Gan, G. L.: Lithium ion and water balance in rats. Toxicol. Appl. Pharmacol. 25:617-620, 1973.
102. Hochman, S. and Gutman, Y.: Lithium: ADH antagonism and ADH independent action in rats with diabetes insipidus. Eur. J. Pharmacol. 28: 100-107, 1974.
103. Singer, I., Rotenburg, D., and Puschett, J. B.: Lithium induced nephrogenic diabetes insipidus in vivo and in vitro studies. J. Clin. Invest. 51: 1081-1091, 1972.
104. Forrest, J. N., Cohen, A. D., Torretti, J., Himmelhoch, J. M., and Epstein, F. H.: On the mechanism of lithium induced diabetes insipidus in man and the rat. J. Clin. Invest. 53:1115-1123, 1974.
105. Davis, J. M. and Fann, W. E.: Lithium. Ann. Rev. Pharmacol. 11:285-302, 1971.
106. Bentley, P. J. and Wasserman, A.: The effects of lithium on the permeability of an epithelial membrane, the toad urinary bladder. Biochem. Biophys. Acta 266:285-292, 1972.
107. Herrera, F. G., Egea, R., and Herrera, A. M.: Movement of lithium across toad urinary bladder. Am. J. Physiol. 220:1501-1508, 1971.
108. Solomon, S.: Action of alkali metals on papillary-cortical sodium gradient of dog kidney. Proc. Soc. Exp. Biol. Med. 125:1183-1186, 1967.
109. Harris, C. A. and Dirks, J. H.: Effects of acute lithium infusion on proximal and distal tubular reabsorption in rats. Fed. Proc. 32:381, 1973.
110. Martinez-Maldonado, M.: Renal concentration and dilution during acute lithium infusion. VI. International Congress of Nephrology, Abstracts of Free Communication, #222, June 8-12, 1975, Florence, Italy.
111. Singer, I. and Franko, E. A.: Lithium-induced ADH resistance in toad bladder. Kidney Int. 3:151-159, 1973.
112. Harris, C. A. and Jenner, F. A.: Some aspects of the inhibition of the action of antidiuretic hormone by lithium ions in the rat kidney and bladder of the toad Bufo Marinus. Br. J. Pharmacol. 44:223-232, 1972.
113. Torp-Pederson, C. and Thorn, N. A.: Acute effects of lithium on the action and release of ADH in rats. Acta Endocrinol. (Kbh) 73:665-671, 1973.
114. Beck, N., Reed, S. W., and Davis, B. B.: Effect of lithium on catecholamine-induced antidiuresis and cyclic AMP. VI. International Congress of Nephrology, Abstracts of Free Communication #54, June 8-12, 1975, Florence, Italy.
115. Dousa, T. P. and Hechter, O.: The effect of NaCl and LiCl on vasopressin sensitive adenyl cyclase. Life Sci. 9:765-770, 1970.
116. Beck, N. P., Reed, S. W., and Davis, B. B.: Effect of lithium on renal concentration of cyclic AMP. Clin. Res. 19:684, 1971.
117. Eknoyan, G., Corey, G. R., Loomis, J., Suki, W. N., and Martinez-Maldanado, M.: Lithium-induced diabetes insipidus: Effect on urinary cyclic

AMP excretion and renal tissue adenylate cyclase activity. Clin. Res. 22: 524, 1974.
118. Geisler, A., Wraae, O., and Olesen, O. V.: Adenyl cyclase activity in kidneys of rats with lithium-induced polyuria. Acta Pharmacol. Toxicol. 31: 203-208, 1972.
119. Dousa, T. P.: Lithium: Interaction with ADH dependent cyclic AMP system of human renal medulla. Clin. Res. 21:282, 1973.
120. Rall, T. W. and Sutherland, E. W.: Formation of a cyclic adenine sehonucleotide by fat particles. J. Biol. Chem. 232:1065-1076, 1958.
121. Murad, F., Brewer, H. B., and Vaugh, M.: Effect of thyrocalcitonin on adenosine 3'-5' cyclic phosphate formation by rat kidney and bone. Proc. Nat. Acad. Sci. U.S. 65:446-453, 1970.
122. Dousa, T. and Rychlick, I.: The effect of parathyroid hormone on adenyl cyclase in rat kidney. Biochem. Biophys. Acta 158:484-486, 1968.
123. Oye, I. and Sutherland, E. W.: Effects of epinephrine and other agents on adenyl cyclase in the cell membrane of avian erythrocytes. Biochem. Biophys. Acta 127:347-354, 1966.
124. Butcher, R. W., Baird, C. E., and Sutherland, E. W.: Effects of lipolytic and antilipolytic substances on adenosine 3',5'-monophosphate levels in isolated fat cells. J. Biol. Chem. 243:1705-1712, 1968.
125. Sutherland, E. W., Rall, T. W., and Menon, T.: Adenyl cyclase. I. Distribution, preparation and properties. J. Biol. Chem. 237: 1220-1227, 1962.
126. Birnbaumer, L. and Rodbell, M.: Adenyl cyclase in fat cells. J. Biol. Chem. 244:3477-3482, 1969.
127. Ramwell, P. W. and Shaw, J. E.: Biological significance of the prostaglandins. Rec. Progress in Hormone Research 26:139-187, 1970.
128. Orloff, J., Handler, J. S., and Bergstrom, S.: Effect of prostaglandin (PGE_1) on the permeability response of toad bladder to vasopressin, theophylline and adenosine 3'-5' monophosphate. Nature (London) 205:397-398, 1965.
129. Lipson, L. C. and Sharp, G. W. G.: Effect of prostaglandin E_1 on sodium transport and osmotic water flow in the toad bladder. Am. J. Physiol. 220:1046-1052, 1971.
130. Granthum, J. J. and Orloff, J.: Effect of prostaglandin E_1 on the permeability response of the isolated collecting tubule to vasopressin, adenosine 3'-5' monophosphate and theophylline. J. Clin. Invest. 47:1154-1161, 1968.
131. Bergstrom, S.: Prostaglandins: Members of a new hormonal system. Science 157:382-391, 1967.
132. Omachi, R. S., Robbie, D. E., Handler, J. S., and Orloff, J.: Effects of ADH and other agents on cyclic AMP accumulation in toad bladder epithelium. Am. J. Physiol. 226:1152-1157, 1974.
133. Bar, H. P., Hechter, O., Schwartz, I. L., and Walter, R.: Neurohypophyseal hormone sensitive adenyl cyclase of toad urinary bladder. Proc. Nat. Acad. Science U.S.A. 67:7-12, 1970.
134. Wong, P. D. W., Bedwani, J. R., and Cuthbert, A. W.: Hormone action and the levels of cyclic AMP and prostaglandins in the toad bladder. Nature 238:27-31, 1972.
135. Higgins, C. B. and Braunwald, E.: The prostaglandins. Biochemical, phys-

iologic and clinical considerations. Am. J. Med. 53:92-112, 1972.
136. Flores, J., Witkum, P. A., Beckman, B., and Sharp, G. W. G.: Stimulation of osmotic water flow in toad bladder by prostaglandin E_1. Evidence for different compartments of cyclic AMP. J. Clin. Invest. 56:256-262, 1975.
137. Johnston, H. H., Herzog, J. P., and Lauler, D. P.: Effects of prostaglandin E_1 on renal hemodynamics, sodium and water excretion. Am. J. Physiol. 213:939-946, 1967.
138. Martinez-Maldonado, M., Tsaparas, N., Eknoyan, G., and Suki, W. W.: Renal action of prostaglandin. Comparison with acetylcholine and volume expansion. Am. J. Physiol. 222:1147-1152, 1972.
139. Gross, J. B. and Bartter, F. C.: Effects of prostaglandin E_1, A_1, F_2 alpha on renal handling of salt and water. Am. J. Physiol. 225:218-224, 1973.
140. Lee, J. B.: Prostaglandin A_1: Antihypertensive and renal effects. Ann. Intern. Med. 74:703-710, 1971.
141. Carr, A. A.: Hemodynamic and renal effects of prostaglandin PGA_1 in subjects with essential hypertension. Am. J. Med. Sci. 259:21-26, 1970.
142. Lee, S. J., Johnson, J. G., Smith, C. J., and Hatch, F. E.: Renal effects of prostaglandin A_1 in patients with essential hypertension. Kidney Int. 1:254-262, 1972.
143. Fichman, M. P., Littenburg, G., Brooker, G., and Horton, R.: Effect of prostaglandin A_1 on renal and adrenal function in man. Circ. Research (Suppl 11) 31: 19-35, 1972.
144. Lee, J. B.: Hypertension, natriuresis and the renal prostaglandins. Ann. Intern. Med. 70:1033-1038, 1969.
145. Vander, A. J.: Direct effects of prostaglandins on renal function and renin release in anaesthetized dogs. Am. J. Physiol. 214:218-221, 1968.
146. Lafferty, J. J., Kannegiesser, H., Lee, J. B., and Parker, C. W.: Metabolic mechanisms of the action of the renal prostaglandins. Adv. Biosci. 9:293-299, 1973.
147. Vane, J. R.: Inhibition of prostaglandin synthesis as a mechanism of action for aspirin-like drugs. Nature 231:232-235, 1971.
148. Smith, J. B. and Willis, A. L.: Aspirin selectively inhibits prostaglandin production in human platelets. Nature 231:235-237, 1971.
149. Anderson, R. J., Berl, T., McDonald, K. M., and Schrier, R. W.: Evidence for an in vivo antagonism between vasopressin and prostaglandin in the mammalian kidney. J. Clin. Invest. 56:420-426, 1975.
150. Berl, T., Horowitz, J., Czaczkes, J. W., Wald, H., and Kleeman, C. R.: Effect of indomethacin (indo) on the action of vasopressin (ADH). Evidence for a role of prostaglandins (PG) in the control of renal water excretion in humans and rats. VI. International Congress of Nephrology, #200, Abstracts of Free Communications, June 8-12, 1975, Florence, Italy.
151. Fichman, M. P., Speckart, P. F., Zia, P., and Lee, A.: Antidiuretic effect of indomethacin (I) in man due to inhibition of vasopressin (ADH) induced prostaglandins (PG) generation. Clin. Research 34:000, 1976.
152. Zia, P., Golub, M., and Horton, R.: Radioimmunoassay for prostaglandin A_1 in human peripheral blood. J. Clin. Endocrinol. Metab. 41:245-252, 1975.
153. Goldberg, D. C., Schoessler, M. A., and Schwartz, I. L.: Intrinsic and

extrinsic inhibition of the reactivity of the toad bladder to vasopressin. Physiologist 6:188, 1963.
154. Kalisker, A. and Dyer, D. C.: In vitro release. Prostaglandin-vasopressin interaction in the renal medulla. Pharmacologist 13:293, 1972.
155. Flores, A. G. A. and Sharp, G. W. G.: Endogenous prostaglandins and osmotic water flow in the toad bladder. Am. J. Physiol. 323:1392-1397, 1972.
156. Schrier, R. W.: Effects of adrenergic nervous system and catecholamines on systemic and renal hemodynamic sodium and water excretion and renin secretion. Kidney Int. 6:291-306, 1974.
157. Schrier, R. W., Berl, T., Harbottle, J. A., and McDonald, K. M.: Catecholamines and renal water excretion. Nephron 15:186-196, 1975.
158. Klein, L. A., Liberman, B., Laks, M., and Kleeman, C. R.: Inter-related effects of antidiuretic hormones and adrenergic drugs on water metabolism. Am. J. Physiol. 221:1657-1665, 1971.
159. Fisher, D. A.: Norepinephrine inhibition of vasopressin antidiuresis. J. Clin. Invest. 47:540-547, 1968.
160. Liberman, B., Klein, L. A., and Kleeman, C. R.: Effects of adrenergic blocking agents on the vasopressin inhibiting action of norepinephrine. Proc. Soc. Exp. Biol. Med. 133:131-143, 1970.
161. Levi, J., Grinblat, J., and Kleeman, C. R.: Effects of isoproterenol on water diuresis in rats with congenital diabetes insipidus. Am. J. Physiol. 221:1728-1732, 1971.
162. Beck, N. P., Reed, S. W., Murdaugh, H. V., and Davis, B.: Effect of catecholamines and their interaction with other hormones on cyclic 3'-5' adenosine monophosphate of the kidney, J. Clin. Invest. 51:939-944, 1972.
163. Kurokawa, K. and Massry, S. G.: Interaction between catecholamines and vasopressin on renal medullary cyclic AMP of rat. Am. J. Physiol. 225: 825-829, 1973.
164. Handler, J. S., Bensinger, R., and Orloff, J.: Effect of adrenergic agents on toad bladder response to ADH, 3'-5' AMP and theophylline. Am. J. Physiol. 215:1024-1031, 1968.
165. McGiff, J. G., Crowshaw, K., Terragno, N. A., Malik, K. U., and Lonigro, A. J.: Differential effect of noradrenaline and renal nerve stimulation on vascular resistance in the dog kidney and the release of prostaglandin E like substance. Clin. Sci. 42:223-233, 1972.
166. Fujimoto, S. and Lockett, M. F.: The intrarenal release of prostaglandin E by nor-adrenaline. IV. International Congress of Pharmacology, Basel, Schwabe and Co., p. 122, 1969.
167. Fujimoto, S. and Lockett, M. F.: The diuretic actions of prostaglandin E_1 and of nor-adrenaline, and the occurance of a prostaglandin E_1 like substance in the renal lymph of cats. J. Physiol. (Lond.) 208:1-19, 1970.
168. Gryglewski, R. J. and Ocetkiewicz, A.: A release of prostaglandins may be responsible for acute tolerance to norepinephrine infusions. Prostaglandins 8:31-42, 1974.
169. Dunham, E. W. and Zimmerman, B. G.: Release of prostaglandin like material from dog kidney during nerve stimulation. Am. J. Physiol. 219: 1279-1285, 1970.

170. Schrier, R. W., Lieberman, R., and Ufferman, R. C.: Mechanism of antidiuretic effect of beta-adrenergic stimulation. J. Clin. Invest. 51:97-111, 1972.
171. Schrier, R. W. and Berl, T.: Mechanism of effect of alpha-adrenergic stimulation with norepinephrine on renal water excretion. J. Clin. Invest. 52:502-511, 1973.
172. Berl, T., Harbottle, J. A., Cadnapaphaphornchai, P., and Schrier, R. W.: Mechanism of suppression of vasopressin during alpha-adrenergic stimulation with nor-epinephrine. J. Clin. Invest. 53:219-227, 1974.
173. Gill, J. R., Jr. and Casper, A. G. T.: Depression of proximal tubular Na+ reabsorption in the dog in response to beta-adrenergic stimulation by isoproterenol. J. Clin. Invest. 50:112-117, 1971.
174. Rodriquez, H. J., Walls, J., Yates, J., and Klahr, S.: Effects of acetazolamide on the urinary excretion of cyclic AMP and on the activity of renal adenyl cyclase. J. Clin. Invest. 53:122-130, 1974.
175. Agus, Z. S., Puschett, J. B., Senesky, D. J., and Goldberg, M.: Mode of action of parathyroid hormone and cyclic adenosine 3'-5' monophosphate on renal tubular phosphate reabsorption in the dog. J. Clin. Invest. 50: 617-626, 1971.
176. Pullman, T. W., Lavender, A. R., and Aho, I.: Direct effects of glucagon on renal hemodynamics and excretion of inorganic ions. Metabolism 16: 358-373, 1967.
177. Levy, M. and Starr, N. L.: The mechanism of glucagon induced natriuresis in the dog. Kidney International 2:76-84, 1972.
178. Bijvoet, O. L. M., Veer, J. J. D. S., DeVries, M. R., and Van Koppen, A. T. J.: Natriuretic effect of calcitonin in man. New Eng. J. Med. 284: 681-688, 1971.
179. Haas, H. G., Dambacher, M. A., Guncaga, J., and Cauffenburger, T.: Renal effects of calcitonin and parathyroid extract in man. J. Clin. Invest. 50:2689-2702, 1971.
180. Kurtzman, N. A. and Rogers, P. W.: The diuretic effect of antidiuretic hormone. Clin. Res. 20:600, 1972.
181. Humphreys, M. H., Friedler, R. M., and Early, L. E.: The mechanism of vasopressin induced natriuresis. Clin. Res. 18:193, 1970.
182. Gill, J. R. and Casper, A. G. T.: Renal effects of adenosine 3',5'-cyclic monophosphate and dibutyryl adenosine 3',5'-cyclic monophosphate. Evidence for a role of adenosine 3',5'-cyclic monophosphate in the regulation of proximal tubular sodium reabsorption. J. Clin. Invest. 50:1231-1241, 1971.
183. Shaw, J. W., Oldham, S. B., Bethune, J. E., and Fichman, M. P.: Parathyroid hormone (PTH) mediated rise in urinary cyclic AMP (UcAMP) during acute extracellular fluid (ECF) expansion natriuresis in man. J. Clin. Endocrinol. Metab. 39:311-315, 1974.
184. Bourgoignie, J., Schmidt, R. W., Hwang, K., Nawar, T., Klahr, S., Chase, L., and Bricker, N. S.: On the nature of the circulating natriuretic factor in chronic uremia. J. Clin. Invest. 51:122, 1972.
185. Gill, J. R., Jr. and Casper, A. G. T.: Effect of renal alpha-adrenergic stimulation on proximal tubular sodium reabsorption. Am. J. Physiol. 273: 1201-1204, 1972.

186. Gill, J. R., Tate, J., and Kelley, G.: Evidence that guanosine 3'-5' cyclic monophosphate but not guanosine 5'-monophosphate increases sodium reabsorption by the proximal tubule. Proc. Am. Soc. Nephrol., p. 26, 1971.
187. Eckstein, J. W. and Abboud, F. M.: Circulatory effects of sympathomimetic amines. Am. Heart J. 63:119-135, 1962.
188. Namm, D. H. and Mayer, S. E.: Effect of epinephrine on cardiac cyclic 3'-5' AMP, phosphorylase kinase and phosphorylase. Mol. Pharmacol. 4: 61-69, 1968.
189. Porter, G. A. and Edelman, I. S.: The action of aldosterone and related corticosteroids on sodium transport across the toad bladder. J. Clin. Invest. 43:611-620, 1964.
190. Sharp, G. W. G., Coggins, C. H., Lichenstein, N. S., and Leaf, A.: Evidence for a mucosal effect of aldosterone on sodium transport in the toad bladder. J. Clin. Invest. 45:1640-1647, 1966.
191. Eggena, P., Schwartz, I. L., and Walter, R.: Action of aldosterone and hypertonicity on toad bladder permeability to water In: Regulation of Body Fluid Volumes by the Kidney, J. H. Cat and Lichardus, B. (eds.) pp. 182-192, Karger, Basel, 1970.
192. Handler, J. S., Preston, A. S., and Orloff, J.: Effect of adrenal steroid hormones on the response of the toad's urinary bladder to vasopressin. J. Clin. Invest. 48:823-833, 1969.
193. Fanestil, D. D., Porter, G. A., and Edelman, I. S.: Aldosterone stimulation of sodium transport. Biochem. Biophys. Acta 135:74-88, 1965.
194. Goodman, D. B. P., Allen, J. E., and Rasmussen, H.: On the mechanism of action of aldosterone. Proc. Nat. Acad. Sci. U.S.A. 64:330-337, 1969.
195. Stoff, J. S., Handler, J. S., and Orloff, J.: The effect of aldosterone on the accumulation of adenosine 3'-5' cyclic monophosphate in toad bladder epithelial cells in response to vasopressin and theophylline. Proc. Nat. Acad. Sci. U.S.A. 69:805-808, 1972.
196. Stoff, J. S., Handler, J. S., Preston, A. S., and Orloff, J: The effect of aldosterone on cyclic nucleotide phosphodiesterase activity in the toad urinary bladder. Life Sci. 13:545-552, 1973.
197. Manganello, V. and Vaughan, M.: An effect of dexamethasone on adenosine 3'-5' monophosphate content and adenosine 3'-5' monophosphate phosphodiesterase activity in cultured hepatoma cells. J. Clin. Invest. 51: 2763-2767, 1967.
198. Kleeman, C. R., Levi, J., and Better, O.: Kidney and adrenocortical hormones. Nephron 15:261-278, 1975.
199. Cutler, R. E., Kleeman, C. R., Koplowitz, J., Maxwell, M. H., and Dowling, J. T.: Mechanism of impaired water excretion in adrenal and pituitary insufficiency. III. The effect of extracellular or plasma volume expansion or both on the impaired diuresis. J. Clin. Invest. 41:1524-1530, 1962.
200. Dingman, J. F., Finkenstaedt, J. T., Laidlow, J. C., Renold, A. E., Jenkins, D., Merrill, J. P., and Thorn, G. W.: Influence of intravenous adrenal steroids on sodium and water excretion in Addisonian subjects. Metabolism 7:608-623, 1958.
201. Garrod, O., Davies, S. A., and Cahill, G., Jr.: The action of cortisone and desoxycortisone acetate on glomerular filtration rate and sodium and water excretion in adrenalectomized dog. J. Clin. Invest. 34:761-776, 1955.

202. Gill, J. R., Gann, D. S., and Bartter, F. C.: Restoration of water diuresis in Addisonian patients by restoration of extracellular fluid volume. J. Clin. Invest. 41:1078-1085, 1962.
203. Kleeman, C. R., Maxwell, M. H., and Rockney, R. E.: Mechanism of impaired water excretion in adrenal and pituitary insufficiency. I. The role of altered glomerular filtration rate and solute excretion. J. Clin. Invest. 37:1799-1812, 1959.
204. Raisz, L. G., McNeely, W. F., Saxon, L., and Rosenbaum, J. D.: Effects of cortisone and hydrocortisone on water diuresis and renal function in man. J. Clin. Invest. 36:767-779, 1957.
205. Kleeman, C. R., Koplowitz, J., Maxwell, M. H., Cutler, R., and Dowling, J. T.: Mechanisms of impaired water excretion in adrenal and pituitary insufficiency in normal subjects and in diabetes insipidus. II. Interrelationship of adrenal cortical steroids and antidiuretic hormone. J. Clin. Invest. 39:1472-1482, 1960.
206. Yunis, S. L., Bercovitch, D. D., Stein, R. M., Levitt, M. F., and Goldstein, M. H.: Renal tubular effects of hydrocortisone and aldosterone in normal hydropenic man. Comments on sites of action. J. Clin. Invest. 43:1668-1676, 1964.
207. Reforzo-Membrives, J., Power, M. H., and Kepler, E. J.: Studies on the renal excretion of water and electrolytes in cases of Addison's disease. J. Clin. Endocrinol. 5:76-85, 1945.
208. Bethune, J. and Nelson, D.: Hyponatremia and hypopituitarism. New Eng. J. Med. 272:771-779, 1965.
209. Cook, C. R. and Steenburg, R. W.: Effect of aldosterone and cortisol on the renal concentrating mechanism. J. Lab. Clin. Med. 82:784-794, 1973.
210. Jick, H., Snyder, J. G., and Moore, E. W.: The effect of aldosterone and glucocorticoid on free water reabsorption. Clin. Sci. 29:25-33, 1965.
211. Ahmed, A. B., George, B. C., Gonzales-Auwert, C., and Dingman, J. F.: Increased plasma arginine vasopressin in clinical adrenocortical insufficiency and its inhibition by glucocorticoids. J. Clin. Invest. 46:111-123, 1967.
212. Dingman, J. F. and Despointes, R. H.: Adrenal steroid inhibition of vasopressin release from the neuro-hypophysis of normal subjects and patients with Addison's disease. J. Clin. Invest. 39:1851-1863, 1960.
213. George, C. P., Messerli, F. H., Genest. J., Nowaczynski, W., Boucher, R., Kuchel, O., and Rojo-Ortega, M.: Diurnal variation of plasma vasopressin in man. J. Clin. Endocrinol. Metab. 41:332-338, 1975.
214. Aubry, R. H., Nankin, H. R., Moses, A. M., and Streeten, D. H. P.: Measurement of the osmotic threshold for vasopressin release in human subjects and its modification by cortisol. J. Clin. Endocrinol. Metab. 25:1481-1492, 1965.
215. Fichman, M. P., Vorherr, H., Kleeman, C. R., and Telfer, N.: Diuretic induced hyponatremia. Ann. Intern. Med. 75:853-863, 1971.
216. Kleeman, C. R., Czaczkes, J. W., and Cutler, R.: Mechanisms of impaired water excretion in adrenal and pituitary insufficiency. IV. Antidiuretic hormones in primary and secondary insufficiency. J. Clin. Invest. 43:1641-1648, 1964.

217. Green, H. M., Harrington, A. R., and Valtin, H.: On the role of antidiuretic hormone in the inhibition of active water diuresis in adrenal insufficiency and the effects of gluco- and mineralo-corticoids in reversing the condition. J. Clin. Invest. 49:1724-1736, 1970.
218. Katz, A. I., Emmanouel, D. S., and Lindheimer, M. D.: Thyroid hormone and the kidney. Nephron 15:223-249, 1975.
219. Bradley, S. E., Stephan, F., Coelho, J. B., and Reville, P.: The thyroid and the kidney. Kidney Int. 8:346-363, 1974.
220. Ismail-Beigi, F. and Edelman, I. S.: The mechanism of the calorigenic action of thyroid hormone. Stimulation of Na^+-K^+ activated adenosine triphosphatase activity. J. Gen. Physiol. 51:710-722, 1971.
221. Katz, A. I. and Lindheimer, M. P.: Renal sodium and potassium activated adenosine triphosphatase and sodium reabsorption in the hypothyroid rat. J. Clin. Invest. 52:796-804, 1973.
222. Marusic, E. and Torretti, J.: Synergistic action of vasopressin and thyroxine on water transport in the isolated toad bladder. Nature 202:1118-1119, 1964.
223. Vaughan, M.: An in vitro effect of triiodothyronine on rat adipose tissue. J. Clin. Invest. 46:1482-1491, 1967.
224. Mandel, L. R. and Kuehl, F. A., Jr.: Lipolytic action of 3,3'5-triiodo-L-thyronine, a cyclic AMP phosphodiesterase inhibitor. Biochem. Biophys. Res. Comm. 28: 13-18, 1967.
225. Krishna, G., Hynie, S., and Brodie, B. B.: Effects of thyroid hormones on adenyl cyclase in adipose tissue and on free fatty acid mobilization. Proc. Nat. Acad. Sci. U.S.A. 59:884-889, 1968.
226. Nelson, T. E. and Stouffer, J. E.: Thyroxine modulation of epinephrine stimulated secretion of rat parotid amylase. Biochem. Biophys. Res. Comm. 48:480-485, 1972.
227. Stouffer, J. E. and Nelson, T. E.: Thyroxine modulation of epinephrine stimulated secretion of rat parotid alpha-amylase. Fed. Proc. 30:1321, 1971.
228. Weiss, I. W., Harris, S. C., Downing, S. J., and Phang, J. M.: Cyclic AMP mediated changes in Na-K ATPase from rat kidney cortex. Clin. Res. 19:488, 1971.
229. DeRobertis, F. R., Michelis, M. F., Bloom, M. E., Mintz, D. H., Field, J. B., and Davis, B. B.: Impaired water excretion in myxedema. Am. J. Med. 51:41-53, 1971.
230. DiScala, V. A. and Kinney, M. J.: Effects of myxedema on the renal diluting and concentrating mechanism. Am. J. Med. 30:325-335, 1971.
231. Crispell, K. R., Parson, W., and Sprinkle, P.: A cortisone resistant abnormality in the diuretic response to ingested water in primary myxedema. J. Clin. Endocrinol. Metab. 14:640-644, 1954.
232. Emmanouel, D. S., Lindheimer, M. D., and Katz, A. I.: Mechanism of impaired water excretion in the hypothyroid rat. J. Clin. Invest. 54:926-934, 1974.
233. Curtis, R. H.: Hyponatremia in primary myxedema. Ann. Int. Med. 44: 376-385, 1956.
234. Goldberg, M. and Reivich, M.: Studies on the mechanism of hyponatremia

impaired water excretion in myxedema. Ann. Intern. Med. 56:120-130, 1962.
235. Pettinger, W. A., Talner, L., and Ferris, T. F.: Inappropriate secretion of antidiuretic hormone to myxedema. New Eng. J. Med. 272:362-364, 1965.
236. Sterling, F. H., Richter, J. S., and Giampetro, A. M.: Inappropriate antidiuretic hormone secretion and myxedema: Hazards in management. Am. J. Med. Sci. 253:697-699, 1967.
237. Levey, G. S. and Epstein, S. E.: Myocardial adenyl cyclase-activation by thyroid hormones and evidence for two adenyl cyclase systems. J. Cin. Invest. 48:1663-1669, 1969.
238. Levey, G. S., Shelton, C. L., and Epstein, S. E.: Decreased myocardial adenyl cyclase activity in hypothyroidism. J. Clin. Invest. 48:2240-2250, 1969.
239. Bleifer, K. H., Belsky, J. L., Saxon, L., and Papper, S.: The diuretic response to administered water in patients with myxedema. J. Clin. Endocrinol. Metab. 20:409-414, 1960.
240. Skowsky, R., Nielsen, T., and Fisher, D.: Arginine vasopressin kinetics in the thyroidectomized sheep. Clin. Res. 22:168a, 1974.
241. Guttler, R. B., Shaw, J. W., Otis, C. L., and Nicoloff, J. T.: Epinephrine induced alterations in urinary cyclic AMP in hyper- and hypothyroidism. J. Clin. Endocrinol. Metab. 41:707-711, 1975.
242. Lin, T., Kopp, L. E., and Tucci, J. R.: Urinary excretion of cyclic-3',5'-adenosine monophosphate in hyperthyroidism. J. Clin. Endocrinol. Metab. 36:1033-1036, 1973.
243. Hlad, C. J., Jr. and Bricker, N. S.: Renal function and I^{131} clearance in hyperthyroidism and myxedema. J. Clin. Endocrinol. Metab. 14:1539-1550, 1954.
244. Ford, R. V., Owens, J. C., Curd, G. W., Jr., Moyer, J. H., and Spurr, C. L.: Kidney function in various thyroid states. J. Clin. Endocrinol. Metab. 21:548-553, 1961.
245. Bradley, S. E.: Renal function, In: The Thyroid, S. C. Werner and S. H. Ingbar (eds.) pp. 615-623, Harper & Row, New York, 2nd edition, 1962.
246. Dix, A. S., Rogoff, J. M., and Barnes, B. O.: Diuresis of hyperthyroidism. Proc. Soc. Exp. Biol. Med. 32:616-618, 1935.
247. Gaunt, R., Cordsen, M., and Liling, M.: Water intoxication in relation to thyroid and adrenal function. Endocrinology 35:105-111, 1944.
248. Liu, S. H.: The effect of thyroid medication in nephrosis. Arch. Intern. Med. 40:73-79, 1927.
249. Hare, K., Phillips, D. M., Bradshaw, J., Chambers, G., and Hare, R. S.: The diuretic action of thyroid in diabetes insipidus. Am. J. Physiol. 144:187-195, 1944.
250. Cutler, R. E., Glatte, H., and Dowling, J. T.: Effect of hyperthyroidism on the renal concentrating mechanism in humans. J. Clin. Endocrinol. Metab. 27:453-460, 1967.
251. Leaf, A., Mamby, A. R., Rasmussen, H., and Marasco, J. P.: Some hormonal aspects of water excretion in man. J. Clin. Invest. 31:914-927, 1952.
252. Epstein, F. H. and Rivera, M. J.: Renal concentrating ability in thyro-

toxicosis. J. Clin. Endocrinol. Metab. 18:1135-1137, 1958.
253. Epstein, F. H., Friedman, L. R., and Levitin, H.: Hypercalcemia, nephrocalcinosis and reversible renal insufficiency associated with hyperthyroidism. New Eng. J. Med. 258:782-785, 1958.
254. Handler, J. R., Peterson, M., and Orloff, J.: Effect of metabolic inhibitors on the response of the toad bladder to vasopressin. Am. J. Physiol. 211: 1175-1180, 1966.
255. Bentley, P. J.: The physiology of the urinary bladder of amphibia. Biol. Rev. 41:275-316, 1966.
256. Bentley, P. J.: The effects of N-ethylmaleimide and glutathione on the isolated rat uterus and frog bladder with special reference to the action of oxytocin. J. Endocrinol. 30:103-113, 1964.
257. Handler, J. S. and Orloff, J.: Cysteine effect on toad bladder response to vasopressin, cyclic AMP, and theophylline. Am. J. Physiol. 206:505-509, 1964.
258. Cobb, F. R. and McManus, T. J.: Inhibition of neurohypophyseal hormone action by ethacrynic acid. In: Abstracts of the 3rd International Congress of Nephrology, Washington, D.C., Sept. 25-30, p. 172, 1966.
259. Ferguson, D. R.: Effects of furosemide and arginine vasopressin on sodium transport and movement of water by the isolated toad bladder. Ibid., p. 188.
260. Jenny, M. J., deSousa, R. C., and Junod, A. F.: Renal effect of various diuretic agents in conditions with hypervasopressinism. Ibid., p. 216.
261. Patak, R. V., Mookerjee, B. K., Bentzel, C. J., Hysert, P. E., and Lee, J. B.: Abolition by indomethacin of the antihypertensive and natriuretic effects of furosemide in normal and hypertensive man. International Congress on Prostaglandins, Florence, Italy, (Abstr.) 181: 1975.
262. Moore, P. F.: Effects of diazoxide and benzothiadiazine diuretics upon phosphodiesterase. Ann. N.Y. Acad. Sci. 150:256-260, 1968.
263. Schultz, G., Senft, G., Losert, W., and Sitt, R.: Biochemische Grundlagen der Diazoxid—Hyperglykamie. Arch. Pharm. Exp. Pathol. 253:372-387, 1966.
264. Brickman, A. S., Massry, S. G., and Coburn, J. W.: Changes in serum and urine calcium during treatment with hydrochlorothiazide: Studies on mechanisms. J. Clin. Invest. 52:945-954, 1972.
265. Crawford, J. D. and Kennedy, G.: Animal physiology: Chlorathiazide in diabetes insipidus. Nature 183:891-892, 1959.
266. Skadhauge, E.: Studies of the antidiuresis induced by natrichloriuretic drugs in rats with diabetes insipidus. Quart. J. Exp. Physiol. 51:297-310, 1966.
267. Earley, L. E. and Orloff, J.: The mechanism of antidiuresis associated with the administration of hydrochlorothiazide to patients with vasopressin-resistant diabetes insipidus. J. Clin. Invest. 41: 1988-1997, 1962.
268. Bartorelli, C., Gargano, N., and Leonetti, G.: Hypotensive and renal effects of diazoxide: A sodium retaining benzothiadiazine compound. Circulation 27:895-903, 1963.
269. Pohl, J. E. F., Thurston, H., and Swales, J. O.: The antidiuretic action of diazoxide. Clin. Sci. 42:145-152, 1972.

270. Senft, G., Munske, K., Schultz, G., and Hoffman, M.: Der Einfluss von Hydrochlorothiazid und anderen sulfonamidierten Diuretica auf die 3',5'-AMP-Phosphodiesterase-Aktivitat in der Ratteniere. Arch. Pharmakol. Exp. Patrol. 254:344-359, 1968.
271. Fichman, M. P., Speckart, P., Zia, P., and Lee, A.: Antidiuretic response to prostaglandin inhibition by indomethacin in nephrogenic diabetes insipidus. Clin. Res. 34:XXX, 1976.
272. Moses, A. M. and Miller, M.: Drug-induced dilutional hyponatremia. New Eng. J. Med. 291:1234-1279, 1972.
273. Arduino, F., Ferraz, F. P. J., and Rodriquez, J.: Antidiuretic action of chlorpropamide in idiopathic diabetes insipidus. J. Clin. Endocrinol. Metab. 26:1325-1328, 1966.
274. Ehrlich, R. M. and Kooh, S. W.: Oral chlorpropamide in diabetes insipidus. Lancet 1:890, 1969.
275. Meinders, A. E., Touber, J. L., and DeVries, L. A.: Chlorpropamide treatment in diabetes insipidus. Lancet 2:544-546, 1967.
276. Kunstadter, R. H., Cabana, E. C., and Oh, W.: Treatment of vasopressin-sensitive diabetes insipidus with chlorpropamide. Am. J. Dis. Child. 117:436-441, 1969.
277. Froyshov, I. and Haugen, H. N.: Chlorpropamide treatment in diabetes insipidus. Acta Med. Scand. 183:397-400, 1968.
278. Reforzo-Membrives, J., Moledo, L. I., Lanaro, A. E., and Megias, A.: Antidiuretic effect of 1-propyl-3-P-chlorobenzene-sulfonylurea (chlorpropamide). J. Clin. Endocrinol. Metab. 28:332-336, 1968.
279. Ehrlich, R. M. and Kooh, S. W.: The use of chlorpropamide in diabetes insipidus in children. Pediatrics 45:236-245, 1970.
280. Vallet, H. L., Prassad, M., and Goldbloom, R. B.: Chlorpropamide treatment of diabetes insipidus in children. Pediatrics 45:246-253, 1970.
281. Hocken, A. G. and Longson, D.: Reduction of free water clearance by chlorpropamide. Br. Med. J. 1:355-356, 1968.
282. Driedger, A. A. and Linton, A. L.: Familial ADH-responsive diabetes insipidus: Response to thiazides and chlorpropamide. Can. Med. Assoc. J. 109:594-597, 1973.
283. Webster, B. and Bain, J.: Antidiuretic effect and complications of chlorpropamide therapy in diabetes insipidus. J. Clin. Endocrinol. Metab. 30:215-227, 1970.
284. Miller, M. and Moses, A. M.: Mechanism of chlorpropamide action in diabetes insipidus. J. Clin. Endocrinol. Metab. 30:488-496, 1970.
285. Wales, J. K. and Fraser, T. R.: The clinical use of chlorpropamide in diabetes insipidus. Acta Endocrinol. (Kbh) 68:725-736, 1971.
286. Bricaire, H., Saltiel, H., and Schaison, G.: Traitement du diabete insipide par le chlorpropamide. Ann. Endocrinol. 30:61-70, 1969.
287. Rado, J. P., Szende, L., and Borbely, L.: Clinical value and mode of action of chlorpropamide in diabetes insipidus. Am. J. Med. Sci. 260:359-372, 1970.
288. Miller, M. and Moses, A. M.: Potentiation of vasopressin action by chlorpropamide in vivo. Endocrinology 86:1024-1027, 1970.
289. Berndt, W. O., Miller, M., Kettyl, W. M., and Valtin, H.: Potentiation of

the antidiuretic action of vasopressin by chlorpropamide. Endocrinology 86:1028-1032, 1970.
290. Zweig, S. M., Ettinger, B., and Earley, L. E.: Mechanism of antidiuretic action of chlorpropamide in the mammalian kidney. Am. J. Physiol. 221: 911-915, 1971.
291. Mendoza, S.: Effect of chlorpropamide on the permeability of the urinary bladder of the toad and the response to vasopressin adenosine 3'-5' monophosphate and theophylline. Endocrinology 84:411-416, 1969.
292. Ingelfinger, J. R. and Hays, R. M.: Evidence that chlorpropamide and vasopressin share a common site of action. J. Clin. Endocrinol. Metab. 29:738-740, 1969.
293. Moses, A. M., Numann, P., and Miller, M.: Mechanism of chlorpropamide induced antidiuresis in man. Evidence for release of ADH and enhancement of peripheral action. Metabolism 22:59-66, 1973.
294. Murase, T. and Yoshida, S.: Mechanism of chlorpropamide action in patients with diabetes insipidus. J. Clin. Endocrinol. Metab. 36:174-177, 1973.
295. Kleeman, C. R., Rubini, M. E., and Lamdin, E.: Studies on alcohol diuresis. II. The evaluation of ethyl alcohol as an inhibitor of the neurohypophysis. J. Clin. Invest. 34:448-455, 1955.
296. Hagen, G. A. and Frawley, T. F.: Hyponatremia due to sulfonylurea compounds. J. Clin. Endocrinol. Metab. 31:570-575, 1970.
297. Garcia, M., Miller, M., and Moses, A. M.: Chlorpropamide induced water retention in patients with diabetes mellitus. Ann. Intern. Med. 75:549-554, 1971.
298. Fine, D. and Shedrovilzky, H.: Hyponatremia due to chlorpropamide. A syndrome resembling inappropriate secretion of antidiuretic hormone. Ann. Intern. Med. 72:83-87, 1970.
299. Weissman, P. N., Shenkman, L., and Gregerman, R. I.: Chlorpropamide hyponatremia: Drug induced inappropriate antidiuretic hormone activity. New Eng. J. Med. 284:65-71, 1971.
300. Cinotti, G. A., Stivati, G., and Ruggiero, F.: Abnormal water retention and symptomatic hyponatremia in idiopathic diabetes insipidus during chlorpropamide therapy. Post Grad. Med. J. 48:107-111, 1972.
301. Fichman, M. P. and Telfer, N.: Unusual hyponatremic syndromes. Clin. Res. 19:195, 1971.
302. Miller, M. and Moses, A. M.: Radioimmunoassay of urinary antidiuretic hormone in man: Response to water load and dehydration in normal subjects. J. Clin. Endocrinol. Metab. 34:537-545, 1972.
303. Robertson, G. L., Mahr, E. A., Athars, S., and Sinha, T.: The development and clinical application of a new method for the radioimmunoassay of arginine vasopressin in human plasma. J. Clin. Invest. 52:2340-2352, 1973.
304. Robertson, G. L. and Mahr, E. A.: Mechanism of chlorpropamide antidiuresis in diabetes insipidus studies with a new assay for plasma vasopressin. Endocrinology 88:A105 1971 (Suppl.).
305. Lozada, E. S., Gouaux, J., Franki, N., Appel, G. B., and Hays, R. M.: Studies of the mode of action of the sulfonylureas and phenylacetamides

in enhancing the effect of vasopressin. J. Clin. Endocrinol. Metab. 34:704-712, 1972.
306. Levey, G. S., Valmer, R. F., Lasseter, K. C., and McCarthy, J.: Effect of tolbutamide on adenyl cyclase in rabbit and human heart and contactility of isolated rabbit atria. J. Clin. Endocrinol. & Metab. 33:371-374, 1971.
307. Brooker, G. and Fichman, M. P.: Chlorpropamide and tolbutamide inhibition of adenosine 3'-5' cyclic monophosphate phosphodiesterase. Biochem. Biophys. Res. Comm. 42:824-828, 1971.
308. Roth, J., Prout, T. E., Goldfine, I. D., and Parcus, M. L.: Sulfonylureas: Effects in vivo and in vitro. Ann. Intern. Med. 75:607-625, 1971.
309. Urakabe, S. and Shirai, D.: Effect of vasopressin, cyclic 3'-5' AMP and chlorpropamide on water permeability of toad urinary bladder. Med. J. Osaka Univ. 21:151-159, 1971.
310. Urakabe, S., Shirai, D., Ando, A., Takamitsu, Y., Orita, Y., and Abe, H.: Effect of the sulfonylureas on the permeability to water and electrical properties of the urinary bladder of the toad. Jap. Circ. J. 34:595-601, 1970.
311. Zgliczynski, S.: Antidiuretic effect of sulfonylureas in diabetes insipidus. Helv. Medica Acta 34:478-485, 1969.
312. Luethi, A. and Studer, H.: Antidiuretic action of chlorpropamide and tolbutamide. Minn. Med. 52:33-36, 1969.
313. Lambert, A. E., Jeanrenaud, B., and Renold, A. E.: Enhancement by caffeine of glucagon-induced and tolbutamide-induced insulin release from isolated foetal pancreatic tissue. Lancet 1:819-820, 1967.
314. University Group Diabetes Program. Diabetes 19:747-830, 1970 (Suppl. 2).
315. Moses, A. M., Howanitz, J., and Miller, M.: Diuretic action of three sulfonylurea drugs. Ann. Intern. Med. 78:541-544, 1973.
316. Rado, J. P. and Borbely, L.: Glybenclamide enhancement of polyuria in patients with pituitary diabetes insipidus. Endokrinologie 59:397-402, 1972.
317. Van Genert, M., Miller, M., and Moses, A. M.: Sulfonylurea induced diuresis. Fed. Proc. 32:737, 1973.
318. Katsuki, S. and Ito, M.: Antidiuretic effects of diguanides. Lancet 2:530-532, 1966.
319. Eisenberg, E.: The mechanism of the diguanide effect on diabetes insipidus. Clin. Res. 18:168, 1970.
320. Villamor, J., Mita, J., and Barreiro, P.: Diabetes insipidus primaria familiar vasopressin resistente y sensible a la biguanida. Rev. Clin. Esp. 121:51-62, 1971.
321. Nusonywitz, M. L. and Forsham, P. H.: The antidiuretic action of acetominophen. Am. J. Med. Sci. 252:429-435, 1966.
322. Nusonywitz, M. L., Wegienka, L. C., Bower, B. F., and Forsham, P. H.: Effect on vasopressin action of analgesic drugs in vitro. Am. J. Med. Sci. 252:424-428, 1966.
323. Shirai, D., Ubake, S., Ando, A., Takamitsu, Y., Orita, Y., and Abe, H.: Effect of urea derivatives and analgesics on the permeability to water

and electrical properties of the urinary bladder of the toad. Jap. Circ. J. 34:603-608, 1970.
324. Bower, B. F., Wegienka, L. C., and Forsham, P. H.: In vitro studies of the mechanism of polyuria induced by dextro propokyphene (Darvon). Proc. Soc. Exp. Biol. Med. 120:155-157, 1965.
325. Spuhler, O. and Zollinger, H. U.: Die chronisch-interstitielle nephritis. Z. Klin. Med. 151:1-50, 1953.
326. McCarthy, W. H. and Keenan, R. L.: Propoxyphene hydrochloride poisoning: report of the first fatality. JAMA 187:460-461, 1964.
327. Singer, I. and Rotenburg, D.: Demeclocycline-induced nephrogenic diabetes insipidus: In vivo and in vitro studies. Ann. Intern. Med. 79:679-683, 1973.
328. Castell, D. O. and Sparks, H. A.: Nephrogenic diabetes insipidus due to demethylchlortetracycline hydrochloride. JAMA 193:237-239, 1965.
329. Roth, H., Becker, K. L., and Shalhoub, R. S.: Nephrotoxicity of demethylchlortetracycline hydrochloride. Arch. Intern. Med. 120:433-435, 1967.
330. Wilson, D. M., Perry, H. O., and Sams, W. M. J.: Selective inhibition of human distal tubular function by demeclocycline. Curr. Ther. Res. 15: 734-740, 1973.
331. Mazze, R. I., Cousins, M. J., and Kosek, J. C.: Dose related methoxyflurane nephrotoxicity in rats. Anesthesiology 36:571-587, 1972.
332. Mazze, R. I., Tridell, J. R., and Cousins, M. J.: Methoxyflurane metabolism and renal dysfunction. Anesthesiology 35:247-252, 1971.
333. Wosilait, W. D. and Sutherland, E. W.: Relationship of epinephrine and glucagon to liver phosphorylase, liver phosphorylase, preparation and properties. J. Biol. Chem. 218:459-468, 1956.
334. Taylor, A. and Reaven, E.: Microtubule content of toad bladder epithelialcells: effect of colchicine and vasopressin. VI. Int. Congress of Nephrology, Abstracts of Free Communication, June 8-12, 1975, Florence, Italy, 143.
335. Dousa, T. P. and Barnes, L. P.: Effect of colchicine and vinblastine on the cellular action of vasopressin in mammalian kidney, a possible role of microtubules. J. Clin. Invest. 54: 252-262, 1974.
336. Anderson, N. B.: Effect of general anaesthetics on sodium transport in the isolated toad bladder. Anesthesiology 23:304-310, 1966.
337. Braunkofer, J. and Zicha, L.: Eröffnet Tegretal neue Therapiemoglichkeiten bei bestimmten neurologischen und endokrinen Krankheitsbildern? Med. Welt. 17:1875-1880, 1966.
338. Frahm, H. and Smejkal, V.: Hemmung der Polysipsie und Polyurie durch Tegretal bei hypophysenoperierten Patienten und bei kranken Diabetes Insipidus. Med. Welt. 20:1529-1533, 1969.
339. Tietz, H. U. and Finkenwirth, M.: Beeinflussung des Diabetes Insipidus durch Tegretal. Monatsschr. Kinderheilkd. 118:237-238, 1970.
340. Rado, J. P.: Clinical use of additive antidiuretic action of carbamazepine and chlorpropamide. Horm. Metab. Res. 5:309, 1973.
341. Rado, J. P.: Combination of carbamazepine and chlorpropamide in the treatment of "hyporesponder" pituitary diabetes insipidus. J. Clin. Endocrinol. Metab. 38:1-7, 1974.

342. Rado, J. P.: Water intoxication during carbamazepine treatment. Br. Med. J. 3:479, 1973.
343. Frahm, M., Smejkal, E., and Kratzenstein, R.: Antidiuretic effect of an anticonvulsant drug (5-carbamyl-5H-dibenzo (BF) azepin = Tegretol) associated with measurable increase of ADH activity in serum of patients suffering from diabetes insipidus and of patients with polyuria and polydipsia following hypophysectomy. Acta Endocr. (Kbh) 138:240, 1969 (Suppl.).
344. Kimura, T., Matsui, K., and Sato, I.: Mechanism of carbamazepine (Tegretol) induced antidiuresis: evidence for release of antidiuretic hormone and impaired excretion of a water load. J. Clin. Endocrinol. Metab. 38: 356-362, 1974.
345. Tietze, H. U., Oetliker, O. H., and Chattas, A.: Carbemyldibenzo-azepine in the treatment of diabetes insipidus in children. IV. International Congress of Endocrinology (Int. Cong. Series 256), Amsterdam, Excerpta Med., p. 131, 1972.
346. Uhlich, E., Loeschke, K., and Eigler, J.: Zur antidiuretischen Wirkung von Carbamazepin bei Diabetes Insipidus. Klin. Wochenschr. 50:1127-1133, 1972.
347. Baisset, A., Cotonat, J., and Dumas, J. L.: Recherches sur l'action antidiuretique de la carbamazépine. Therapie 28:663-69, 1973.
348. Bonnici, F.: Antidiuretic effect of clofibrate and carbamazebine in diabetes insipidus. Studies on free water clearance and response to a water load. Clin. Endocrinol. 2:265-275, 1973.
349. Schaison, G.: Diversiste des traitments actuels du diabete insipide. Presse Med. 79:561-563, 1971.
350. Uhlich, E., Loeschke, K., Eigler, J., and Holbach, R.: Clofibrat bei Diabetes Insipidus. Klin. Wochenschr. 49:436-437, 1971.
351. deGennes, J. L., Bertrand, C., Bigorie, B., and Truffert, J.: E'tudes preliminaires de l'action antidiuretique du clofibrate (ou atromide S) dans le diabéte insipide pitressosensible. Ann. Endocrinol. (Paris) 31:300-308, 1970.
352. de Gennes, J. L., Desbois, J. C., and Marie, J.: Etude therapeutique du clofibrate au cours des diabetes insipides pitresso-sensibles de l'enfant. Ann. Pediatr. (Paris) 17:754-759, 1970.
353. Moses, A. M., Howanitz, J., Van Gemert, M., and Miller, M.: Clofibrate induced antidiuresis. J. Clin. Invest. 53:535-542, 1973.
354. Baisset, A., Cotonat, J., and Dumas, J. C.: Recherches sur l'action antiuretique du clofibrate. Therapie 28:651-661, 1973.
355. Fine, R. N., Clark, R. R., and Shore, N. A.: Hyponatremia and vincristine therapy. Syndrome possibly resulting from inappropriate antidiuretic hormone secretion. Am. J. Dis. Child. 112:256-259, 1966.
356. Oldham, R. K. and Pomeroy, T. C.: Vincristine induced syndrome of inappropriate secretion of antidiuretic hormone. South. Med. J. 65:1010-1012, 1972.
357. Cutting, H. O.: Inappropriate secretion of antidiuretic hormone secondary to vincristine therapy. Am. J. Med. 51:269-271, 1971.
358. Nicholson, R. G. and Feldman, W.: Hyponatremia in association with

vincristine therapy. Can. Med. Assoc. J. 106:356-357, 1972.
359. Rosenthal, S. and Kaufman, S.: Vincristine neurotoxicity. Ann. Int. Med. 80:733-737, 1974.
360. Haggard, M. E., Feinback, D. J., Holcomb, T. M., Sutow, W. W., Vietti, T. J., and Windmiller, J.: Vincristine in acute leukemia of childhood. Cancer 22:438-444, 1968.
361. Slater, L. M., Warner, R. A., and Serpick, A. A.: Vincristine neurotoxicity with hyponatremia. Cancer 23:122-125, 1969.
362. Steele, T. M., Serpick, A. A., and Block, J. B.: Antidiuretic response to cyclophosphamide in man. J. Pharm. Exp. Ther. 185:245-253, 1973.
363. DeFronzo, R. A., Braine, H., Colvin, M., and Davis, P. J.: Water intoxication in man after cyclophosphamide therapy: Time course and relation to drug activation. Ann. Intern. Med. 78:861-869, 1973.
364. Schwartz, W. B., Bennett, W., Curelop, S., and Bartter, F. C.: A syndrome of renal sodium loss and hyponatremia probably resulting from inappropriate secretion of antidiuretic hormone. Am. J. Med. 23:529-542, 1957.
365. Bartter, F. C. and Schwartz, W. B.: The syndrome of inappropriate secretion of antidiuretic hormone. Am. J. Med. 42:790-806, 1967.
366. Rufener, C., Nordmann, J., and Rouiller, C.: Effect of vincristine on the rat posterior pituitary in vitro. Neurochirurgie 18:137-141, 1972.
367. Robertson, G. L., Bhoopalam, N., and Zelkowitz, L. J.: Vincristine neurotoxicity and abnormal secretion of antidiuretic hormone. Arch. Intern. Med. 132:717-720, 1973.
368. Zedeck, M. S., Mellett, L. B., and Cafruny, E. J.: The diuretic effects of cyclophosphamide and nor-nitrogen mustard: relationship to antidiuretic hormone. J. Pharm. Exp. Ther. 153:550-561, 1966.
369. Levin, S. R., Booker, J., Smith, D., and Grodsky, G.: Inhibition of insulin secretion by diphenylhydantoin in the isolated perfused pancreas. J. Clin. Endocrinol. Metab. 30:400-401, 1970.
370. Avery, S., Clark, C. M., Jr., Trygstad, C., and Bell, N. M.: Effects of cyclic adenosine 3'5' monophosphate (AMP) and dibutyryl cyclic AMP in antidiuretic hormone deficient and antidiuretic hormone resistant diabetes insipidus. J. Clin. Invest. 50:3a, 1971 (Abstr.).
371. Bode, H. H., Harley, B. M., and Crawford, J.: Restoration of normal drinking behavior by chlorpropamide in patients with hypodipsia and diabetes insipidus. Am. J. Med. 51:304-313, 1971.
372. Orloff, J. and Burg, M. B.: Vasopressin-resistant diabetes insipidus. In: The Metabolic Basis of Inherited Disease. J. B. Stanbury, Wyngaarden, J. B., and Frederickson, D. S. (eds.) pp. 1247-1261, McGraw Hill, New York, 1960.
373. Bell, N. H., Clark, C. M., Jr., Avery, S., Sinha, C., Trygstad, L., and Allen, D. O.: Demonstration of defects in the formation of and response to cyclic adenosine 3'-5' monophosphate in vasopressin resistant diabetes insipidus. Clin. Res. 20:586, (Abstr.).
374. Dousa, T. P. and Valtin, H.: Cellular action of antidiuretic hormone in mice with inherited vasopressin resistant urinary concentrating defects. J. Clin. Invest. 54:753-762, 1974.

375. Harrington, J. T. and Cohen, J. J.: Clinical disorders of urinary concentration and dilution. Arch. Intern. Med. 131:810-825, 1973.
376. Epstein, F. H.: Disorders of renal concentrating ability. Yale J. Biol. Med. 39:186-195, 1966.
377. Sherwood, L. M., O'Riordan, J. L. M., Aurbach, G. D., and Potts, J. T.: Production of parathyroid hormone by non-parathyroid tumors. J. Clin. Endocrinol. Metab. 27:140-146, 1967.
378. Thomas, A. N., Loken, H. F., Gordon, G. S., and Goldman, L.: Hypercalcemia of metastatic breast cancer. Surg. Forum 11:70-71, 1960.
379. Brereton, H. D., Halushka, P. V., Alexander, R. W., Mason, D. M., Keiser, H. R., and DeVita, V. T.: Indomethacin responsive hypercalcemia in a patient with renal-cell adenocarcinoma. New Eng. J. Med. 291:83-85, 1974.
380. Fichman, M. and Bethune, J.: Effects of neoplasms on renal electrolyte function. Ann. N.Y. Acad. Sci. 230:448-472, 1974.
381. Moses, A. M. and Spencer, H.: Hypercalcemia in patients with malignant lympoma. Ann. Intern. Med. 59:531-536, 1963.
382. Kronfeld, S. J. and Reynolds, T. B.: Leukemia and hypercalcemia, report of a case and review of the literature. New Eng. J. Med. 271:399-401, 1964.
383. James, D. G., Siltzbach, L. E., Sharma, O. P., and Carstairs, L. P.: A tale of two cities. A comparison of sarcoidosis in London and New York. Arch. Intern. Med. 123:187-191, 1969.
384. Tumulty, P. A. and Howard, J. E.: Irradiated Ergosterol poisoning. JAMA 119:233-236, 1942.
385. Burnett, C. H., Commons, R. R., Albright, F., and Howard, J. E.: Hypercalcemia without hypercalciuria or hypophosphatemia, calcinosis or renal insufficiency. A syndrome following prolonged intake of milk or alkalae. New Eng. J. Med. 240:787-794, 1949.
386. Baxter, J. D. and Bondy, P. K.: Hypercalcemia of thyrotoxicosis. Ann. Intern. Med. 65:429-442, 1966.
387. McIntosh, D. A., Peterson, E. W., and McPharl, J. J.: Autonomy of parathyroid function after renal homotransplantation. Ann. Intern. Med. 65:900-907, 1966.
388. Kurtzman, N. A. and Boonjavern, S.: Physiology of antidiuretic hormone and the interrelationship between the hormone and the kidney. Nephron 15:167-185, 1975.
389. Beck, N., Singh, H., Reed, S. W., Murdaugh, H. V., and Davis, B.: Pathogenic role of cyclic AMP in the impairment of urinary concentrating ability in acute hypercalcemia. J. Clin. Invest. 54:1049-1055, 1974.
390. Epstein, F. and Whittom, R.: The mode of inhibition by calcium of cell-membrane adenosine-triphosphatase activity. Biochem. J. 99:232-238, 1966.
391. Manitius, A., Levitin, H., Beck, D., and Epstein, F. H.: On the mechanism of impairment of renal concentrating ability in hypercalcemia. J. Clin. Invest. 39:693-697, 1960.
392. Bennett, C. M.: Urine concentration and dilution in hypokalemic and hypercalcemic dogs. J. Clin. Invest. 49:1447-1457, 1970.
393. Black, D. A.: Potassium metabolism. In: Clinical Disorders of Fluid and

Electrolyte Metabolism, M. Maxwell and C. R. Kleeman (eds.) pp. 121-161, McGraw-Hill, New York, 1972.
394. Darrow, D. C.: The retention of electrolyte during recovery from severe dehydration due to diarrhea. J. Ped. 28:515-540, 1946.
395. Schwartz, W. B. and Relman, A. S.: Metabolic and renal studies in chronic potassium depletion resulting from overuse of laxatives. J. Clin. Invest. 32:258-271, 1953.
396. Stamey, T. A.: The pathogenesis and implications of electrolyte imbalance in ureterosigmoidostomy. Surg. Gynec. Obstet. 103:736-758, 1956.
397. Morrison, A. B., Rawson, A. J., and Fitts, W. T.: The syndrome of retracting watery diarrhea and hypokalemia in patients with a non-insulin secreting islet-cell tumor. Am. J. Med. 32:119-127, 1962.
398. Roy, A. D. and Ellis, H.: Potassium secreting tumors of the large intestine. Lancet 1:759-760, 1959.
399. Banyajati, L., Keoplug, M., Biesel, W. R., Gangarosa, E. J., Sprinz, H., and Sitprija, V.: Acute renal failure in asiatic cholera: Clinical pathologic correlations with acute tubular necrosis and hypokalaemic nephropathy. Ann. Intern. Med. 52:960-975, 1960.
400. Conn, J. W.: Primaryaldosteronism: A new clinical syndrome. J. Lab. Clin. Med. 45:6-17, 1955.
401. Davis, W. G., Newsome, H. H., Wright, D. W., Hammond, W. G., Easton, J., and Bartter, F. C.: Bilateral adrenal hyperplasia as a cause of primary aldosteronism with hypertension, hypokalemia, and suppressed renin activity. Am. J. Med. 42:642-647, 1967.
402. Barraclough, M. A., Bacchus, B., Brown, J. J., Davies, D. L., Lever, A. F., and Robertson, J. I. S.: Plasma renin and aldosterone secretion in hypertensive patients with renal or renal artery lesions. Lancet 2:1310-1313, 1965.
403. Gill, J. R., George, J. M., Solomon, A., and Bartter, F. C.: Hyperaldosteronism and renal sodium loss reversed by drug treatment for malignant hypertension. New Eng. J .Med. 270:1088-1092, 1964.
404. Conn, J. W., Cohen, E. L., Lucas, C. P., McDonald, W. J., Mayor, G. H., Blough, W. H., Eveland, W. C., Booktein, J. J., and Lapides, J.: Primary reninism, hypertension, hyperreninemia and secondary aldosteronism due to renin producing juxta-glomerular cell tumors. Arch. Intern. Med. 130: 682-696, 1972.
405. Brown, J. J., Davies, D. L., Lever, A. F., and Robertson, J. I. S.: Variation in plasma renin concentration in several physiological and pharmacological states. Can. Med. Assoc. J. 90:206-210, 1964.
406. Newton, M. A., Sealey, J. E., Ledingham, J. G. G., and Laragh, J. H.: High blood pressure and oral contraceptives. Changes in plasma renin and renin-substrate and in aldosterone excretion. Am. J. Obstet. Gynec. 101:1037-1045, 1968.
407. Biglieri, E. G., Herron, M. W., and Brust, N.: 17-Hydroxylation deficiency in man. J. Clin. Invest. 45:1946-1954, 1966.
408. Bagshawe, K. D.: Hypokalemia, carcinoma, and Cushing's syndrome. Lancet 2:284-287, 1960.

409. Conn, J. W., Rovner, D. R., and Cohen, E. L.: Licorice induced pseudoaldosteronism. JAMA 205:492-496, 1968.
410. Bartter, F. C., Pronove, P., Gill, J. R., Jr., and MacCardle, R. C.: Hyperplasia of the juxtaglomerular apparatus with hyperaldosteronism and hypokalemic alkalosis. Am. J. Med. 38:811-828, 1962.
411. Morris, R. C., Jr.: Renal tubular acidosis mechanisms, classification, and implications. New Eng. J. Med. 281:1405-1413, 1969.
412. Klastersky, J., Vanderkelen, B., Daneau, D., and Muthieu, M.: Carbenicillen and hypokalemia. Ann. Intern. Med. 78: 774-775, 1973.
413. Shils, M. E.: Experimental production of magnesium deficiency in man. Ann. N.Y. Acad. Sci. 162:847-855, 1969.
414. Muggia, F. M., Heinemann, H. O., Farhangi, M., and Osserman, E. F.: Lysozymuria and renal tubular dysfunction in monocytic and myelomonocytic leukemia. Am. J. Med. 47:351-366, 1969.
415. Liddle, G. W., Bledsoe, T., and Coppage, W. S.: A familial disorder simulating primary aldosteronism but with negligible aldosterone secretion. Trans. Assoc. Am. Phys. 76:199-213, 1963.
416. Fichman, M. P., Crane, M. G., and Bethune, J. E.: Hypokalemia with normal blood pressure, aldosterone and renin levels secondary to a renal or adrenal tumor. Am. J. Med. 48:509-514, 1970.
417. Wigley, R. D.: Potassium deficiency in anorexia nervosa with reference to renal tubular vacuolation. Brit. Med. J. 2:110-113, 1960.
418. Relman, A. S. and Schwartz, W. B.: Effects of electrolyte disorders on renal structure and function. In: Renal Disease D. A. R. Black (ed.), 2nd edition, Blackwell Scientific Publications Ltd., Oxford, England, pp. 754-774, 1967.
419. Manitius, A., Levitin, H., Beck, D., and Epstein, F. H.: On the mechanism of impairment of renal concentrating ability in potassium deficiency. J. Clin. Invest. 39:684-692, 1960.
420. Gottschalk, C. W., Mylle, N. F., Jones, R., and Winter, W.: Osmolality of renal tubular fluid in potassium depleted rodents. Clin. Sci. 29:249-260, 1965.
421. Schultze, R. O.: Recent advances in the physiology and pathophysiology of potassium excretion. Arch. Intern. Med. 131:885-897, 1973.
424. Paulson, L. G., Taylor, A., Mintz, D. H., Field, J. B., and Davis, B. B.: priate secretion of antidiuretic hormone with lithium carbonate. New Eng. J. Med. 292:390-392, 1975.
423. Perillie, P. E. and Epstein, F. H.: Sickling phenomenon produced by hypertonic solutions: A possible explanation for the hyposthenuria of sicklemia. J. Clin. Invest. 42:570-580, 1963.
424. Paulson, L. G., Haylor, A., Mintz, D. H., Field, J. B., and Davis, B. B.: Effect of vasopressin on renal cyclic AMP generation in potassium deficiency and patients with sickle hemoglobin. Metab. (Clin. Exp.) 19: 694-700, 1970.
425. Dousa, T. P. and Wilson, D. M.: Effects of demethylchlortetracycline on cellular action of antidiuretic hormone in vitro. Kidney Int. 5:279-284, 1974.

426. DeTroyer, A. and Demanet, J. C.: Correction of antidiuresis by demeclocycline. New Eng. J. Med. 293:915-917, 1975.
427. Cherrill, D. A., Stote, R. M., Birge, J. R., and Singer, I.: Demeclocycline treatment in the syndrome of inappropriate antidiuretic hormone secretion. Ann. Intern. Med. 83:654-656, 1975.
428. Carone, F. A. and Epstein, F. M.: Nephrogenic diabetes insipidus caused by amyloid disease. Am. J. Med. 28:539-544, 1960.
429. Shearn, M. A. and Tu, W. H.: Nephrogenic diabetes insipidus and other defects of renal function in Sjogren's syndrome. Am. J. Med. 39:312-318, 1965.
430. Massry, S. G., Schainuck, L. I., Goldsmith, C., and Shreiner, G. E.: Studies on the mechanism of diuresis after relief of urinary tract obstruction. Ann. Int. Med. 66:149-158, 1967.
431. Stahl, W.: Effect of mannitol on the kidney. New Eng. J. Med. 272:381-386, 1965.
432. Suki, W., Rector, F. C., and Seldin, D. W.: The site of action of furosemide and other sulfonamide diuretics in the dog. J. Clin. Invest. 44:1458-1469, 1965.
433. Goldberg, M.: Ethacrynic acid: Site and mode of action. Ann. N.Y. Acad. Sci. 139:443-452, 1966.
434. Lockett, M. P. and Gwynne, H. L.: Release of oxytocin contributes to the natriuretic action of aminophylline in rats. J. Pharm. Pharmacolo. 20: 688-696, 1968.
435. Luzecky, M. H., Burman, K. D., and Schultz, E. R.: The syndrome of inappropriate secretion of antidiuretic hormone associated with amitriptyline administration. South. Med. J. 67:495-497, 1974.
436. Self, J.: Water intoxication induced by oxytocin administration. Am. J. Med. Sci. 252:573-574, 1966.
437. Kennedy, R. M. and Earley, L. E.: Profound hyponatremia resulting from a thiazide induced decrease in urinary diluting capacity in a patient with primary polydipsia. New Eng. J. Med. 282:1185-1186, 1970.
438. Fuisz, R. E., Lauler, D. P., and Cohen, P.: Diuretic induced hyponatremia and sustained antidiuresis. Am. J. Med. 33:783-791, 1962.
439. Friedman, M., Freed, C., and Rosenman, R.: Effect of potassium administration on 1) peripheral vascular reactivity and 2) blood pressure of the potassium-deficient rat. Circulation 5:415-418, 1952.
440. Brown, J. J., Chinn, R. H., Lever, A. F., and Robertson, J. I. S.: Renin and angiotensin as a mechanism of diuretic-induced antidiuresis in diabetes insipidus. Lancet 1:237-239, 1969.
441. Michelekis, A. M., Caudle, J., and Liddle, G. W.: The effects of norepinephrine, epinephrine and cyclic AMP on renin release. Proc. Soc. Exp. Biol. Med. 130:748-753, 1969.
442. Bourgoigne, J. J., Catanzaro, F. J., and Perry, H. M.: Renin-angiotensin-aldosterone system during chronic thiazide therapy of benign hypertension. Circulation 37:27-35, 1968.
443. Urakabe, S., Handler, J. S., and Orloff, J.: Release of cyclic AMP by toad urinary bladder. Am. J. Physiol. 228:954-958, 1975.

444. Gorman, R. R.: Prostaglandin endoperoxides: Possible new regulators of cyclic nucleotide metabolism. J. Cyclic Nucleotide Res. 1:1-9, 1975.
445. Rajerison, R., Marchetti, J. J., Roy, C., Bockaert, J., and Jard, S.: The vasopressin sensitive adenylate cyclase of the rat kidney. Effect of adrenalectomy and corticosteroids on hormonal receptor-enzyme coupling. J. Biol. Chem. 249:6390-6400, 1974.
446. Broadus, A. E., Kaminsky, N. I., Hardman, J. G., Sutherland, E. W., and Liddle, G. W.: Kinetic parameters and renal clearances of plasma adenosine 3'-5' monophosphate and guanosine 3'-5' monophosphate in man. J. Clin. Invest. 49:2222-2236, 1970.
447. Coulson, R. and Bowman, R. H.: Excretion and degradation of exogenous adenosine 3'-5' monophosphate by isolated perfused rat kidney. Life Sci. 14:545-566, 1974.
448. Levine, S. D., Levine, R., Worthington, R., and Hays, R. M.: Selective inhibition of osmotic water flow in toad urinary bladder by general anesthetics. In: Proc. 5th International Biophysics Congress, Copenhagen, 1975.
449. Hays, R. M.: Antidiuretic hormone and water transfer. Kidney International 9:223-230, 1976.
450. Imbert, M., Cambardes, D., Montegut, M., Clique, A., and Morel, F.: Vasopressin dependent adenylate cyclase in single segments of rabbit kidney tubule. Pflügers Arch. 357:173-186, 1975.

Copyright 1977, Spectrum Publications, Inc.
Clinical Aspects of Cyclic Nucleotides

5

Extracellular Cyclic AMP in Human Hypertension

PAVEL HAMET

Adenosine 3',5'-monophosphate (cyclic AMP) is involved in the secretion and/or the action of many blood pressure-regulating substances, such as catecholamines, steroid hormones, the renin–angiotensin system and prostaglandins (1); various abnormalities of these substances have been demonstrated in various forms of human hypertension (2, 3). Cyclic AMP is also thought to be involved in the regulation of smooth muscle contraction (4, 5). Previous investigations showed that infusions of catecholamines are associated with an increase in the plasma level of this nucleotide (6, 7), and we have recently observed that endogenous catecholamine secretion resulting from insulin-induced hypoglycemia is also accompanied by a striking rise in the plasma cyclic AMP level (8).

The question then arises: What is the role of cyclic AMP in the pathophysiology of hypertension? As a first approach to answering this question we have undertaken measurements of extracellular cyclic AMP in different types of human hypertension.

METHODOLOGY

At the present time the most widely used methods for measurement of cyclic AMP in biologic fluids are competitive binding assays using either a preparation of protein kinase (9) or an antibody (10) as the binding molecule; correlation between these assays is excellent (11). The measurement of cyclic AMP levels in urine presents relatively few problems, since the levels are high enough to allow substantial dilution, thereby overcoming the nonspecific influence of many substances, such as NaCl on binding equilibrium. Greater difficulty is encountered when measuring plasma levels, since the low levels of cyclic AMP in plasma necessitate the concentration of samples, which results in high concentrations of interfering substances. In our experience, rather extensive purification of plasma is required for reliable results. Different types of purification procedures may be used, but they must be accompanied by careful attention to "blanks" induced by the purification procedure itself. It is also imperative to treat representative samples with phosphodiesterase to estimate effects of contaminants on binding equilibrium or on the separation of "bound" and "free" nucleotide (12, 13). As with any other competitive assay, it is important to measure samples at different dilutions for verification of parallelism with the "standard curve" (14). Useful modifications of the original protein-binding assay described by Gilman (9) have been reported by Brostrom and Kon (15).

The control subjects and the patients in the present study were investigated in a uniform state of hydration and sodium and potassium balance. Foodstuffs containing methylxanthines were excluded from the diet, and tobacco smoking was not allowed (11, 12). Patients with hypertension were investigated and assigned to different categories using strictly defined criteria (16, 17).

ESSENTIAL HYPERTENSION

Essential hypertension is a disease characterized by an increased blood pressure of uncertain etiology (18). For the purpose of our studies, the patients were subdivided into three groups: patients having (I) labile hypertension, (II) stable hypertension and (III) low-renin hypertension.

Labile Hypertension

Labile hypertension is characterized by increased plasma renin activity, rapid pulse rate, and elevated cardiac output, accompanied by inappropriately "normal" peripheral vascular resistance. This group predominately includes young patients whose blood pressure normalized spontaneously when hospitalized. These patients frequently show a characteristic symptom complex which includes hyperhidrosis of extremities, anxiety, dermographism and positive "cold pressor test".

Stable Hypertension

Stable hypertension includes patients of all ages with "fixed" high blood pressure, normal cardiac output and normal plasma renin activity but with increased peripheral vascular resistance.

Low Renin Hypertension

Low renin hypertension is a category defined by an abnormally low response of plasma renin activity to stimulation by posture and sodium depletion (17).

In our initial work (11), we demonstrated (Fig. 1) that the urinary excretion of cyclic AMP increases in response to isoproterenol or upright position in patients with labile "hyperkinetic" hypertension but decreases or does not change in age-matched control subjects (20). The excretion of cyclic AMP in the recumbent position without any stimulation was indistinguishable in control subjects and in patients with labile or stable hypertension. Treatment with propranolol seemed to normalize the abnormal postural increase in urinary cyclic AMP (21) (Fig. 1).

Plasma levels of cyclic AMP (12) were found to increase in response to upright posture (about 30 percent at the end of a 4-hr period) in control subjects and in patients with essential hypertension. Comparable levels of cyclic AMP were found in control subjects (18.6 ± SE 1.1nM) and in patients with essential hypertension (20.4 ± 2.6nM); propranolol administration diminished the postural increase of cyclic AMP in control subjects, but it produced an opposite effect in patients with labile hypertension (12). This latter effect is difficult to interpret at the present time, but it points to the possibility that the "normalization" of urinary excretion of

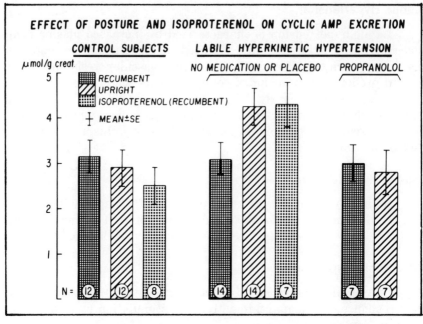

1 Cyclic AMP excretion as measured by radioimmunoassay. The effect of upright posture and isoproterenol infusion is demonstrated in control subjects and in patients with labile "hyperkinetic" hypertension. Propranolol administration (40 mg four times per day for 4 days) reverses the effect of upright posture on cyclic AMP level in patients. (Data from 11, 21).

cyclic AMP by this drug was only an apparent one. Further studies of the metabolic fate and the kidney handling of cyclic AMP in hypertension are needed.

To explore further the abnormally high levels of cyclic AMP under conditions of β-adrenergic stimulation in patients with labile hypertension, we compared effects of isoproterenol infusion in these patients and in control subjects. In collaboration with Messerli, Kuchel, Tolis, Fraysse and Genest, we demonstrated that infusion of isoproterenol (20 ng • kg^{-1} • min^{-1}) produces a significantly greater increase in pulse rate (75 percent) and in plasma cyclic AMP levels (80 percent) in patients than in control subjects (55 and 39 percent, respectively). However, this apparent β-adrenergic hyperresponsiveness in the patients was accompained by a smaller

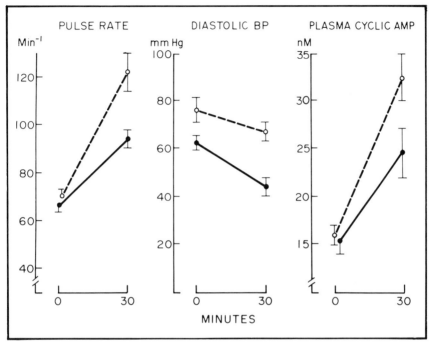

2 Effects of isoproterenol infusion (20 ng · min^{-1} · kg^{-1}) on pulse rate, diastolic blood pressure and plasma cyclic AMP in control subjects (solid line) and in patients with labile hypertension (broken line). (Data from 22, 23).

decrease in diastolic blood pressures than was seen in control subjects (22, 23) (Fig. 2).

In another collaborative study (with Lowder, Hardman, and Liddle), we have demonstrated that insulin-induced hypoglycemia has a striking effect on plasma levels of cyclic AMP (8). The plasma level of cyclic AMP increased by about fourfold in control subjects. This response was absent in adrenalectomized (cortisol-treated) patients, and it was almost completely abolished by propranolol in normal subjects. Thus, the effect of hypoglycemia on plasma cyclic AMP appears to be mediated by β-adrenergic activity. We have recently studied effects of insulin-induced hypoglycemia in "stable" essential hypertension and in hypertensive patients with "low plasma renin activity" (24). As illustrated in Fig. 3, the re-

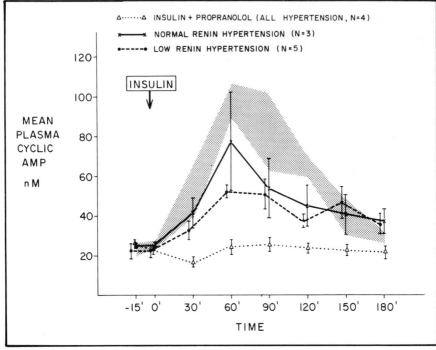

3 Effect of insulin-induced hypoglycemia (0.15 to 0.20 U/kg) upon plasma cyclic AMP in patients with "normal" (solid line) and "low" (dashed line) renin hypertension. The shaded area represents the mean ± SEM of values observed in normal subjects. The dotted line shows the lack of the effect of hypoglycemia when insulin injection was preceded and followed by infusion of propranolol (15 mg during 180 min). (Date from 8, 24).

sponse of plasma cyclic AMP to hypoglycemia is significantly diminished in hypertensive patients with low plasma renin activity.

It can be, therefore, tentatively concluded that patients with labile hypertension and those with low plasma renin activity form opposite ends of the physiopathological spectrum—one with abnormally high and the other abnormally low responsiveness to β-adrenergic stimulation. The categorization of patients with essential hypertension is helpful when choosing the most appropriate therapeutic regimen. Patients with labile hypertension respond favorably to β-blocking agents (19), and those with low renin activity are successfully treated with spironolactone or diuretics (24). The possible relation between an abnormality in adrenergic responsiveness and mineralocorticoid excess (26, 27) in patients with low renin activity remains to be evaluated.

SECONDARY HYPERTENSION

To date, we have studied two types of hypertension of known etiology.

Pheochromocytoma

One patient with a pheochromocytoma had plasma levels of cyclic AMP, which increased from control values of 20 to 25 to a peak of 130 to 260nM during episodes of hypertensive crisis (12). This was accompanied by a 10-fold increase in plasma norepinephrine and a 15-fold rise in urinary epinephrine. With the exception of periods of acute blood pressure elevation, plasma levels of cyclic AMP in this and other patients with pheochromocytoma were within normal limits (unpublished observations). This observation suggests again that β-adrenergic stimulation may induce an acute rise of plasma cyclic AMP levels.

Renovascular Hypertension

The hypertension secondary to stenosis of the renal artery is thought to be dependent on renin secretion by the affected kidney (28, 29). Plasma renin activity in blood from renal veins, particularly the ratio between stenotic and nonstenotic sides, is a useful diagnostic tool in renovascular hypertension (30). Since cyclic AMP was suggested to be a mediator of catecholamine-induced renin release (31), it was of interest for us to study extracellular levels of cyclic AMP in this type of hypertension. Our observations of normal urinary excretion of cyclic AMP (20) and increased plasma levels (12) in patients with "significant" renal artery stenosis (defined by the ratio of plasma renin activity in renal veins) led us to consider an abnormal renal clearance of cyclic AMP. We then measured plasma cyclic AMP in the aorta and renal veins of patients submitted to diagnositc angiographic studies (32, 33). As demonstrated in Fig. 4, cyclic AMP in renal veins in patients with clinically "nonsignificant" renal artery stenosis is significantly lower than that in blood from the aorta. The difference, averaging

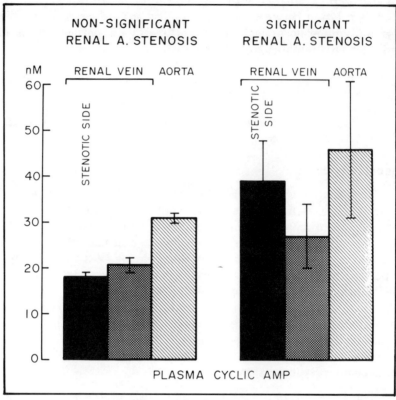

4 Plasma levels of cyclic AMP in renal veins and aorta of patients with "nonsignificant" and "significant" renal artery stenosis. Significant renal artery stenosis is that exceeding 60 percent and resulting in an increase of plasma renin activity in ipsilateral vein to 1.5-fold that in the contralateral vein. (Data from 32, 33).

40 percent, exceeds that amount contained in the filtration fraction of normal blood flow. Since it has been established that cyclic AMP in man reaches urine only via glomerular filtration (34), present observations suggest degradation of cyclic AMP within the kidney in addition to the urinary clearance. This observation is in accordance with several experimental studies in animals (35, 36), demonstrating the importance of the kidney in plasma clearance of cyclic AMP.

In contrast to patients with nonsignificant renal artery stenosis, those with significant stenosis of the renal artery have significantly higher cyclic AMP in the venous outflow from the compromised kidney, than from the contralateral kidney. In addition, the arterio-venous difference across the affected kidney is less than in patients with nonsignificant renal artery stenosis (averaging about 20 percent). This may point to either a decreased removal of the nucleotide or an increase in its production by the renal parenchyma. However, the relevance of an increased plasma level of cyclic AMP in the outflow from the stenotic side in which plasma renin activity is also increased has not as yet been defined.

The secretion of renin is thought to be mediated by cyclic AMP. The parallelism of extracellular cyclic AMP and plasma renin activity demonstrated in our studies supports this concept. Perhaps an exception to this general observation is content in our recent study (8, 17) using insulin-induced hypoglycemia as a stimulus for both renin and cyclic AMP: in adrenalectomized (cortisol-treated) patients, cyclic AMP did not rise, while plasma renin activity rose significantly and similarly to that of control subjects. It should be appreciated, however, that the intracellular regulating activity of cyclic AMP may not be reflected at the extracellular level.

CONCLUSIONS

Extracellular cyclic AMP was measured in three groups of patients with essential hypertension. Cyclic AMP rose to higher levels in plasma and urine of patients with *labile hypertension* than in control subjects in response to β-adrenergic stimulation; cyclic AMP levels in patients with *low renin hypertension* rose less than in patients with *stable hypertension* and control subjects following the β-adrenergic stimulation associated with insulin-induced hypoglycemia.

In patients with renovascular hypertension, we observed increased levels of peripheral plasma cyclic AMP and a decreased arterio-venous difference across the affected kidney.

Plasma renin activity and cyclic AMP levels in plasma behave in a parallel fashion in hypertensive subjects. The precise significance of this observation has not been defined.

ACKNOWLEDGMENTS

The author wishes to express his gratitude for collaboration and guidance to Dr. J. Genest and Dr. O. Kuchel of the Clinical Research Institute in Montreal and to Dr. G. W. Liddle and Dr. J. G. Hardman of the School of Medicine at Vanderbilt University. The technical assistance of Jacques Fraysse and Christine E. Baird are gratefully appreciated.

REFERENCES

1. Robison, G. A., Butcher, R. W., and Sutherland, E. W.: *Cyclic AMP*, Academic Press, New York and London, (1971).
2. Genest, J., and Koiw, E., Eds., *Hypertension '72*, Springer Verlag, Heidelberg, (1972).
3. Onesti, G., Kim, D. E., and Moyer, J. H., Eds., *Hypertension: Mechanisms and Management*, Grune & Stratton, New York, (1973).
4. Schultz, G., Hardman, J. G., and Sutherland, E. W.: Cyclic nucleotides and smooth muscle function in *Asthma, Physiology, Immunopharmacology, and Treatment*. (Eds, K. Frank Austen, and L. M. Lichenstein) Academic Press, New York, p. 123 (1973).
5. Bär, H-P.: Cyclic nucleotides and smooth muscle in *Advances in Cyclic Nucleotide Research*, Vol. *4*, (Eds., Paul Greengard, and G. Alan Robison) Raven Press, New York, p. 195 (1974).
6. Broadus, A. E., Kaminsky, N. I., Northcutt, R. C., Hardman, J. G., Sutherland, E. W., and Liddle, G. W.: Effects of glucagon on adenosine 3',5'-monophosphate and guanosine 3',5'-monophosphate in human plasma and urine. *J. Clin. Invest.*, *49*:2237 (1970).
7. Ball, J. H., Kaminsky, N. I., Hardman, J. G., Broadus, A. E., Sutherland, E. W., and Liddle, G. W.: Effects of catecholamines and adrenergic-blocking agents on plasma and urinary cyclic nucleotides in man. *J. Clin. Invest.*, *51*:2124 (1972).
8. Hamet, P., Lowder, S. C., Hardman, J. G., and Liddle, G. W.: Effect of hypoglycemia on extracellular level of cyclic AMP in man. *Metabolism: 24*:1139 (1975).
9. Gilman, A. G.: A protein binding assay for adenosine 3',5'-cyclic monophosphate. *Proc. Natl. Acad. Sci. U.S.A.*, *67*:305 (1970).
10. Steiner, A. L., Parker, C. W., and Kipnis, D. M.: Radioimmunoassay for cyclic nucleotides. I. Preparation of antibodies and iodinated cyclic nucleotides. *J. Biol. Chem.*, *247*:1106 (1972).
11. Hamet, P., Kuchel, O., and Genest, J.: Effect of upright posture and isoproterenol infusion on cyclic adenosine monophosphate excretion in control subjects and patients with labile hypertension. *J. Clin. Endocrinol. Metab.*, *36*:218 (1973).
12. Hamet, P., Kuchel, O., Fraysse, J., and Genest, J.: Plasma adenosine 3',5'-cyclic monophosphate in human hypertension. *Can. Med. Assoc. J.*, *111*: 323 (1974).

13. Hamet, P., Stouder, D. A., Ginn, H. E., Hardman, J. G., and Liddle, G. W.: Studies of the elevated extracellular concentration of cyclic AMP in uremic man. *J. Clin. Invest.*, *56*:339 (1975).
14. Yalow, R. S.: Radioimmunoassay; practices and pitfalls. *Circ. Res.*, *32*:1-116 (1973).
15. Brostrom, C. O. and Kon, C.: An improved protein binding assay for cyclic AMP. *Anal. Biochem.*, *58*:459 (1974).
16. Cuche, J.-L., Kuchel, O., Barbeau, A., Langlois, Y., Boucher, R., and Genest, J.: Autonomic nervous system and benign essential hypertension in man. I. Usual blood pressure, catecholamines, renin and their interrelationships. *Circ. Res.*, *35*:281 (1974).
17. Lowder, S. C., Frazer, M. G., and Liddle, G. W.: Effect of insulin-induced hypoglycemia upon plasma renin activity in man. *J. Clin. Endocrinol. Metab.* *41*:97 (1975).
18. Pickering, G.: High Blood Pressure, 2nd Edition, Grune & Stratton, New York, (1968).
19. Kuchel, O., Cuche, J.-L., Hamet, P., Barbeau, A., Boucher, R., and Genest, J.: Catecholamines, cyclic adenosine monophosphate and renin in labile hypertension, in *Mechanisms of Hypertension*. Excerpta Medica, Amsterdam, p. 160 (1973).
20. Hamet, P., Kuchel, O., and Genest, J.: L'excrétion de l'AMP cyclique dans l'hypertension artérielle. *Union Med. Can.*, *102*:805 (1973).
21. Hamet, P., Kuchel, O., Cuche, J.-L., Boucher, R., and Genest, J.: Effect of propranolol on cyclic AMP excretion and plasma renin activity in labile essential hypertension. *Can. Med. Assoc. J.*, *109*:1099 (1973).
22. Messerli, F. H., Kuchel, O., Tolis, G., Hamet, P., Fraysse, J., and Genest, J.: Plasma cyclic AMP response to isoproterenol and glucagon in labile (borderline) hypertension (LTH). *Clin. Res.*, *23*:197A (1975).
23. Messerli, F. H., Kuchel, O., Hamet, P., Tolis, G., Guthrie, Jr., G. P., Fraysse, J., Nowaczynski, W., and Genest, J.: Plasma adenosine 3',5'-cyclic monophosphate response to isoproterenol and glucagon in borderline (labile) hypertension. *Circ. Res. 38*, Suppl. II: 42, 1976.
24. Lowder, S. C., Hamet, P., and Liddle, G. W.: Contrasting effects of hypoglycemia on plasma renin activity and cyclic AMP in low-renin and normal-renin essential hypertension. *Circ. Res.*, *38*:105, 1976.
25. Douglas, J. G., Hollifield, J. W., and Liddle, G. W.: Treatment of low renin essential hypertension. Comparison of spironolactone and a hydrochlorothiazide-triamterene combination. *JAMA*, *227*:518 (1974).
26. Melby, J. C., Dale, S. L., Grekin, R. J., Gaunt, R., and Wilson, T. E.: 18-hydroxy-11-deoxycorticosterone (18-OH-DOC) secretion in experimental and human hypertension in *Hypertension: Mechanisms and Management.*, (Eds., Gaddo Onesti, Kwan Eun Kim, and John H. Moyer) Grune & Stratton, New York, p. 523 (1973).
27. Liddle, G. W., and Sennett, J. A.: Steroids and the syndrome of low-renin hypertension. Symp. 23, *Steroids and Hypertension* (1), Mexico City (1974).
28. Page, I. H., and McCubbin, J. W., Eds.,: *Renal Hypertension*, Year Book Publ., Chicago (1968).

29. Streeten, D. H. P., Anderson, G. H., Freiberg, J. M. and Dalakos, T. G.: Use of an angiotensin II antagonist (Saralasin) in the recognition of "Angiotensinogenic" hypertension. *N. Engl. J. Med.*, *292*:657 (1975).
30. Michelakis, A. M., Foster, J. H., Liddle, G. W., Rhamy, R. K., Kuchel, O., and Gordon, R. D.: Measurement of renin in both renal veins. Its use in diagnosis of renovascular hypertension. *Arch. Intern. Med.*, *120*:444 (1967).
31. Michelakis, A. M., Caudle, J., and Liddle, G. W.: In vitro stimulation of renin production by epinephrine, norepinephrine and cyclic AMP. *Proc. Soc. Exp. Biol. Med.*, *130*:748 (1969).
32. Tolis, G., Kuchel, O., Hamet, P., Fraysse, J., Trachewsky, D., Messerli, F., Boucher, R., and Genest, J.: Renal vein (RV) plasma cyclic AMP (cAMP) in hypertension (HBP) associated with renal artery stenosis (RAS). *Clin. Res.*, *22*:548A (1974).
33. Kuchel, O., Messerli, F., Tolis, G., Hamet, P., Fraysse, J., and Genest, J.: Renal vein plasma 3',5'-adenosine cyclic monophosphate in renovascular hypertension, Can. Med. Assoc., in press (1976).
34. Broadus, A. E., Kaminsky, N. I., Hardman, J. G., Sutherland, E. W., and Liddle, G. W.: Kinetic parameters and renal clearances of plasma adenosine 3',5'-monophosphate and guanosine 3',5'-monophosphate in man. *J. Clin. Invest.*, *49*:2222 (1970).
35. Blonde, L., Wehmann, R. E., and Steiner, A. L.: Plasma clearance rates and renal clearance of ^{3}H-labeled cyclic AMP and ^{3}H-labeled cyclic GMP in the dog. *J. Clin. Invest.*, *53*:163 (1974).
36. Coulson, R., and Bowman, R. H.: Excretion and degradation of exogenous adenosine 3',5'-monophosphate by isolated perfused rat kidney. *Life Sci.*, *14*:545 (1974).

Copyright 1977, Spectrum Publications, Inc.
Clinical Aspects of Cyclic Nucleotides

6

Cyclic Nucleotides in Shock and Trauma

CHU-JENG CHIU

In hemorrhagic shock, trauma, and other low-flow states, massive hormonal, metabolic, and autonomic nervous responses occur to maintain homeostasis and to counter the effects of such insults that otherwise may lead to the "rude unhinging of the machinery of life", as shock was defined by Samuel D. Gross in 1872. Since cyclic nucleotides are now known to mediate many hormonal and metabolic stimuli, elucidation of changes in cyclase-cyclic nucleotide system should shed more light on the cellular processes involved, and possibly lead to an improved therapy. The information available to date, particularly in the clinical field, is still limited, and indeed has so far only scratched the surface of the problem. However, in view of the potential importance of such investigations, the current knowledge in this area will be reviewed and discussed in perspective.

THE EFFECTS OF HEMORRHAGIC SHOCK AND TRAUMA ON CYCLIC NUCLEOTIDES

Experimental Studies and the Mechanisms Postulated for the Changes in Cyclic Nucleotides in Shock

Rutenburg, et al. (1) studied the effects of canine hemorrhagic shock on liver adenosine 3',5'-monophosphate (cyclic AMP) levels. Laparotomy and anesthesia with intravenous pentobarbital (30 mg/kg) for 8 hr produced minimal, insignificant elevation of liver cyclic AMP levels. After 1 hr of hemorrhagic shock (B.P. 30 to 35 mm Hg), the liver cyclic AMP decreased 29 percent and reached a level less than 50 percent that of control ($p < 0.05$) following the retransfusion of the shed blood at the late refractory phase of shock. The liver slice cyclic AMP response to epinephrine was likewise reduced; 44 percent less than normal after 1 hr of shock and 67 percent less ($p < 0.001$) in late refractory shock. The possibility that such impaired cyclase responsiveness may be present in vivo following hemorrhagic shock was examined by Rutenburg, Polgar and Bell et al. (2). Not only the shocked rabbit liver contained less cyclic AMP, the maximal response to intravenous injection of glucagon was also significantly reduced. Measurement of adenylate cyclase in the particulate liver cell preparations in vitro revealed less enzymatic activity in shocked rabbits, and their responses to the addition of several concentrations of glucagon were reduced, while their response to the addition of NaF, a nonspecific stimulator of enzymatic activity, was not altered. Thus the results suggest damage to the hormone receptor site in shock, without an intrinsic malfunction of the adenylate cyclase or its catalytic site. The same authors also observed similar impairments of the adenylate cyclase cyclic AMP system in shocked canine adrenal cortical cells to ACTH stimulation (2).

The functional depression of adenylate cyclase activity in shock may be a manifestation of alteration in membrane structure caused by cellular ischemic damage. Such damage may also change the membrane permeability, resulting in the leakage of cyclic nucleotides to the extracellular space. McArdle, Chiu, and Hinchey (3) observed more than fourfold increase in the canine plasma cyclic AMP within one hour of hemorrhagic shock (B.P. 35 to 40 mm Hg). It remained elevated until 3 hr later when the shed blood was reinfused, at which time the plasma level returned toward normal value. Such response appears to be suppressed by pretreating the dogs with acepromazine maleate, an alpha-adrenergic blocking agent. Since plasma catecholamine level is elevated in shock, and the catecholamines are known to stimulate cyclic AMP formation in many tissues, the effects of epinephrine infusion in dogs were studied. Rapid, sustained four- to fivefold increase in plasma cyclic AMP occurred during constant epinephrine infusion, as was reported earlier in man by Ball, et al. (4). In order to ascertain the source of this cyclic AMP, simultaneous sampling of blood from aorta, femoral, hepatic, and portal veins were made following a single pulse injection of epinephrine. Most rapid, significant rise in cyclic AMP was noted in the portal venous blood, with later and lesser increase in the hepatic venous blood, then in the femoral venous and aortic blood. Kinetic studies by Issekutz (5) using ($8-^3$H) cyclic AMP demonstrated that the plasma level of cyclic AMP was controlled by the rate of cyclic AMP appearance, both of them were found to be increased during catecholamine infusion. The rate of disappearance of cyclic AMP from plasma, on the other hand, was thought to be the result of mass-action effect of the concentration.

McArdle et al. (3) also found that the intestinal mucosal cyclic AMP content steadily decreased during hemorrhagic shock, which paralleled the decrease in tissue ATP (adenosine triphosphate) level. It has been argued that decreased ATP content per se cannot reduce cyclic AMP formation, since the amount of ATP required as a substrate to be converted into cyclic AMP is only a small fraction of the cellular ATP. However, since the ATP available for this purpose may be compartmentalized (6), and since we know little about the fate of this particular ATP pool, the possibility that the decrease in cellular cyclic AMP may be due to the decreased availability of ATP cannot be readily excluded at present. Alpha-adrenergic blockade inhibited changes in the intestinal mucosal ATP as well as cyclic AMP, while epinephrine infusion in normal dogs did not significantly affect the mucosal ATP and cyclic AMP levels, even though the plasma cyclic AMP was elevated.

Since the intestine is the well known "shock organ" in the dog, prone to develop severe hemorrhagic necrosis following shock, the question is whether splanchnic organs are also the main source of plasma cyclic AMP in shock in other species of animals. Prudhoe et al. (7) found in shocked pigs a significantly higher plasma cyclic AMP level ($p < 0.0005$) in the portal vein than in the systemic vein. These authors, however, were impressed by the hyperglycemia and increased portal plasma insulin, which appeared to correlate with the plasma cyclic AMP level ($p < 0.0005$). It was therefore postulated that the increased portal venous cyclic AMP was a reflection of pancreatic insulin release during hemorrhagic shock (8). Obviously further studies are needed to elucidate the pathophysiology involved, including the possible alteration of phosphodiesterase activity in shock (9) and the release of glucagon and other hormones within the portal circulation.

Clinical Studies

Few clinical patients suffer from "pure" hemorrhagic shock, and no information is available from such patients on their cyclic nucleotide changes. Observations made in patients who underwent surgical trauma, however, revealed many interesting parallels with those found in experimental hemorrhagic shock. Eight normal human subjects studied by Chiu, McArdle, and Hinchey (unpublished data) had plasma cyclic AMP level of 21.8 ± 3.1 nmole/liter. Seven patients who had general anesthesia and minor surgical procedures had no. or minor transient elevation of plasma cyclic AMP, whereas 12 patients who received open heart surgery had markedly elevated plasma cyclic AMP (79.6 ± 13.5 nmole/liter) during the operations, declining gradually for several days in the postoperative period. Gill, et al. (10) found elevation of plasma cyclic AMP in six of their seven general surgical patients. Cyclic AMP level fell to normal within 6 hr except in one patient who developed pneumonia, in whom the plasma cyclic AMP showed a prolonged elevation. Thus tissue injury and stress related hormonal and metabolic responses appear to increase plasma cyclic AMP.

Speculations on the roles of increased plasma cyclic AMP:

The biological significance and the homeostatic roles, if any, of the elevated plasma cyclic AMP in shock and trauma are not known. Gill et al. (9) correlated the temporal sequence of elevated plasma cortisol level with the increase in plasma cyclic AMP level in their surgical patients. The plasma cyclic AMP levels were always at a peak before those of cortisol in each case. In the one patient who developed complications, both plasma cyclic AMP and cortisol levels remained elevated. One nonresponder showed normal cyclic

AMP and cortisol levels. Although afferent nervous impulses from traumatized tissue to the hypothalamus are known to stimulate ACTH release (11), more recently it has been shown that animals with denervated pituitary glands can also show a cortisol response to stress, suggesting the presence of a humoral factor or factors (12, 13). Cyclic AMP is known to elevate plasma cortisol level in man (14) and in pigs (15) by stimulating ACTH release from the pituitary. Thus, increased plasma cyclic AMP in shock and trauma may play an essential role in stress reactions, but the data available to date in support of this hypothesis remain presumptive and not conclusive. For example, it has been shown that at least 1,000-fold increase from normal in plasma cyclic AMP level would have to be achieved with exogenous cyclic AMP to exert physiologic effects comparable with those seen after the administration of hormones (16). The increase in plasma cyclic AMP observed in shock and trauma is far below such magnitude. It is likely instead that plasma cyclic AMP levels in these conditions will be found to be useful as an overall index reflecting the severity of insult and the prognosis (17).

Cyclic nucleotides play many important biologic roles as an intracellular second messenger and modulator. It has been shown, for example, to be involved in various immunologic responses (18). It signals cell differentiation (19) and affects vascular smooth muscle function (20). Whether the altered cellular cyclase-cyclic nucleotide system plays any role in the known immunologic depression following shock and trauma (21), or stimulation of cellular proliferation and differentiation in the reparative process, or the altered vascular responsiveness in refractory shock, all still remain speculative at present, but illustrates some of the most fascinating areas for future investigation.

CYCLIC NUCLEOTIDES IN OTHER LOW FLOW STATES

Endotoxic and Septic Shock

Although the escherichia coli endotoxin was shown to activate hepatic adenylate cyclase by Gunpel et al. (22), no detailed clinical study in man of the changes in cyclic nucleotides in endotoxin and septic shock has been performed. One crucial aspect in the clinical course of septic patients is the patient's resistance to infection, and the impaired neutrophilic leukocyte function observed in sepsis (23) may be at least partly related to cyclic AMP, since the hormonal control of the neutrophil lysosomal enzyme release appears to be mediated by cyclic AMP (24). Many other metabolic abnormalities observed in endotoxin shock may also be associated with cyclic nucleotide changes, but no firm data is available at present.

Myocardial Infraction and Cardiogenic Shock

The importance of cyclic AMP in mediating the cardiac metabolism and contractility has been studied extensively in the last several years (25, 26). Clinically, Rabinowitz, Kligerman, and Parmley (27) reported slight increase in the plasma cyclic AMP level during the first 24 hr of acute myocardial infarction in the 35 patients who survived the attack. Nine nonsurvivors had abnormally high plasma cyclic AMP, five of them had cardiogenic shock, and seven had heart failure. Plasma cyclic AMP levels were found to correlate inversely with cardiac stroke work index and appeared to be of prognostic value in these patients. The source of this increased plasma cyclic AMP is not known, but the authors implied leakage from the cardiac muscle stimulated by endogenous catecholamine released within the myocaridum as the possible source (28). Subsequent canine experimental studies by the same authors (29) did show a significant early increase in the myocardial cyclic AMP content following the cross-clamping of the canine aortic root. Pretreatment with propranolol abolished such increase. However, no significant change in plasma cyclic AMP was noted when localized coronary occlusion was induced.

Both massive myocardial infarction and experimental cross-clamping of the ascending aorta produce not only myocardial ischemia but also a generalized low flow state. In order to separate these two factors, Chiu, McArdle, and MacRae (unpublished data) placed four dogs on total cardiopulmonary bypass before normothermic myocardial ischemia was induced by cross-clamping the aortic root. Adequate blood flow and pressure to the rest of the body were maintained with the pump-oxygenator. The aortic cross-clamp was released after 30 min of anoxic arrest. Coronary sinus blood cyclic AMP levels obtained before and immediately after the ischemic episode were not significantly different. It may be that the catecholamines released by extracardiac tissue in response to systemic low-flow state in cardiogenic shock is more important in this regard than the endogenous catecholamines released within the myocardium (30). Furthermore, from the data obtained in hemorrhagic shock, the possibility that the splanchnic organs are also the main source of plasma cyclic AMP in cardiogenic shock warrants further investigation.

THERAPEUTIC APPROACHES

Decreased intracellular cyclic AMP content in the liver and the intestinal mucosa, depressed responsiveness of adenylate cyclase to hormonal stimuli, and leakage of cyclic AMP to the extracellular space and plasma were observed in shock and trauma (Fig. 1). Such changes in cyclic nucleotides may be amenable to therapeutic manipulations.

Earlier, Levine and Vogel (31) reported on the cardiovascular effects of exogenous cyclic AMP in normal conscious dogs, and Levine, Dixon, and Franklin (32) described similar effects in man. In 26 human volunteers, 10 mg/kg intravenous dose of cyclic AMP induced prompt cardioacceleration and increased cardiac output. With 0.5 mg/kg/min infusion of cyclic AMP, progressive and sustained elevation of systolic blood pressure was also noted. Beta-adrenergic and cholinergic blockades with propranolol and atropine respectively did not inhibit the hemodynamic responses to exogenous cyclic AMP. When cyclic AMP was injected during marked hypotension induced by trimethaphan (Arfonad) and glyceryl trinitrate, prompt increase in heart rate occurred without any change in blood pressure. Such inotropic and chronotropic actions were consistent with those observed in vitro in the isolated perfused heart preparations (25), and may be of value in the treatment of shock state.

Since the dibutyryl derivative of cyclic AMP (db-cAMP) was thought to be more readily permeable through the membrane into the cell (33), Levine (34) further studied the effects of exogenous db-cAMP. In five human subjects, single intravenous injections of db-cAMP produced hyperglycemia and increased cardiac rate, and the hemodynamic response was reported to be greater with intravenous infusion of 0.2 mg/kg/min of db-cAMP than with the infusion of 0.3 to 0.5 mg/kg/min of cyclic AMP. Pretreatment with or simultaneous injections of aminophylline or caffeine, methyl xanthine inhibitors of phosphodiesterase, with db-cAMP did not affect the subjects' responses to the latter.

In contrast to these observations, canine experiments of Mac-

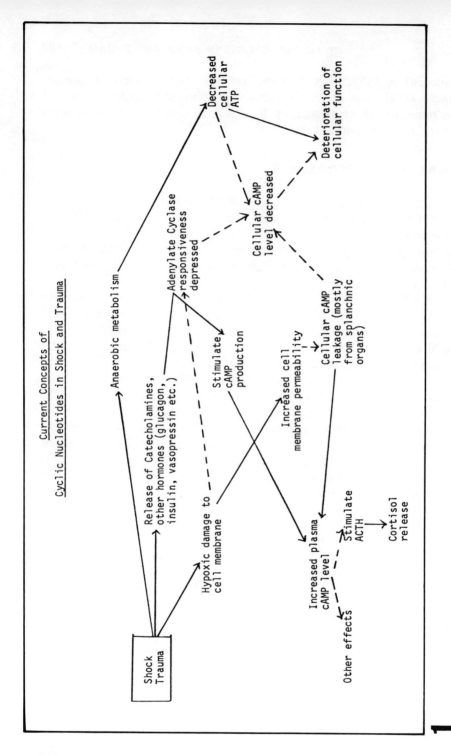

Rae, Chiu, and Hinchey (35) failed to document any increase in cardiac output, heart rate, blood pressure and cardiac contractility (dp/dt) following the bolus intravenous injections of 2 mg/kg of db-cAMP, although hyperglycemia was noted. In dogs subjected to a modified Wigger's hemorrhagic shock (B.P. 40 mm Hg), bolus intravenous injections of 2 mg/kg of db-cAMP also failed to elicit hemodynamic changes during the 3-hr hypotensive period or 30 min after the reinfusion of shed blood. Single injections of epinephrine, on the other hand, produced dramatic increase in blood pressure and reflex bradycardia. Only when a massive dose of 10 mg/kg of db-cAMP was injected, transient hypotension due to vasodilatation occurred. These results differ from that reported by Levine et al. but are in agreement with others who found only small and variable cardiovascular effects following the administration of exogenous cyclic AMP (16). Such contradictions may imply, as suggested by Rasmussen, Goodman, and Tanenhouse (36) and others, that cyclic AMP is one of the modulators of hormone action rather than the sole intracellular mediator. Other uncontrolled relevant factors in the experiments, such as the calcium flux, phosphodiesterase, and cyclic GMP changes, may have caused different experimental results. It should be added, furthermore, that the role and significance of myocardial depression in hemorrhagic shock still remain controversial (37).

There are other ways to manipulate the cyclase-cyclic nucleotide system. Hohl, Jolley, and Smith (38), for example, reported on the improvement in cardiac contractility and hemodynamic parameters with phosphodiesterase inhibitor PDI [3-bromo-5, 7-dimethylpyrazolo (1,5a) pyrimidine]. Direct evidence for cyclic AMP involvement was not available and the efficacy of this agent in shocked animals has not been studied.

Corticosteroids have been used in the treatment of shock patients, presumably for its "membrane-stabilizing" effects. Issekutz (5) reported 50 percent decrease in plasma level and turnover rate of cyclic AMP in normal dogs treated with methylprednisolone. Such treatment greatly potentiated not only the hyperglycemic effect (fourfold) of epinephrine but also the rise of plasma cyclic AMP (8.5-fold).

Goldfarb and Glenn (39) observed a 50 percent reduction in tissue cyclic AMP during shock, which was thought to destabilize the lysosomes and enhance the release of acid hydrolases. Prior treatment with glucocorticoids restored cyclic AMP to nonshocked levels and reduced lysosomal release in shock. Glucocorticoids were also found to noncompetitively inhibit cyclic AMP phosphodiesterase both in vitro and in vivo in supraphysiologic doses. Whether steroids may also prevent the impaired responsiveness of the membrane-bound adenylate cyclase to hormonal stimuli in shock has not been clarified.

It is obvious that investigations on the role of cyclic nucleotides in shock and trauma are still in its infancy. The possible roles of guanosine 3',5'-monophosphate (cyclic GMP) (40), for example, have not been clarified in shock, and the clinical data available in this field are few and fragmentary. However, preliminary experimental and clinical observations presented here strongly suggest the potential importance of such investigations, which may lead to better management of those critically ill patients.

SUMMARY

Massive endocrine, metabolic, and autonomic nervous responses in shock and trauma, and the cellular deterioration that occur in these conditions, cause alterations in cyclic nucleotides with important consequences. Depression of cellular cyclic AMP, particularly in the splanchnic organs, reduced responsiveness of the membrane-bound adenylate cyclase to hormonal stimuli, and leakage of cellular cyclic AMP, which results in high plasma cyclic AMP level, are some of the changes found to occur in shock and trauma. The pathophysiologic and therapeutic implications, as well as the areas for future investigations, are discussed.

REFERENCES

1. Rutenburg, A. M., Bell, M. L., Butcher, R. W., Polgar, P., Dorn, B. D., Egdahl, R. H.: Adenosine 3',5'-monophosphate levels in hemorrhagic shock. Ann. Surg. 174:461, 1971.
2. Rutenburg, A. M., Polgar, P., Bell, M., Egdahl, R. H.: Adenosine 3',5'-monophosphate metabolism in the liver in experimental hemorrhagic shock. Surgery 74:660, 1973.
3. McArdle A. H., Chiu, C. J., Hinchey, E. J.: Cyclic AMP response to epinephrine and shock. Arch. Surg. 110:316, 1975.
4. Ball, J. H., Kaminsky, N. I., Hardman, J. G.: Effects of catecholamines and adrenergic blocking agents on plasma and urinary cyclic nucleotides in man. J. Clin. Invest. 51:2124, 1972.
5. Issekutz, T.B.: Estimation of cyclic AMP turnover in normal and methylprednisolone-treated dogs: Effect of catecholamines. Am. J. Physiol. 229: 291, 1975.

6. Lindl, T., Hein-Sawaya, C. B., Cramer, H.: Compartmentation of an ATP substrate pool for histamine and adrenaline sensitive adenylate cyclase in rat superior cervical ganglia. Biochem. Pharmacol. 24:947, 1975.
7. Prudhoe, K., Wright, P. D., Latner, A. L., Johnston, I. D. A.: The effect of hemorrhagic shock upon plasma glucose, insulin and cyclic AMP levels in the pig. Brit. J. Surg. 62:158, 1975.
8. Zawalich, W. S., Karl, R. C., Ferrendelli, J. A., Matschinsky, F. M.: Factors governing glucose induced elevation of cyclic 3'5' AMP levels in pancreatic islets. Diabetolagia 2:231, 1975.
9. Senft, G., Schultz, G., Munske, K., Hoffman, M.: Effects of gluco-corticoids and insulin on 3',5'-AMP phosphodiesterase activity in adrenolectomized rats. Diabetolagia 4:330, 1968.
10. Gill, G. V., Prudhoe, K., Cook, D. B., Latner, A. L.: Effect of surgical trauma on plasma concentrations of cyclic AMP and cortisol. Brit. J. Surg. 62:441, 1975.
11. Hume, D. H., Bell, C. C., Barker, F.: Direct measurement of adrenal secretion during operative trauma and convalescence. Surgery 52:174, 1962.
12. Witorsch, R. N., Brodish, A.: Evidence for acute ACTH release by extrahypothalamic mechanisms. Endocrinology 90:1160, 1972.
13. Greer, M. A., Allen, C. F., Gibbs, F. P., Gullickson, C.: Pathways at the hypothalamic level through which traumatic stress activates ACTH secretion. Endocrinology 86:1404, 1970.
14. Angeli, A., Boccuzzi, G., Grajria, R., Bisbocci, D., Ceresa, F.: Early variations of plasma cortisol after pulse intravenous injections of dibutyryl cyclic AMP in normal subjects. Acta Endocrinol. 72:752, 1973.
15. Cook, D. B., Gill, G. V., Jackson, I. M. D., Smart, G. A.: Inhibition by dexamethasone of adrenocorticotrophin and cortisol release induced by intravenous infusion of ATP and dibutyryl cyclic AMP in piglets. J. Endocrinol. 60:65, 1974.
16. Henion, W. F., Sutherland, E. W., Posternak, T. H.: Effects of derivatives of adenosine 3',5'-monophosphate on liver slices and intact animals. Biochem. Biophys. Acta 65:558, 1962.
17. Sehgal, C. R., Kraft, A. R., Romero, C., Saletta, J. D.: Cyclic AMP as a determinant of the course of acute alcoholic pancreatitis. Surg. Forum 26:448, 1975.
18. Parker, C. W.: cAMP and the immune response. In "Advances in nucleotide research" 4:1, 1974. P. Greengard and G. A. Robinson, eds. Raven Press (N.Y.).
19. Gross, J. D.: Periodic cAMP signals and cell differentiation. Nature 255: 522, 1975.
20. Seidel, C. L., Schnarr, R. L., Sparks, H. V.: Coronary artery cAMP content during adrenergic receptor stimulation. Am. J. Physiol. 229:265, 1975.
21. MacLean, L. D., Meakins, J. L., Taguchi, K., Duignan, J. P., Dhillon, K. S., Gordon, J.: Host resistance in sepsis and trauma. Ann. Surg. 182:207, 1975.
22. Gunpel, L.: Effects of endotoxin on hepatic adenylate cyclase activity. Cir. Shock 1:31, 1974.
23. Cole, W. Q., Cook, J. J., Grogan, J. B.: Invitro neutrophil function and lysosomal enzyme levels in patients with sepsis. Surg. Forum 26:79, 1975.
24. Ignarro, L. J., Paddock, R. J., George, W. J.: Hormonal control of neutro-

phil lysosomal enzyme release: effect of epinephrine on adenosine 3',5'-monophosphate. Science 183:855, 1974.
25. Ahren, K., Hjalmarson, A., Isaksson, O.: Inotropic and metabolic effects of dibutyryl cyclic adenosine 3',5'-monophosphate in the perfused rat heart. Acta Physiol. Scand. 82:79, 1971.
26. Entman, M. L.: The role of cAMP in the modulation of cardiac contractility. In "Advances in nucleotide research", 4:195, 1974. P. Greengard & G. A. Robinson, eds., Raven Press (N.Y.).
27. Rabinowitz, B., Kligerman, M., Parmley, W. W.: Plasma cyclic adenosine 3'5'-monophosphate (AMP) levels in acute myocardial infarction. Am. J. Cardiol. 34:7, 1974.
28. Wollenberger, A., Krause, E. G., Heier, G.: Stimulation of 3',5'-cyclic AMP formation in dog myocardium following arrest of blood flow. Biochem. Biophys. Res. Commun. 36:664, 1969.
29. Rabinowitz, B., Parmley, W. W., Kligerman, M., Norman, J., Fujimura, S., Chiba, S., Matloff, J. M.: Myocardial and plasma levels of adenosine 3',5'-cyclic phosphate: studies in experimental myocardial ischemia. Chest 68:69, 1975.
30. Videbaek, J., Christensen, N. J., Sterndoft, B.: Serial determination of plasma catecholamines in myocardial infarction. Circulation 46:846, 1972.
31. Levine, R. A., Vogel, J. A.: Cardiovascular and metabolic effects of cyclic adenosine 3',5'-monophosphate in unanesthetized dogs. J. Pharmacol. Exp. Ther. 151:262, 1966.
32. Levine, R. A., Dixon, L. M., Franklin, R. B.: Effects of exogenous adenosine 3',5'-monophosphate in man. I. Cardiovascular response. Clin. Pharmacol. Therap. 9:168, 1968.
33. Ryan, W., Durick, M. A.: Adenosine 3',5'-monophosphate and N^6-2'C-dibutyryl adenosine 3',5'-monophosphate transport in cells. Science 177:1002, 1972.
34. Levine, R. A.: Effects of exogenous adenosine 3',5'-monophosphate in man. III. Increased response and tolerance to the dibutyryl derivative. Clin. Pharmac. Therap. 11:238, 1970.
35. MacRae, M. L., Chiu, C. J., Hinchey, E. J.: Effects of exogenous cyclic adenosine monophosphate in hemorrhagic shock. Surgery 78:254, 1975.
36. Rasmussen, H., Goodman, D. B. P., Tenehouse, A.: The role of cyclic AMP and calcium in cell activation. C. R. C. Crit. Rev. Biochem. 1:95, 1971.
37. Zweifach, B. W., Fronek, A.: The interplay of central and peripheral factors in irreversible hemorrhagic shock. Prog. Cardiovasc. Dis. 18:147, 1975.
38. Hohl, M. K., Jolley, W. B., Smith, L. L.: Alteration of cardiac contractility and hemodynamics by phosphodiesterase inhibition. Surg. Forum 25:158, 1974.
39. Goldfarb, R. D., Glenn, T. M.: Effects of synthetic corticosteroids on shock-induced alterations in cyclic nucleotide metabolism and lysosomal enzyme release. Proceedings of World Microcirculation Congress, Toronto, 1975.
40. Goldberg, N. D., O'Dea, R. F., Haddox, M. K.: Cyclic GMP. In "Advances in cyclic nucleotide research", Vol. 3, Greengard, P. and Robinson, G. A., eds., Raven Press (NY), p. 155, 1973.

7

Role of Cyclic Nucleotides in the Normal Lung and in Bronchial Asthma

RICHARD J. SOHN
ALEKSANDER A. MATHÉ
LADISLAV VOLICER

> ... for the turbid blood returns from the ambit of the body, widowed elsewhere of particles, to which a new humour from the subclavian vein is added to be perfected by further action of Nature. This happens in order that it may be arranged and prepared into the nature of particles of flesh, bone, nerve, etc., while it enters the myriad vessels of the lungs. It is conducted into divers very small threads. Thus a new form, situation, and motion is prepared for the particles of blood, from which flesh, bone and spirits may be formed ...

From this beginning by Malphigius (1661) (1), the establishment of the lung as the principal gas exchange organ has evolved. However, it was nearly 300 years later before Schild's investigations (1936) (2) on the release of histamine led to the realization that the lung is also important in the synthesis, release, and metabolism of many endogenous substances, especially those involved in allergic and immunologic reactions. The release of these mediators is modulated by the sympathetic and parasympathetic nervous systems. An imbalance in this regulation could be one of the manifestations of the pathophysiologic state (3–5). In both the normal and the pathophysiologic states, the cyclic nucleotides play an important role acting as second messengers for the various mediators. It is this role which will be discussed here.

This chapter will be divided into two sections: the first dealing with the role of cyclic nucleotides and their related enzymes in the normal function of the lung; and the second, with the relationship between the cyclic nucleotides and the release of mediators, in bronchial asthma.

CYCLIC NUCLEOTIDES IN THE NORMAL LUNG

Neurotransmitters and the Cyclic Nucleotides

The lung has been shown to contain all the enzymes necessary for the synthesis, metabolism, and expression of both adenosine 3′,5′-monophosphate (cyclic AMP) (6, 7) and guanosine 3′,5′-monophosphate (cyclic GMP) (8, 9).

Murad et al. (10) described the stimulation of adenylate cyclase by adrenergic agents. It was later shown that the receptors which cause the enzyme stimulation by catecholamines are β-receptors (11). Beta receptors have been classified as β-1 and β-2 receptors (12). Burges and Blackburn (13) demonstrated that the β-receptor associated with adenylate cyclase in the rat lung is of the β-2 type (the ratio of relative potencies required to produce 50% of the maximal response being: isoproterenol:epinephrine:norepinephrine::33:6.2:1) rather than the β-1 type (example: rat heart: isoproterenol:epinephrine:norepinephrine::3.9:0.6:1). Similarly, the β-2 agonists, salbutamol and soterenol were more effective in the lung than in the heart. It was also demonstrated that an antagonist of the β-2 receptors, butoxamine, was more effective than practolol, a β-1 antagonist, in inhibiting the stimulation of adenylate cyclase by agonists. These findings were confirmed by Lefkowitz (14) using a canine lung preparation.

At the present, it is not clear whether the phosphodiesterase (PDE) is also regulated by β-adrenergic receptors. Hitchcock (15) found both a low affinity (PDE I) and a high affinity (PDE II) soluble cyclic AMP phosphodiesterase present in the guinea pig lung. Both forms required Mg^{++} and could be inhibited by theophylline. However, only PDE I was inhibited by Ca^{++}. Adrenergic drugs inhibited PDE I (isoproterenol = epinephrine > norepinephrine > methoxamine at 0.1mM) but only norepinephrine inhibited PDE II and that to a lesser extent than the former. However, the fact that high concentrations of the catecholamines were needed to elicit a response together with the failure of propranolol, a β-receptor blocker, to antagonize their effects casts doubts on the role of the catecholamines in phosphodiesterase regulation.

Investigations of the effects of catecholamines on cyclic AMP levels support the findings with adenylate cyclase. Using isolated perfused rat lung, Collins et al. (16) demonstrated that isoproterenol, norepinephrine, and epinephrine caused increases in cyclic AMP levels from a mean of 14 pmoles/gm to 42, 45, and 27, respectively, using a drug concentration of $5 \times 10^{-5}M$. Pretreatment with propranolol or chlorpromazine blocked the action of isoproterenol, whereas aminophylline, potentiated the effects of the catecholamines. Similar results have been obtained in lung slices from several species (17, 18, 19, 6).

The effects of the catecholamines on lung cyclic AMP appear before birth. Palmer (19) has shown that in the fetal rabbit lung, the initial response to epinephrine occurred by day 21 and that of norepinephrine by day 25. The greatest response to the catecholamines occurred 4 days postpartum and dropped to adult levels by 8 days. The activities of the cyclic nucleotide-dependent protein kinases also underwent a change as the lung developed from the fetal to the adult stage. Kuo (20) has shown that reciprocal changes occur in the cyclic nucleotide dependent protein kinases. Whereas the fetal lung produced the highest cyclic GMP and the lowest cyclic AMP stimulation of the protein kinases (2.5 times and 1.5 times basal, respectively), the situation was reversed in the adult lung (1.5 times and 2.5 times basal, respectively). This reciprocal change caused a decrease in the cyclic GMP to cyclic AMP dependent protein kinase activity ratio from 3 to 0.5. These changes were only found in the soluble fraction from the lung homogenates. The nucleotide-dependent and basal activities in the particulate fraction and the basal activity in the soluble fraction from the lung did not exhibit much change as the animal developed from fetus to adult.

The data presented indicate the presence of an adenylate cyclase-cyclic AMP-protein kinase-phosphodiesterase system in the lung. The adenylate cyclase appears to be coupled to a β-2 receptor and both cyclase activity and the levels of cyclic AMP are elevated by adrenergic agents. This stimulation is abolished by specific β-2 antagonists whereas β-1 antagonists have little effect. There appear to be two soluble phosphodiesterases present, one of which is inhibited by most adrenergic agents while the other only responds to norepinephrine. The high concentrations needed and the failure of β-antagonists to block the inhibition raises doubts as to the physiologic role of catecholamines in PDE regulation. Whether the cyclic nucleotide system is also affected by an α-adrenergic receptor mechanism is still not clear. Although no evidence has been found in the normal lung, some data have been obtained in the diseased lung and will be discussed in this chapter.

Just as the sympathetic nervous system appears to affect the cyclic AMP system through the β-adrenoceptors, there is some evidence that the parasympathetic nervous system via cholinergic agents and muscarinic receptors may be related to the cyclic GMP system. Kuo and Kuo (21, 22) demonstrated that acetylcholine increased cyclic GMP levels threefold in rat lung slices and that this increase was largely blocked by atropine, a muscarinic antagonist, and not by hexamethonium, a nicotinic antagonist. It was also shown that isoproterenol antagonized the increase in cyclic GMP induced by acetylcholine without affecting basal cyclic GMP levels. Similarly, acetylcholine antagonized the isoproterenol induced increase in cyclic AMP without affecting basal cyclic AMP levels. Futhermore, the antagonistic action of isoproterenol on the acetylcholine-induced increase of cyclic GMP was blocked by a β-adrenergic blocker, propranolol, whereas the action of acetylcholine on the isoproterenol-induced increase of cyclic AMP was blocked by atropine (Table 1).

Stoner et al. (23) working with guinea pig lung slices, confirmed the findings that acetylcholine increased cyclic GMP levels and that this effect was blocked by atropine. It was also observed that acetylcholine increased cyclic AMP levels two to three-fold and that this increase could be blocked by atropine, findings in variance with those of Kuo and Kuo (22). Guanylate cyclase could not be stimulated by either acetylcholine or other cholinergic agents (23, 24). However, physostigmine increased enzyme activity in the same experiments. This increase, approximately 50 percent, was the same either in the presence or absence of acetylcholine. The authors felt that the apparent lack of response of the enzyme to acetylcholine might have been due to inappropriate enzyme preparation or assay conditions (23, 24). When β-adrenergic agonists and antagonists

Table 1

Reciprocal Regulation by Isoproterenol and Acetylcholine of Intracellular Cyclic AMP and GMP Levels in Rat Lung Slices[22].

Addition	Cyclic AMP	Cyclic GMP†	Cyclic AMP to cyclic GMP ratio
	(pmoles/mg protein)		
None (control)	6.2 ± 0.4	5.1 ± 0.2	1.2
Isoproterenol, 0.1μM	12.5 ± 0.8	5.2 ± 0.4	2.4
Isoproterenol, 1μM	19.6 ± 0.8	4.6 ± 0.2	4.3
Acetylcholine, 0.1μM	6.0 ± 0.5	12.6 ± 0.2	0.5
Acetylcholine, 1μM	5.8 ± 0.4	18.2 ± 0.6	0.3
Isoproterenol, 0.1μM + acetylcholine, 0.1μM	10.5 ± 0.2	9.1 ± 0.2	1.2
Isoproterenol, 0.1μM + acetycholine, 1μM	5.8 ± 0.3	12.5 ± 1.2	0.5
Isoproterenol, 1μM + acetycholine, 0.1μM	14.2 ± 0.2	4.9 ± 0.2	2.9
Isoproterenol, 1μM + acetycholine, 1μM	7.5 ± 0.6	8.2 ± 0.5	0.9
Isoproterenol, 1μM + acetylcholine, 1μM, + propranolol 1μM	7.0 ± 0.8	19.2 ± 0.6	0.4
Isoproterenol, 1μM + acetycholine, 1μM + atropine, 1μM	18.2 ± 1.2	6.2 ± 1.1	2.9

*Each value represents mean ± SE of triplicate incubation.
Reprinted from J.F. Kuo and W.-N. Kuo, Biochem. Biophys. Res. Commun. 55: 660-665, 1973 with the permission of the Editors.

were tested, it was found that they exerted no effect upon the basal cyclic GMP levels.

These results, although less conclusive than those concerning the β-receptor and cyclic AMP, seem to indicate that the parasympathetic nervous system is associated with cyclic GMP by the muscarinic receptor. In addition, the work of Kuo and Kuo (22, 23) presents some evidence of an interplay between the sympathetic-parasympathetic systems in the mutual regulation of these separate nucleotide systems providing a basis for extending the cyclic nucleotide yin-yang hypothesis (25) to the lung. In this context, it is of interest that the sympathetic-parasympathetic prejunctional balance, that is the control of catecholamine release by acetylcholine, and vice versa, has been found in heart, lung, and other organs (26–28).

Effects of Chemical Mediators on the Cyclic Nucleotides in the Lung

Besides the adrenergic and cholinergic agonists and antagonists, other substances are known to affect the cyclic nucleotide systems in the healthy lung. In most cases, these substances are either mediators which are released during the immunologic and allergic responses of the lung, or substances which affect the release or synthesis of these mediators.

Prostaglandins

It has been shown that exogenous PGE_1 increases cyclic AMP levels (29, 22, 23) while PGE_2 has a smaller effect and PGA_2 and $PGF_{2\alpha}$ have no effect in lung fragments (19, 23) and whole perfused lung (30). The effects of prostaglandins on cyclic GMP in lung slices are not clear. Kuo and Kuo (22) found that PGE_1 did not change cyclic GMP levels itself, but antagonized the increase in the nucleotide level elicited by acetylcholine. Similarly, Stoner et al. (23) noticed only a slight reduction in cyclic GMP levels upon the addition of either PGE_1, PGE_2, or $PGF_{2\alpha}$ to guinea pig lung slices. In perfused guinea pig lung studies, it was demonstrated that PGE_1 markedly increased cyclic AMP levels but had no effect on cyclic GMP. $PGF_{2\alpha}$, at the same concentrations as the PGE_1 (1 and 10 μg/ml), increased cyclic GMP by 40 percent but had no effect on cyclic AMP.

Substances which affect endogenous prostaglandins have been shown to have an effect upon cyclic nucleotide levels. Stoner et al. (31) using bradykinin or acetylcholine (1 to 100 μg/ml) elicited increases in the levels of both cyclic nucleotides in guinea pig lung slices. Promethazine, propranolol, and atropine did not affect either the bradykinin-induced increase or the basal levels of the nucleotides. When the slices were pretreated with either indomethacin or aspirin, the increase in cyclic AMP was abolished. However, the basal levels of the cyclic nucleotides and the bradykinin stimulation of cyclic GMP were not affected by the pretreatment. The authors suggested that bradykinin causes an increase in cyclic GMP levels which leads to increased prostaglandin synthesis and release.

The effects of prostaglandins on adenylate cyclase have also been investigated. Weinryb et al. (32) showed that PGE_1, PGE_2, and $PGF_{2\alpha}$ at $10^{-4}M$ all stimulated enzyme activity in guinea pig lung. $PGF_{1\alpha}$ did not stimulate at the same concentration. White (33) confirmed the increase of adenylate cyclase activity produced by PGE_1 and PGE_2 using rat and monkey lung preparations. In addition, it was shown that PGE_3 and PGA_1 had approximately the same effect as PGE_2 while the prostaglandins of the F-series produced no change in activity.

To summarize, stimulation of the adenylate cyclase-cyclic AMP system occurs with PGE_1, is sometimes observed with PGE_2, and generally does not occur with $PGF_{2\alpha}$. Prostaglandins of the E-type do not affect the guanylate cyclase-cyclic GMP system and the data with regard to $PGF_{2\alpha}$ are inconclusive.

Histamine

Weinryb et al. (32) have demonstrated that exogenous histamine ($10^{-4}M$) increased adenylate cyclase activity in guinea pig lung. Palmer (18, 19) and others (16, 34, 35) have shown that histamine also increased cyclic AMP levels and that this effect was inhibited by histamine antagonists. When either tripelenamine, an H_1 antagonist (16, 19) or metiamide, an H_2 antagonist (33), was employed, the histamine-induced increase in cyclic AMP levels was antagonized. This would indicate that exogenous histamine can act on cyclic AMP levels and that this effect can be blocked by histamine receptor blockers. This question will be pursued more thoroughly when discussing the effects of histamine blockers on the sensitized and anaphylactic lung. From these studies and investigations on other organs, it appears that stimulation of the H_1 receptor leads to increased cyclic GMP levels, while stimulation of the H_2 receptor results in higher cyclic AMP levels.

Cyclic Nucleotides in Bronchial Smooth Muscle

All of the above experiments were carried out using various lung preparations. Studies on the lung are complicated by the presence of different cell types (smooth muscle, Type I cells, Type II cells, pneumocytes, alveolar macrophages, mast cells, etc.). Since the lung is such a heterogeneous organ, the data obtained represent the cumulative and/or sequential effects of all the various cell types. Some investigators desiring to diminish this complexity have used the relatively homogeneous smooth muscle tissue of the trachea and bronchi. An added advantage of this system is that it is possible to measure muscle contraction or relaxation and thereby obtain a tension parameter with which to correlate the biochemical and pharmacological findings.

Working with tracheo-bronchial preparations, Moore et al. (36) and others (37–41) have demonstrated that dibutyryl cyclic AMP produced relaxation of the tracheal and bronchial smooth muscle. This effect was not antagonized by β-blocking agents such as propranolol. It was also noted that cyclic AMP itself was incapable of producing relaxation at concentrations equal to that of its dibutyryl derivative. This observation can be explained by the relatively poor membrane permeability for cyclic AMP. However, if cyclic AMP is incubated in the presence of theophylline, a sufficient intracellular cyclic AMP concentration occurs to produce relaxation. As with dibutyryl cyclic AMP, this response was unaffected by β-blocking agents. Furthermore, theophylline by itself was capable of inducing relaxation. Other substances that have been tested and found to produce relaxation are ACTH (39), whose effect could be blocked by the β-adrenergic antagonist, sotalol; various cyclic AMP analogs (42); 8-bromo-cyclic GMP (43, 44); and hydrocortisone (45).

Having demonstrated that ACTH produced relaxation of tracheal smooth muscle, Andersson et al. (41) measured the intracellular cyclic AMP content after stimulation by ACTH. It was observed that the cyclic nucleotide levels increased approximately twofold and that this increase correlated well with the decrease in muscular tension. Similarly, hydrocortisone was shown to produce an increase in cyclic AMP in the tracheal muscle (45). In the same experiment, changes in either adenylate cyclase or phosphodiesterase activity after treatment with hydrocortisone could not be found. Treating the animals with reserpine and exposing them to β-adrenergic antagonists only partially blocked the hydrocortisone induced relaxation. The investigators felt that the drug, in addition to working through the β-adrenergic receptor, had a direct effect upon cyclic nucleotide metabolism.

In order to determine if substances which relax tracheal smooth muscle increase cyclic AMP and those which contract it increase cyclic GMP, Murad (46) and Murad and Kimura (47) examined a series of hormones and drugs. In general, it was found that the hypothesis was correct, but exceptions do exist. Epinephrine and PGE_1, both of which relax tracheal smooth muscle, produced increases in cyclic AMP and had no effect upon cyclic GMP. Propranolol antagonized the epinephrine-induced increase in cyclic AMP but did not affect that produced by PGE_1. Phenoxybenzamine had no effect on either agent. Acetylcholine and carbachol, which increased cyclic GMP levels, also increased cyclic AMP levels. Their effects could be blocked by either atropine or propranolol, suggesting that their action might be secondary to catecholamine release. The choline esters, histamine, and theophylline increased cyclic GMP. The effect of the choline esters could be blocked by atropine and that of histamine by either atropine or diphenylhydramine. Serotonin, angiotensin II, and ACTH had no effect on cyclic AMP levels. (The latter results conflict with those obtained by Andersson et al. (41).) Prostaglandin $F_2\alpha$ was shown to have no effect on either cyclic nucleotide.

Andersson et al. (48), in a study that coupled measurements of changes in tension to changes in tissue levels of intracellular cyclic nucleotides, showed that in the presence of either carbachol or histamine, cyclic GMP levels increased (Fig. 1). Interestingly, levels of cyclic AMP after either decreasing (carbachol) or remaining constant (histamine) for approximately 1 min, also exhibited an increase. The effects of carbachol on the cyclic nucleotides and on muscular tension could be antagonized by atropine and those of histamine by isoproterenol. The carbachol induced contraction was less sensitive to β-adrenoceptor stimulation. It was not affected by used concentrations of isoproterenol.

Few investigations have been carried out with the enzymes obtained from the bronchial or tracheal smooth muscle preparations. However, those indicate that the adenylate cyclase responds similarly to that of the whole lung preparation or to that of other smooth muscles. Vulliemoz et al. (49) demonstrated that the catecholamines stimulate dog and guinea pig bronchial adenylate cyclase and produced bronchial smooth muscle relaxation in the same order of potency (isoproterenol > epinephrine > norepinephrine).

In summary, it appears that substances which cause relaxation of tracheal and bronchial smooth muscle generally increase cyclic AMP levels, while those that cause contraction increase cyclic GMP levels. However, the data also indicate that this may be a rather simplistic interpretation. For example, theophylline, which produces relaxation, and histamine and carbachol, which produce contraction, all cause increases in the levels of both cyclic nucleotides.

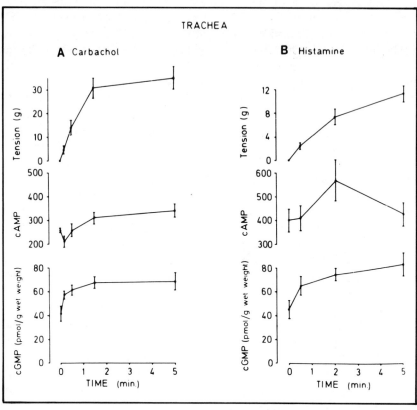

1 Effects of carbachol 8 × 10⁻⁷ gm/ml (A) and histamine 3 × 10⁻⁶ gm/ml (B) on tension and cyclic nucleotide levels in bovine tracheal muscle. Mean ± SE, N = 6-8. Reprinted from R. Andersson, K. Nilsson, J. Wikberg, S. Johansson, E. Mohme-Lundholm and L. Lundholm, Advances in Cyclic Nucleotide Research Vol. 5, Editors, G. I. Drummond, P. Greengard and G. A. Robinson, Raven Press, New York, 1975, pp. 491-518, with the permission of the editors.

This seems to indicate that it is not the absolute change in the levels of either cyclic nucleotide, but the change in the ratio or the sequence of the change in the cyclic nucleotide levels which is important for contraction. It would also indicate that other events, independent of changes in cyclic nucleotides, play a role in changes of the smooth muscle tone. Temporal studies should help to clarify whether the changes are occurring simultaneously or sequentially. It is conceivable that different effects might be achieved depending upon which cyclic nucleotide level is changed first.

BRONCHIAL ASTHMA

> ... Asthmatical persons can endure nothing violent or unaccustomed ... Whatsoever therefore makes the blood to boyl, or raises it into an effervescence, as violent motion of the body or mind, excess of extern cold or heat, the drinking of Wine, Venery, yea sometimes meer heat of the Bed doth cause Asthmatical assaults to such as are predisposed ...
>
> Thomas Willis (1684) (50)

One of the major pathophysiologic manifestations of the asthmatic disease is increased tone of the airways. Generally, the residual volume increase; the forced expiratory volume in 1 sec (FEV_1) and the forced vital capacity (FVC) decrease; and the FEV_1/FVC ratio changes little. One of the principal sites of airway obstruction in asthma is at the level of the small airways (less than 2 mm) that normally constitute only 10 to 30 percent of the total tracheobronchial tree resistance (51). When these airways are partially obstructed, regional variations in both gas distribution and perfusion ensue, with changes in the ventilation/perfusion ratios (51, 52).

It is assumed that during an acute asthmatic attack, the lung releases mediators that act to produce the above pathophysiologic changes. The principal chemical mediators implicated are histamine, slow reacting substance of anaphylaxis (SRS-A), eosinophil chemotactic factor (ECF-A), platelet activating factor (PAF), probably the prostaglandins, and possibly kallikrein and bradykinin. Of these, histamine and ECF-A and partially SRS-A are preformed and stored whereas the others are synthesized at the time of attack.

Due to the inherent problems involved in studying asthma in humans, various models of the asthmatic state have been employed. Among these are actively sensitized lungs from several species, passively sensitized human lung tissue and that of other animals, control and antigen sensitized mast cells from the peritoneal and thoracic cavity, and human leukocytes. In the appropriate model systems, it is assumed that the sensitized state is similar to that of the asthmatic in remission while a system undergoing anaphylactic challenge serves as an approximation of an asthmatic attack. It should be understood that these models are only approximations and, consequently, not necessarily exactly representing the phenomena occurring in asthmatic lung *in vivo*. This section is divided into two subsections to simplify presentation. The first will present the effects of various agents on cyclic nucleotides in sensitized and anaphylactic models. The second will examine the effect of the cyclic nucleotides on the release of mediators.

Cyclic Nucleotides in Asthma

Szentivanyi (53) has proposed the beta adrenergic theory of asthma to explain pathogenesis of asthma. In essence, the theory states that in the healthy state, a balance exists between the alpha and beta adrenergic mechanism maintaining homeostatic conditions within the lung. In bronchial asthma, which the author describes as ". . . a unique pattern of bronchial hyperreactivity to a broad spectrum of immunological, psychic, infectious, chemical, and physical stimuli . . .," this balance is disrupted. Upon mediator release, due to any of the above triggering events, there is a release of adrenergic neurotransmitters to counterbalance the mediator effect. However, a reduced functioning of the beta adrenergic system, due to defective beta receptors, prevents the balance from being restored. While this theory is capable of explaining some of the experimental data and clinical observations, other results are difficult to reconcile with the proposed defective beta receptor. Rather than trying to provide a unifying explanation for all types of asthma and all the phenomena, a more profitable approach appears to be to explore the immunologic phenomena in this disease.

Mathé and his coworkers (54, 55) have looked at the effect of epinephrine on the cyclic AMP system in control and sensitized guinea pigs (See Fig. 2). It was found that epinephrine infused at concentrations of 0.01 to 10 ug/ml caused a smaller increase in cyclic AMP levels in sensitized than in control perfused guinea pig lungs. These results indicated an altered receptor sensitivity. However, when the effect of epinephrine on adenylate cyclase activity was determined, an apparent paradox was discovered (See Fig. 3). At low concentrations ($10^{-9}M$ to $10^{-6}M$), epinephrine caused maximal stimulation (approximately 175% of the base value) in sensitized lungs but had a minimal effect in healthy controls. At higher concentrations ($10^{-5}M$ to $10^{-3}M$), the activity in the healthy con-

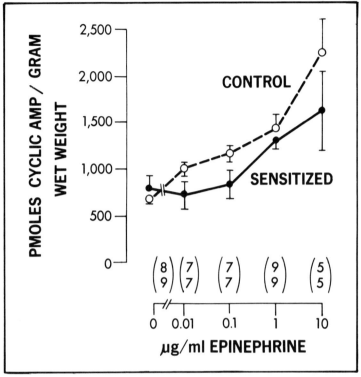

2 Cyclic AMP levels in control and sensitized guinea pig lungs. Control and sensitized lungs were perfused for 4 minutes with (−)-epinephrine in final concentrations of 0.01 to 10 ug/ml. Results expressed as mean ± SE of pmoles cyclic AMP per gram wet weight. Number of animals per group given in parentheses (upper line the sensitized and lower line the control animals). Reprinted from A. A. Mathé, S. K. Puri, R. J. Sohn and L. Volicer, Pharmacology (in press), with the permission of the authors.

trols further increased while sensitized adenylate activity decreased. The basal level of activity and the NaF stimulated response were the same in both the sensitized and the healthy controls. It thus appears that the sensitivity of adenylate cyclase to epinephrine has been increased. Phosphodiesterase activities were found to be equal in both healthy and sensitized lungs. Similarly, epinephrine inhib-

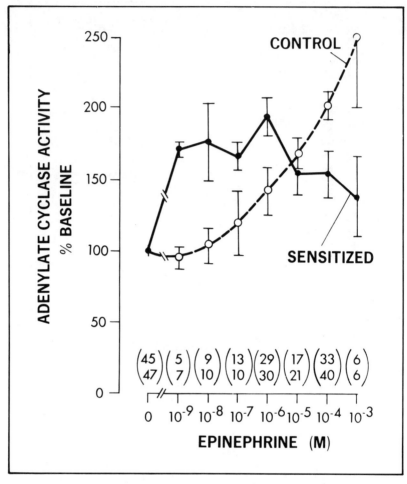

3 Adenylate cyclase activity in control and sensitized guinea pig lung. Homogenates from control and sensitized guinea pig lungs were incubated with (—)-epinephrine. Results are expressed as mean percent of baseline ± SE. Number of animals per group given in parentheses (upper line the sensitized and lower line the control animals). Student's t-test showed that stimulation by epinephrine was significantly different between the two groups at all epinephrine concentrations expect 10^{-5}M. Reprinted from A. A. Mathé, S. K. Puri, R. J. Sohn and L. Volicer, Pharmacology (in press) 1976, with the permission of the authors.

ited both phosphodiesterase preparations to the same extent. It was suggested that this discrepancy was due to the difference in preparations—whole perfused lung for measuring cyclic AMP levels versus crude homogenate for AC activity—the act of homogenization might have either increased the access of epinephrine to the enzyme receptor sites or disrupted the relationship between the receptor site and an inhibitory factor.

Mathé et al. (35) have also shown that the levels of both cyclic nucleotides increased differentially in the control, sensitized and anaphylactic lungs after administration of exogenous histamine. Pyrilamine antagonized the histamine stimulated increase in cyclic GMP and partially antagonized the cyclic AMP increase in anaphylactic lungs. Burimamide blocked the effect of histamine, as well as that of anaphylaxis on cyclic AMP levels, but had no effect on cyclic GMP levels (35). In a study on control and sensitized guinea pig lungs (34), metiamide was also able to block the increase in cyclic AMP levels produced in both states by exogenous histamine. Thus it appears that the H_1 and the H_2 receptors act differently in promoting changes in cyclic nucleotide levels. The H_1 receptor being more closely linked to the cyclic GMP system and the H_2 with that of cyclic AMP.

Infusion of exogenous histamine into control and sensitized guinea pig pulmonary circulation caused an increase in the lung tissue levels of cyclic AMP. This effect was less pronounced in lungs from antigen-sensitized animals and also in control animals pretreated with histamine (34). In the same and other experiments, determination of histamine "leakage" indicated that more endogenous histamine was appearing in the lung outflows from perfused sensitized than from control animals. It was suggested that the feedback by which cyclic AMP inhibited histamine release was impaired in antigen sensitized guinea pigs due to a hyposensitivity of the H_2 receptor caused by the histamine "leakage." In *Bordetella pertussis* sensitized mice, however (56–58), there was a larger increase in cyclic AMP levels in sensitized than in nonvaccinated mice by exogenous histamine. In this model, cyclic GMP levels also increased in the presence of histamine, but only in the sensitized animals. The discrepancy between these results and those of Mathé et al. (34) might be due to the different model systems employed: Bordetella pertussis sensitized mice versus ovalbumin sensitized guinea pigs.

Polson et al. (56) also examined the effect of propranolol on the histamine-induced increases in the cyclic nucleotides in the Bordetella pertussis model. It was found that this β-adrenergic antagonist was capable of partially blocking the effect of histamine on cyclic AMP levels but was ineffectual against the increase in cyclic GMP. It was also established that cholinergic agents, such as methacholine, increased cyclic GMP levels in both sensitized and nonvaccinated mice, the effect being greater in the sensitized animals. Atropine blocked the effect of metacholine but only partially antagonized the increase in cyclic GMP levels due to exogenously administered histamine. Polson et al. (57) hypothesized that the hypersensitivity to histamine caused by *Bordetella pertussis* might be due to a relationship between increased accumulation of cyclic GMP and an imbalance in the autonomic nervous system.

It has been demonstrated that the levels of cyclic AMP and cyclic GMP in guinea pig lungs are higher in the anaphylactic than in the sensitized and control animals (34, 35). Since the increase in cyclic AMP is greater than that of the cyclic GMP, the ratio between the two nucleotides changes from 3.4 to 13.3. These data can be explained by a change in cyclic AMP levels produced by the release of mediators. In essence, during anaphylaxis, there is a series of changes in the cyclic AMP levels. The adrenergic/cholinergic agents cause a change in the cyclic AMP/cyclic GMP system, thereby producing modulation of mediator release, and the release of mediators produces a further change in the cyclic AMP/cyclic GMP system. In general, mediator release appears to cause increases in the levels of the cyclic nucleotides.

In addition to studying changes in lung cyclic AMP levels, urinary levels of cyclic AMP have also been investigated. Normally, 2 to 7 μmoles of cyclic AMP is excreted in the 24-hr urine (7). Bernstein et al. (59) and Fireman (60) have shown that epinephrine, given subcutaneously, increased cyclic AMP excretion twofold in healthy subjects but caused no change in that of the asthmatics. If glucagon was administered, the urinary excretion of cyclic AMP increased in both groups. Coffey and Middleton (61) demonstrated that administration of methylprednisolone increased the urinary cyclic AMP excretion and that asthmatics on steroids had normal excretory levels. They suggested that these data indicate that corticoids restore the normal beta adrenergic receptor responsiveness. This effect might be due to the altered uptake and disposition of epinephrine, as demonstrated in heart and lung (62–64), which leads to a larger epinephrine effect.

Several clinical studies have compared the effect of adrenergic agents on healthy subjects and asthmatics in remission and undergoing attack. Logsdon et al. (65) found a diminished response of adenylate cyclase to isoproterenol in leukocytes from asthmatic patients. These findings were confirmed by Parker and Smith (66) and Alston et al. (67), who also showed that the degree of responsiveness to the catecholamines varies with the degree of airway obstruction and the severity of the asthmatic attack. It was demonstrated that normal response to isoproterenol can be restored by treating the leukocytes with α-blockers. Patel et al. (68) confirmed Logsdon's findings and also showed that, when asthmatics were in remission, leukocytes incubated with isoproterenol showed a greater increase in adenylate cyclase activity than similarly treated leukocytes from active asthmatics. This, together with a higher basal activity in leukocytes from active asthmatics, and the data of Alston et al. (67) suggested to the investigators that in the active state, the β-adrenoreceptors are maximally stimulated by endogenous catecholamines and the effect of the α-adrenoceptor may become more prominent upon the addition of exogenous catecholamines.

Some recent data raise doubts as to the interpretation of the aforementioned clinical studies. Nelson (69) showed that following ephedrine administration, there was a smaller rise in blood-free fatty acids, lactate, and glucose in response to intravenous epinephrine. Gillespie et al. (70) were unable to find differences in adenylate cyclase activities from isoproterenol treated leukocytes from healthy and asthmatic subjects. Morris et al. (71) have shown that if leukocytes from healthy subjects were stimulated twice with epinephrine, *in vivo*, with a 60-min separation between stimulations, there was no difference in cyclic AMP levels. However, if cells from these subjects were taken and stimulated with epinephrine *in vitro* 60 minutes or more after a prior *in vivo* treatment, they exhibited a decreased response. This indicated that prior *in vivo* treatment could affect subsequent *in vitro* studies. Kalisker and his coworkers (72, 73) have shown that B-lymphocytes from healthy subjects and asthmatics who had not undergone therapy with sympathomimetics did not exhibit different responses to isoproterenol. If cells from both groups were given a prior exposure to isoproterenol, they both showed a subsequent diminished response. Similarly, oral ingestion of a β-agonist caused subsequent diminished cyclic AMP response to isoproterenol. It has also been demonstrated (74) that leukocytes from asthmatics on bronchodilator therapy showed an impaired cyclic AMP response to isoproterenol. However, when taken off the therapy for 5 days, leukocytes from the same patients showed normal response. In light of these findings, decreased response of cyclic AMP to sympathomimetics in leukocytes from asthmatics might be related to previous bronchodilator therapy.

In summary, the basal levels of the cyclic nucleotides are not different in healthy and sensitized lung preparations and leukocytes. Cyclic AMP and GMP levels are elevated during anaphylaxis. Since this increase appears to be relatively larger for cyclic AMP, the cyclic AMP/cyclic GMP ratio increases. Lower urinary excretion of cyclic AMP has been observed in asthmatic subjects after epinephrine administration, a possible indication that the effect of asthma on the cyclic AMP system may be systemic. In the sensitized and anaphylactic states as compared to healthy controls, the ability of exogenous agents to affect the cyclic nucleotide levels and adenylate cyclase is modified. Most notably, the effects of epinephrine and histamine appear to be changed. Similarly, cholinergic agents produce an altered and, in this case greater, increase in cyclic GMP levels in the *Bordetella pertussis* sensitized animals. The response of the cyclic AMP system to epinephrine in the sensitized animals is somewhat puzzling. Epinephrine causes a smaller increase in cyclic AMP levels in the sensitized than in the control state, even though the cell-free adenylate cyclase enzyme preparation is hyperresponsive. These paradoxical effects, we have hypothesized, are due to changes in the "normal" compartmentalization and organization of the tissue. Several clinical studies previously reported that asthmatics exhibit a decreased leukocyte adenylate cyclase response to isoproterenol. However, more recent studies have shown that this effect is probably due to prior bronchodilator therapy and that, in fact, there is no difference in responses between leukocytes from healthy subjects, acute asthmatics, or asthmatics in remission.

Cyclic Nucleotides and Release of Mediators

Among the several models proposed to explain the immunologic release of mediators from the lung, two appear to have gained the most acceptance. The first one, by Lichtenstein (75), consists of two stages: an activation stage (Ca^{++} independent) and a histamine release stage (Ca^{++} dependent). The second one, by Kaliner and Austen (76), consists of five stages: (1) a Ca^{++} dependent activation of serine esterase, (2) an autocatalytic esterase activation

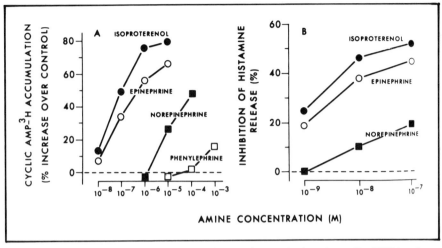

4 The effects of adrenergic agonists on leukocyte cyclic AMP levels and histamine release. Reprinted from H. R. Bourne, L. M. Lichtenstein and K. L. Melmon, J. Immunol. 108:695-705, 1972, with the permission of the authors.

step, (3) a 2-deoxyglucose inhibitable, energy requiring stage, (4) a Ca^{++}-dependent, EDTA inhibitable step, and (5) a cyclic AMP suppressible release of histamine and SRS-A. With respect to cyclic AMP, both models propose that at one or more stages, the immunologic release of mediators can be suppressed by cyclic AMP.

Histamine Release

Lichtenstein and his coworkers and Austen and his coworkers have provided much of the evidence for the role of cyclic AMP in the inhibition of histamine release. Lichtenstein and Margolis (77) showed that catecholamines and the methylxanthines inhibit the release of histamine. These results were confirmed and expanded by Assem and Schild (78) in passively sensitized human lung, Ishizaka et al. (79) in monkey lung sensitized with human IgE, and Orange et al. (80) in passively sensitized human lung. Subsequent experiments with lung or leukocyte preparations showed that the ability of catecholamines to increase cyclic AMP levels correlated well with the potency with which they inhibited histamine release (81). This order was typical of β-adrenergic agonists (isoproterenol > epinephrine > norepinephrine > phenylephrine) (81–83) (Fig. 4).

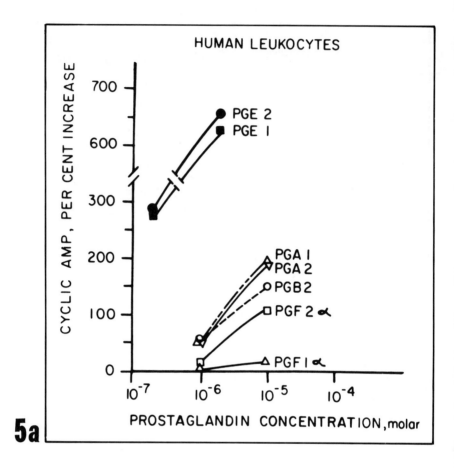

Confirmation of the β-adrenergic action of these agents was obtained when it was found that propranolol blocked their effect whereas the α-blocker, phentolamine, did not (81, 84). Similarly, a positive correlation was found between the ability of prostaglandins to inhibit the release of histamine and to raise the levels of cyclic AMP (82, 85) (Fig. 5). These biochemical findings were supported by morphologic evidence showing that agents that increased cyclic AMP caused a decrease in the extent of mast cell degranulation (86).

A more direct indication that increased cyclic AMP leads to inhibition of histamine release was obtained using dibutyryl cyclic AMP (84, 87–90) and cholera enterotoxin (91–93). The delay pe-

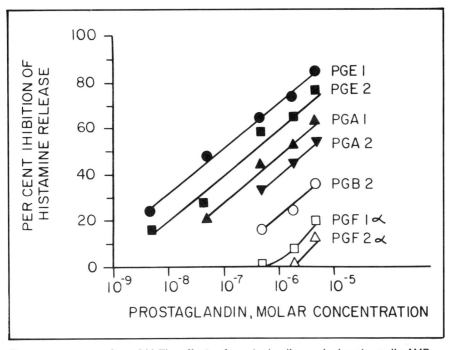

(*a* and *b*) The effects of prostaglandins on leukocyte cyclic AMP levels and histamine release. Reprinted from L. M. Lichenstein, E. Gillespie, H. B. Bourne and C. S. Henney, Prostaglandins 2:519-528, 1972, with the permission of the editors.

riod necessary for the enterotoxin to stimulate adenylate cyclase and to inhibit the release of histamine (in general 30 to 60 min) was similar. Dibutyryl cyclic GMP had no effect on the immunogenic release of histamine (89). Exogenous histamine also caused a rise in cyclic AMP levels accompanied by a concomitant decrease in histamine release (94). Five different classes of antihistamines which block H_1 receptors were found to have no effect on the inhibition of histamine release by histamine. However, this inhibition could be blocked by burimamide and metiamide, antihistamines of the H_2 type (34, 35, 90). It thus appears that there exists a negative feedback system for histamine which acts through a H_2 receptor.

Histamine, the catecholamines, and the prostaglandins appear to work at the activation, that is the Ca^{++}-independent stage of the Lichtenstein model (75). If added at the second stage, they are incapable of preventing the release of histamine. In the model proposed by Kaliner and Austen (76), the cyclic AMP inhibitable step is the last stage before the release of histamine. These differences may be partially reconciled by the different experimental systems employed—human leukocytes for the former, and passively sensitized human lung tissues for the latter.

While inhibition of histamine release is caused by an increase in cyclic AMP levels, enhancement of release may be caused by a decrease in these levels. There is some evidence to indicate that this phenomenon is mediated via adrenergic control. Treatment of lung tissue with phenylephrine or with natural catecholamines in combination with β-blockers produces a decrease in cyclic AMP levels and an increase in the release of histamine (80, 96). Indeed, it appeared that all agents (low concentrations of PGE_1 and $PGF_{2\alpha}$, low concentrations of norepinephrine, imidazole, etc.) which reduced cyclic AMP levels increased mediator release (97). If either acetylcholine, carbachol, or 8-bromo-cyclic GMP was incubated with human lung fragments before challenge, the amount of histamine released during anaphylaxis would be greater than that of the anaphylactic control (see Fig. 6). Atropine was able to block the responses produced by the cholinergic agents, but was ineffectual against phenylephrine. While phenylephrine caused a decrease in cyclic AMP, the cholinergic agents did not produce a consistent change. When isoproterenol was added in combination with carbachol, the cyclic AMP levels rose and histamine release was inhibited (96, 97).

In summary, it has been shown that in both normal and asthmatic lungs catecholamines cause an increase in cyclic AMP levels via a β_2 adrenoreceptor. However, the response of the cyclic AMP system to the catecholamines may be less in the sensitized state. Substances which produce increases in cyclic AMP levels (i.e., dibutyryl cyclic AMP, catecholamines, prostaglandins, phosphodiesterase inhibitors, etc.) can cause an inhibition of histamine release both during anaphylaxis and in the sensitized state. Exogenously administered histamine can also cause an increase in cyclic AMP levels and a decrease in anaphylactic histamine release. This effect is apparently controlled by a H_2 receptor. In order for the above substances to be effective in blocking histamine release, they must be administered before anaphylactic challenge. Some evidence has been found for both an α-adrenergic effect upon cyclic AMP and a possible cholinergic effect on cyclic GMP. Phenylephrine or

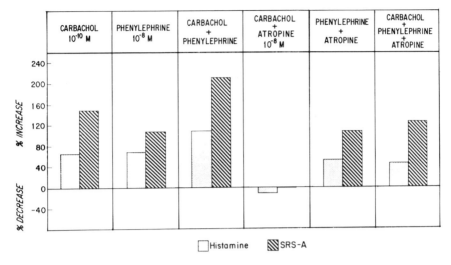

6 The effect of atropine on cholinergic and α adrenergic-induced enhancement of the immunologic release of histamine and SRS-A from human lung tissue. Atropine was added to the incubation medium 1 minute before the addition of carbachol, phenylephrine, or both. Antigen was added 3 minutes after the agonists. Reprinted from M. Kaliner, R. P. Orange, K. F. Austen, J. Exp. Med. 136:566-567, 1972, with the permission of the editors.

natural catecholamines in combination with β-blockers cause a decrease in cyclic AMP levels and a concomitant increase in histamine release during anaphylactic challenge. An increase in histamine release without a decrease in cyclic AMP levels can be caused by cholinergic agents or 8-bromo-cyclic GMP. This effect can be blocked by atropine.

Release of Other Mediators

In addition to histamine, SRS-A, ECF-A, PAF, the prostaglandins and possibly other substances are also released during anaphylaxis (98, 99). In comparison to the number of investigations concerned with the role of cyclic AMP in histamine release, relatively little has been done to study the effect of cyclic AMP on other mediators. Mathé *et al.* (100) have shown that indomethacin-treated sensitized guinea pigs did not have significantly lower lung

Table 2

Effect of Epinephrine on Tissue Levels of Cyclic AMP and Release of PGE, $PGF_{2\alpha}$ and Histamine from Sensitized and Anaphylactic Guinea Pig Lungs.

Treatment		Cyclic AMP	PGE	$PGF_{2\alpha}$	Histamine
I	NaCl	1.8 ± 0.2	3.5 ± 0.9	4.7 ± 1.0	3.7 ± 0.8
II	Ovalbumin	8.1 ± 1.3	8.5 ± 2.1	14.3 ± 3.4	30.0 ± 6.1
III	Epinephrine, 5 min				
	Epi 1 + NaCl	2.3 ± 0.4	3.5 ± 0.3	4.1 ± 0.6	1.1 ± 0.2*
	Epi 10 + NaCl	6.4 ± 1.2*	2.6 ± 0.4	4.2 ± 0.2	1.6 ± 0.3*
	Epi 1 + Ovalbumin	11.1 ± 2.0	3.9 ± 0.4†	5.9 ± 0.7†	14.9 ± 2.3†
	Epi 10 + Ovalbumin	9.2 ± 3.2	4.8 ± 1.1†	6.4 ± 1.5†	17.6 ± 4.0†
IV	Epinephrine, 3 min				
	NaCl + Epi 1	2.1 ± 0.8*	3.7 ± 0.4	4.3 ± 0.9	4.4 ± 0.2
	NaCl + Epi 10	5.1 ± 0.8*	3.1 ± 0.8	4.5 ± 0.8	3.9 ± 1.3
	Ovalbumin + Epi 1	16.7 ± 1.9†	8.4 ± 2.1	16.7 ± 5.3	39.0 ± 4.9
	Ovalbumin + Epi 10	14.5 ± 1.8†	4.9 ± 0.9	11.4 ± 2.4	27.9 ± 6.3

Lungs from ovalbumin sensitized guinea pigs were perfused with Tyrode's solution via the pulmonary artery. Perfusates were collected for 5 minutes and 0.9 percent NaCl (I) or 100 mg ovalbumin in 0.9 percent NaCl (II) was infused during the second minute of collection. Epinephrine (1 or 10 µg/ml), was infused during the entire 5-min period (III) or only the last 3 min (IV). The results are expressed as means ± SE; cyclic AMP as nmoles/gm lung wet weight, prostaglandins as ng released and histamine as % released. N = 5 animals/group. More interesting statistical differences from sensitized controls (I NaCl) and anaphylactic controls (II Ovalbumin) are indicated by * and †, respectively. Reprinted from A.A. Mathé, S.-S. Yen, R.J. Sohn and Hedqvist, P., Biochem. Pharm. (in press) 1976 with permission of authors.

levels of cyclic AMP than the nontreated controls. However, there was a lower histamine and prostaglandin "spontaneous" release during perfusion. In contrast, indomethacin-treated anaphylactic animals had lower cyclic AMP levels than the anaphylactic controls. These indomethacin-treated animals also had decreased outflows of the prostaglandins and histamine. Diisopropylfluorophosphate was found to have a similar effect to that of indomethacin. The effect of epinephrine and anaphylaxis on cyclic AMP and the release of prostaglandins and histamine in whole perfused lung were also studied (See Table 2) (100). The anaphylactic reaction (ovalbumin infusion), elevated the cyclic AMP levels and also increased the outflows of PGE, $PGF_{2\alpha}$, and histamine over the sensitized control state (NaCl infusion).

Infusion of epinephrine, 1 or 10 µg/ml, over a period of 5 min in sensitized nonchallenged lungs suppressed the spontaneous release of histamine without influencing that of the prostaglandins. The higher epinephrine concentration also increased the tissue levels of cyclic AMP. In anaphylactic lungs, both epinephrine concentrations reduced the release of histamine and prostaglandins but had no clear-cut effect on cyclic AMP accumulation.

Infusion of epinephrine, 1 or 10 µg/ml, over a period of 3 min had no effect on the release of prostaglandins and histamine in sensitized nonchallenged lungs. However, the higher epinephrine concentration raised the cyclic AMP levels. In anaphylactic lungs, the lower concentration of epinephrine had an additive action on the cyclic AMP accumulation. The higher epinephrine concentration had an additive effect on the cyclic AMP rise caused by the anaphylactic release and caused a decrease in PGE. It would thus appear that the relationship between cyclic AMP, histamine and the prostaglandins is complex. A primary effect of cyclic AMP on the release of mediators followed by a secondary effect of the mediators on cyclic AMP levels is suggested by these results. More studies are necessary to establish the sequential relationship of these changes.

Wasserman et al. (93) have shown that increases in cyclic AMP in passively sensitized human fragments inhibited the release of ECF-A. They demonstrated that both dibutyryl cyclic AMP and treatment with cholera enterotoxin caused inhibition. Carbachol ($10^{-9}M$ to $10^{-11}M$) and 8-bromo-cyclic GMP ($10^{-6}M$ to $10^{-9}M$) enhanced release of ECF-A. It thus appears that as with histamine, the release of ECF-A is under both adrenergic and cholinergic control.

As with ECF-A, studies on SRS-A have revealed a regulatory mechanism similar to that of histamine. Ishizaka et al. (79) demonstrated that the release of SRS-A was inhibited by epinephrine, isoproterenol and theophylline and that β-receptor antagonists block this effect. In addition, isoproterenol and theophylline exert a synergisitic effect upon the inhibition of SRS-A release. Furthermore, concentrations of dibutyryl cyclic AMP of $10^{-5}M$ to $10^{-3}M$ also produced an inhibition. Studies with acetylcholine, carbachol and 8-bromo-cyclic GMP all showed enhanced release of SRS-A (96) without causing changes in cyclic AMP levels. Isoproterenol ($5 \times 10^{-7}M$) in combination with carbachol ($10^{-12}M$ to $10^{-9}M$)

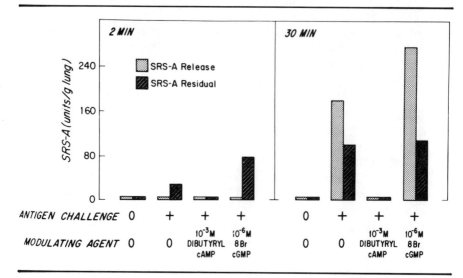

7 Cyclic nucleotide modulation of SRS-A generation from passively sensitized human lung fragments. Reprinted from R. A. Lewis, S. I. Wasserman, E. J. Goetzl and K. F. Austen, J. Exp. Med. 140:1133-1146, 1974, with the permission of the editors.

blocked SRS-A release and increased cyclic AMP levels (97). An α-adrenergic effect has been demonstrated by employing phenylephrine and propranolol. When these two agents were used together or when phenylephrine was used alone, there was an increased release of SRS-A accompained by a decrease in cyclic AMP. While atropine blocked the effects of the cholinergic agents, it was not effective in antagonizing the effect of phenylephrine.

In addition to the effect of the cyclic nucleotides upon SRS-A release, Lewis et al. (101) have demonstrated that the cyclic nucleotides also affected the generation of SRS-A (see Fig. 7). They found that preincubation of human lung fragments with $10^{-3}M$ dibutyryl cyclic AMP prevented the generation of SRS-A as well as its release for at least 30 min. Preincubation with $10^{-3}M$ 8-bromo-cyclic GMP increased tissue SRS-A at both 2 and 30 min. However, despite these increases, no SRS-A was released by 2 min. At 30 min, the amount released was both proportionally and absolutely greater.

From the data available, it would appear that the relationship of cyclic nucleotides to SRS-A and ECF-A is similar to their relationship to histamine. Cyclic AMP analogs and agents which increase endogenous cyclic AMP cause an inhibition of the release of these compounds. Agents that cause a decrease in cyclic AMP by activating α-adrenergic receptors or by acting directly enhance the release of SRS-A. No evidence has been reported on the effect of α-adrenergic agonists on ECF-A. Similarly, stimulation of the cholinergic system or the presence of high levels of cyclic GMP cause an enhancement of anaphylatic mediator release. In addition, it has been shown that both cyclic AMP and cyclic GMP have an effect upon the generation of SRS-A, distinct from their action on its release. Preincubation with dibutyryl cyclic AMP before challenge prevented SRS-A generation whereas cyclic GMP enhanced its production several-fold. The role of cyclic AMP with respect to the prostaglandins is more complex. Substances which inhibit prostaglandin synthesis can also cause decreases in the level of cyclic AMP. However, depending upon the dose and time of administration (with respect to anaphylactic challenge) changes in cyclic AMP and prostaglandin levels could occur independently. Further studies are necessary to establish the sequential relationship between the primary and secondary changes in the cyclic nucleotide levels and mediator release.

SUMMARY

In both the healthy and asthmatic lung, neurotransmitters of the sympathetic and parasympathetic nervous systems interact with the cyclic nucleotides. Adrenergic agonists increase adenylate cyclase activity without changing the activity of the particulate or soluble phosphodiesterases. This results in an increase in the cyclic AMP level, an increase which can be blocked by β_2 antagonists. On the other hand, there is some evidence indicating that stimulation of the α adrenergic receptors leads to a decrease in the cyclic AMP level. Cholinergic agents increase lung cyclic GMP by acting via muscarinic receptors. This action can be antagonized by atropine and by isoproterenol. Similarly, the isoproterenol induced rise in cyclic AMP is antagonized by acetylcholine.

Chemical agents that are released during anaphylaxis have also been shown to affect the cyclic nucleotide systems in the healthy lung. The prostaglandins, notably PGE_1 and PGE_2, increase the lung cyclic AMP levels, while $PGF_{2\alpha}$ increases that of cyclic GMP. Substances that have a primary effect on endogenous prostaglandin production can secondarily affect the levels of the cyclic nucleotides. Hence, indomethacin and aspirin prevent the rise of cyclic AMP induced by factors which release prostaglandins. Histamine increases both cyclic AMP and cyclic GMP levels. The histamine-induced increase of cyclic AMP can be partially antagonized by H_1 antagonists and is completely blocked by H_2 antagonists. In contrast, effect of histamine on cyclic GMP levels is antagonized by H_1 antagonists.

Antigen sensitization of experimental animals alters the effect of adrenergic and cholinergic agents on cyclic nucleotides. Epinephrine and histamine cause smaller increases of lung cyclic AMP levels in sensitized than in control animals. On the other hand, carbachol produces a higher lung cyclic GMP level in the sensitized than in the healthy state. In man, urinary cyclic AMP excretion exhibits less of an increase following epinephrine administration to asthmatics than to healthy subjects.

Substances which increase cyclic AMP levels cause a decrease in the release of the mediators and anaphylaxis. Conversely, substances which decrease the cyclic AMP level, enhance mediator release. While data on the cyclic GMP system are not yet definite, it appears that increasing the cyclic GMP level enhances mediator release, and decreasing it has the opposite effect.

REFERENCES

1. Malpighius, M.: De pulmonibus observationes anatomiae. Bologna 1661 Translated by J. Young. Proc. Roy. Soc. Med. (Pt 1) 23:1-11, 1929-1930.
2. Schild, H.: Histamine release and anaphylactic shock in isolated lungs of guinea-pigs. Quart. J. Exp. Physiol. 26:165-179, 1936.
3. Eppinger, H. and Hess, L.: Vagotonia: A Clinical Study in Vegatative Neurology II. Nervous and Mental Disease Publishing Co., New York, 1917.
4. Williams, H. L.: Trans. Amer. Acad. Opthalmol. Otolaryng 123:1950.
5. Gellhorn, E. and Loofburrow, G. N.: Emotions and Emotional Disorders. Harper & Row, New York, 1963.
6. Kaliner, M. A., Orange, R. P., Koopman, W. J., Austen, K. F. and Laraia, P. J.: Cyclic adenosine 3'5' monophosphate in human lung. Biochem. Biophys. Acta 252:160-164, 1971.
7. Butcher, R. W. and Sutherland, E. W.: Adenosine 3'5' phosphate in bio-

logical materials. I Purification and properties of cyclic 3'5' nucleotide phosphodiesterase and use of this eyzmes to characterize adenosine 3'5' phosphate in human-urine. J. Biol. Chem. 237:1244-1250, 1962.
8. Kuo, J. F.: Guanosine 3'5' monophosphate-dependent protein kinases in mammalian tissue. Proc. Natl. Acad. Sci. 71:4037-4041, 1974.
9. Chrisman, T. D., Garbers, D. L. and Hardman, J. G.: Soluble and particulate guanylate cyclases from rat lung. Fed. Proc. 33:1250, 1974.
10. Murad, F., Chi, Y. M., Rall, T. W. and Sutherland, E. W.: Adenyl Cyclase III. The effect of catecholamines and choline esters on formation of adenosine 3'5' phosphate by preparations from cardiac muscle and liver. J. Biol. Chem. 237:1233-1238, 1962.
11. Robinson, G. A., Butcher, R. W. and Sutherland, E. W.: Cyclic AMP. Academic Press, New York, 1971.
12. Lands, A. M., Arnold, A., McAuliff, J. P., Luduena, F. P. and Brown, T. G.: Differentiation of receptor systems activated by sympathomimetric amines. Nature 214:597-598, 1967.
13. Burges, R. A. and Blackburn, K. J.: Adenyl cyclase and the differentiation of β-adrenoreceptors. Nature New Biol. 235:249-250, 1972.
14. Lefkowitz, R. J.: Heterogeneity of adenylate cyclase-coupled β-adrenergic receptors. Biochem. Pharm. 24:583-590, 1975.
15. Hitchcock, M.: Adenosine 3'5' cyclic monophosphate phosphodiesterase in guinea pig lung-properties and effect of adrenergic drugs. Biochem. Pharm. 22:959-969, 1973.
16. Collins, M., Palmer, G. C., Baca, G. and Scott, H. R.: Stimulation of cyclic AMP in the isolated perfused rat lung. Res. Commun. Chem. Path. Pharm. 6:805-812, 1973.
17. Sutherland, E. W., Robinson, G. A. and Butcher, R. W.: Some aspects of the biological role of adenosine 3'5' monophosphate (cyclic AMP). Circulation 37:279-305, 1968.
18. Palmer, G. C.: Characteristics of the hormonal induced cyclic adenosine 3'5' monophosphate response in the rat and guinea pig lung *in vitro*. Biochem. Biophys. Acta 252:561-566, 1971.
19. Palmer, G. C.: Cyclic 3'5' adenosine monophosphate response in the rabbit lung-adult properties and development. Biochem. Pharm. 21:2907-2914, 1972.
20. Kuo, J. F.: Reciprocal changes in levels of guanosine 3'5' monophosphate dependent and adenosine 3'5' monophosphate dependent protein kinases in developing guinea pig lungs. J. Cyclic Nucleotide Res. 1:151-157, 1975.
21. Kuo, W.-N. and Kuo, J. F.: Regulation of cyclic GMP and cyclic AMP levels in rat lung and other tissues by various agents as determined by double-prelabeling with radioactive guanine and adenine. Fed. Proc. 773A, 1973.
22. Kuo, J. F. and Kuo, W.-N.: Regulation of β-adrenergic receptor and muscarinic cholinergic receptor activation of intracellular cyclic AMP and cyclic GMP levels in rat lung slices. Biochem. Biophys. Res. Commun. 55:660-665, 1973.
23. Stoner, J., Manganiello, V. C. and Vaughan, M.: Guanosine cyclic 3'5' monophosphate and guanylate cyclase activity in guinea pig lung: Effect

of acetylcholine and cholinesterase inhibitors. Molec. Pharm. 10:155-161, 1974.
24. Krishnan, N. and Krishna, G.: Hormone Induced Modulation of Cyclic GMP in the Lung. Advances in Nucleotide Research, Vol. 5, Drummond, G. I., Greengard, P. and Robinson, G. A. (Eds.) p. 820. Raven Press, New York, 1975.
25. Goldberg, N. D., Haddox, M. K., Hartle, D. K. and Hadden, J. W.: The Biological Role of Cyclic 3'5' Guanosine Monophosphate. 5th International Congress of Pharmacology, San Francisco, Vol. 5, 146-169, Karger, (Basel), 1972.
26. Muscholl, E.: Muscarinic Inhibition of the Norepinephrine Release from Peripheral Sympathetic Fibers. Pharmacology and the Future of Man. Proc. 5th International Congress of Pharmacology, San Francisco, Vol. 4, 440-457, Karger, (Basel), 1973.
27. Kosterlitz, H. W. and Lees, G. M.: Interrelationships Between Adrenergic and Cholinergic Mechanisms. Catecholamines. Blaschko, H. and Muscholl, E. (Eds.), p. 762-812, Springer-Verlag, Berlin, 1972.
28. Mathé, A. A., Tong, E. Y. and Tisher, P. W.: Sympathetic-parasympathetic balance in the lung: Release of norepinephrine by nerve stimulation. Fed. Proc. 35:600, 1976.
29. Butcher, R. W. and Baird, C. E.: Effects of prostaglandins on adenosine 3'5' monophosphate levels in fat and other tissues. J. Biol. Chem. 243: 1713-1717, 1968.
30. Mathé, A. A.: Prostaglandins and the Lung. The Prostaglandins, Vol. 3, P. W. Ramwell (Ed.), Plenum Press, New York, 1976 (in press).
31. Stoner, J., Manganiello, V. C. and Vaughan, M.: Effects of bradykinin and indomethacin on cyclic GMP and cyclic AMP in lung slices. Proc. Natl. Acad. Sci. 70:3830-3833, 1973.
32. Weinryb, I., Michel, I. M. and Hess, S. M.: Adenylate cyclase from guinea pig lung: Further characterization and inhibitory effects of substrate analogs and cyclic nucleotides. Arch. Biochem. Biophys. 154:240-249, 1973.
33. White, G.: Stimulation of rat and monkey lung adenyl cyclase by various prostaglandins. Fed. Proc. 33:590, 1974.
34. Mathé, A. A., Sohn, R. J. and Volicer, L.: Cyclic adenosine monophosphate and release of histamine in healthy, sensitized and anaphylactic guinea pig lung. Fed Proc. 34:798, 1975.
35. Mathé, A. A., Volicer, L. and Puri, S. K.: Effect of anaphylaxis and histamine, pyrilamine, and burimamide and cAMP and cGMP in guinea pig lung. Res. Commun. Chem. Path. Pharm. 8:635-651, 1974.
36. Moore, P. F., Iorio, L. C. and McManus, J. M.: Relaxation of the guinea pig tracheal chain preparation by N^6O^2 dibutyryl 3'5' cyclic adenosine monophosphate. J. Pharm. Pharmacol. 20:368-372, 1968.
37. Middleton, E. and Finke, S. R.: Effect of adrenergic blockade on tracheal smooth muscle response to histamine, mecholyl, anaphylaxis, and catecholamines. J. Clin. Invest. 47:69a, 1968.
38. Guirgis, H. M.: The effect of adrenergic drugs on the isolated tracheal tube *in vitro* and on the bronchial resistance to inflation *in vivo*. Arch.

Internatl. de Pharmacodynamine et de Thérapie 182:147-160, 1969.
39. Svedmyr, N., Andersson, R., Bergh, N. P. and Malmberg, R.: Relaxing effect of ACTH on human bronchial muscle *in vitro*. Scand. J. Resp. Dis. 51:171-176, 1970.
40. Mathé, A. A., Åström, A. and Persson, N. Å.: Some broncho-constricting and broncho-dilating responses of human isolated bronchi: Evidence for the existence of α-adrenoceptors. J. Pharm. Pharmacol. 23:905-910, 1971.
41. Andersson, R., Bergh, P. and Svedmyr, N.: Metabolic actions in human bronchial muscle associated with ACTH induced relaxation. Scand. J. Resp. Dis. 53:125-128, 1972.
42. Rubin, B., O'Keefe, E. H., Waugh, M. H., Kotler, D. G., DeMaio, D. A. and Horowitz, Z. P.: Activities *in vitro* of 8-substituted derivatives of adenosine 3'5' cyclic monophosphate on guinea pig trachea and rat portal vein. Proc. Soc. Expt. Biol. & Med. 137:1244-1248, 1971.
43. Szaduykis-Szadurski, L. and Berti, F.: Smooth muscle relaxing activity of 8-bromo-guanosine 3'5' monophosphate. Pharm. Res. Commun. 4:53-61, 1972.
44. Szaduykis-Szadurski, L., Weiman, G. and Berti, F.: Pharmacological effects of cyclic nucleotides and their derivatives on vascular smooth muscle. Pharm. Res. Commun. 4:63-69, 1972.
45. Andersson, R. G. G. and Kövesi, G.: The effect of hydrocortisone on tension and cyclic AMP metabolism in tracheal smooth muscle. Separatum Experientia 30:784-785, 1974.
46. Murad, F.: Beta-blockade of epinephrine-induced cyclic AMP formation in heart, liver, fat and trachea. Biochem. Biophys. Acta 304:181-187, 1973.
47. Murad, F. and Kimura, H.: Cyclic nucleotide levels in incubations of guinea pig trachea. Biochem. Biophys. Acta 343:275-286, 1974.
48. Andersson, R., Nilsson, K., Wikberg, J., Johansson, S., Mohme-Lundholm, E., and Lundholm, L.: Cyclic Nucleotides and the Contraction of Smooth Muscle. Advances in Cyclic Nucleotide Research, Vol. 5. Drummond, G. I., Greengard, P. and Robinson, G. A. (Eds.) p. 491-518, Raven Press, New York, 1975.
49. Vulliemoz, Y., Verosky, M., Nahas, G. G. and Triner, L.: Adenyl cyclase-phosphodiesterase system in bronchial smooth muscle. Pharmacologist 13:256, 1971.
50. Willis, T.: "Practice of Physick, Pharmaceutice Rationalis or the Operations of Medicines in Human Bodies" 2nd ed. Section 1, Chapter 12, Dring, T. (Ed.) p. 78-85, London, 1684.
51. Bates, D. V., Macklem, P. T. and Christie, R. V.: Respiratory Function in Disease. W. B. Saunders Co., Philadelphia, 1971.
52. Gross, N. J.: Bronchial Asthma. Harper & Row, New York, 1974.
53. Szentivanyi, A.: The beta adrenergic theory of the atopic abnormality in bronchial asthma. J. Allergy 42:203-232, 1968.
54. Mathé, A. A., Puri, S. K. and Volicer, L.: Sensitized guinea pig lung: Altered adenylate cyclase stimulation by epinephrine. Life Sci. 15:1917-1924, 1974.
55. Mathé, A. A., Puri, S. K., Sohn, R. J. and Volicer, L.: Effect of epineph-

rine on cyclic AMP levels and adenylate cyclase and phosphodiesterase activities in control and antigen sensitized guinea pig lungs. Pharmacology, in press, 1976.
56. Polson, J. B., Krzanowski, J. J. and Szentivanyi, A.: Histamine induced changes in pulmonary guanosine 3'5' cyclic monophosphate and adenosine 3'5' cyclic monophosphate levels in mice following sensitization by Bordetella pertussis and/or propranolol. Res. Commun. Chem. Path. Pharm. 9: 243-251, 1974.
57. Polson, J. B., Krzanowski, J. J., jr. and Szentivanyi, A.: Elevation of Cyclic GMP Levels Caused by Histamine and Methacholine in mice Simulating Atopic Disease. Advances in Cyclic Nucleotide Research, Vol. 5, Drummond, G. I., Greengard, P. and Robinson, G. A. (Eds.) p. 815, Raven Press, New York, 1975.
58. Ortez, R. A.: Histamine induced accumulation of lung cyclic adenosine 3'5' monophosphate in pertussis vaccinated mice. Fed. Proc. 33:586A, 1974.
59. Bernstein, R. A., Linarelli, L., Facktor, M. A., Friday, G. A., Drash, A. L. and Fireman, P.: Decreased urinary adenosine 3'5' monophosphate in asthmatics. J. Lab. Clin. Med. 80:772-779, 1972.
60. Fireman, P.: Metabolic abnormalities in asthma-decreased urinary cyclic AMP. Internatl. Arch. Allergy 45:123-127, 1973.
61. Coffey, R. G. and Middleton, E., jr.: Effects of glucocorticosteroids on the urinary excretion of cyclic AMP and electrolytes by asthmatic children. J. Allergy & Clin. Immunol. 54:41-53, 1974.
62. Brodie, B. B., Davies, J. T., Hynie, S., Khrishna, G. and Weiss, B.: Interrelationships of catecholamines with other endocrine systems. Pharmacol. Rev. 18:273, 1966.
63. Iversen, L.: Catecholamine uptake processes. Br. Med. Bull. 29:130-135, 1973.
64. Gillespie, J. S.: Uptake of noradrenaline by smooth muscle. Br. Med. Bull. 29:136-141, 1973.
65. Logsdon, P. J., Middleton, E. jr. and Coffey, R. G.: Stimulation of leukocyte adenyl cyclase by hydrocortisone and isoproterenol in asthmatic and non-asthmatic subjects. J. Allergy Clin. Immunol. 50:45-56, 1972.
66. Parker, C. W. and Smith, J. W.: Alteration in cyclic adenosine monophosphate metabolism in human bronchial asthma. J. Clin. Invest. 52:48-59, 1973.
67. Alston, W. C., Patel, K. R. and Kerr, J. W.: The response of leukocyte adenyl cyclase to isoproterenol and the effect of alpha blocking drugs in extrinsic bronchial asthma. Br. Med. J. 1:90-93, 1974.
68. Patel, K. R., Alston, W. C. and Kerr, J. W.: The relationship of leukocyte adenyl cyclase activity and airways response to beta blockade and allergen challenge in extrinsic asthma. Clin. Allergy. 4:311-322, 1974.
69. Nelson, H. S.: The effect of ephedrine on the response to epinephrine in normal man. J. Allergy Clin. Immunol. 51:191-198, 1973.
70. Gillespie, E., Valentine, M. D. and Lichtenstein, L. M.: The beta adrenergic theory of asthma. Fact or Fantasy? J. Allergy Clin. Immunol. 51: 93-94, 1973.

71. Morris, H. G., DeRoche, G. B. and Caro, C. M.: Response of Leukocyte and Cyclic AMP to Epinephrine Stimulation *In Vivo* and *In Vitro*. Advances In Cyclic Nucleotide Research Vol. 5, Drummond, G. I., Greengard, P. and Robinson, G. A. (Eds.) p. 812, Raven Press, New York, 1975.
72. Kalisker, A. and Middleton, E. jr.: Beta adrenergic receptor desensitization in lymphocytes from normal and asthmatic subjects. Fed. Proc. 34: 984, 1975.
73. Kalisker, A.: (personal communication).
74. Conolly, M. E., Greenacre, J. K. and Dollery, C. T.: β-Adrenoreceptor Function (βKF) in Asthma. Advances in Cyclic Nucleotide Research, Vol. 5, Drummond, G. I., Greengard, P. and Robinson, G. A. (Eds.) p. 815, Raven Press, New York, 1975.
75. Lichenstein, L. M.: The immediate allergic response. *In vitro* separation of antigen activation, decay, and histamine release. J. Immunol. 107: 1122-1130, 1971.
76. Kaliner, M. and Austen, K. F.: A sequence of biochemical events in the antigen-induced release of chemical mediators from sensitized human lung tissue. J. Exp. Med. 138:1077-1094, 1973.
77. Lichtenstein, L. M. and Margolis, S.: Histamine release *in vitro*: Inhibition by catecholamines and methylxanthines. Science 161:902-903, 1968.
78. Assem, E. S. K. and Schild, H. O.: Inhibition by sympathomimetic amines of histamine release induced by antigen in passively sensitized human lung. Nature (Lond.) 224:1028-1029, 1969.
79. Ishizaka, T., Ishizaka, K., Orange, R. P. and Austen, K. F.: Inhibition of antigen-induced release of histamine and SRS-A from monkey lung tissues mediated by human IgE. J. Immunol. 106:1267-1273, 1971.
80. Orange, R. P., Austen, W. G. and Austen, K. F.: Immunologic release of histamine and SRS-A from human lung. I. Modulation by agents influencing cellular levels of cyclic 3'5' adenosine monophosphate. J. Exp. Med. 134:136s-148s, 1971.
81. Bourne, H. R., Lichenstein, L. M. and Melmon, K. L.: Pharmacologic control of allergic histamine release *in vitro*: Evidence for an inhibitory role of 3'5' adenosine monophosphate in human leukocytes. J. Immunol. 108:695-705, 1972.
82. Tauber, A. I., Kaliner, M., Stechschulte, D. J. and Austen, K. F.: Immunologic release of histamine and slow reacting substance of anaphylaxis from human lung. V. Effects of prostaglandins on release of histamine. J. Immunol. 111:27-32, 1973.
83. Schmutzler, W., Poblete-Freundt, G. and Derwall, R.: Studies on the role of cAMP in the regulation of anaphylactic histamine release from the guinea pig lung. Agents & Actions 3:192-193, 1973.
84. Lichtenstein, L. M. and DeBernardo, R.: The immediate allergic responses: *in vitro* action of cyclic AMP-active and other drugs on the two stages of histamine release. J. Immunol. 107:1131-1136, 1971.
85. Lichtenstein, L. M., Gillespie, E., Bourne, H. B. and Henney, C. S.: The effects of a series of prostaglandins on the *in vitro* models of the allergic response and cellular immunity. Prostaglandins 2:519-528, 1972.
86. Kimura, Y., Inoue, Y. and Honda, H.: Further studies on rat mast cell

degranulation by IgE-anti-IgE and the inhibitory effects of drugs related to cyclic AMP. Immunology 26:983-988, 1974.
87. Orange, R. P., Kaliner, M. A., LaRaia, P. L. and Austen, K. F.: Immunological release of histamine and slow-reacting substance of anaphylaxis from human lung II. Influence of cellular levels of cyclic AMP. Fed. Proc. 30:1725-1729, 1971.
88. Loeffler, L. J., Lovenberg, W. and Sjoerdsma, A.: Effects of dibutyryl 3'5' cyclic adenosine monophosphate, phosphodiesterase inhibitors, and prostaglandin E_1 on compound 48/80 induced histamine release from rat peritoneal mast cells. Biochem. Pharm. 20:2287-2297, 1971.
89. Renoux, M. L., DeMontis, G. and Marcelli, D.: Adenosine 3'5' monophosphate and guanosine 3'5' monophosphate: Effect on anaphylactic histamine release by rat mast cells. Biomed. 21:410-413, 1974.
90. Mathé, A. A., Levine, L., Yen, S.-S., Sohn, R. J. and Hedqvist, P.: Release of prostaglandins and histamine from guinea pig lung. 2nd Internatl. Conference on Prostaglandins, Florence, p. 188, 1975.
91. Lichenstein, L. M., Henney, C. S., Bourne, H. R. and Greenough, W. B.: Effects of cholera toxin on *in vitro* models of immediate and delayed hypersensitivity: Further evidence for the role of cyclic AMP. J. Clin. Invest. 52:691-697, 1973.
92. Bourne, H. B., Lehrer, R. I., Lichenstein, L. M., Weissman, G. and Zurier, R.: Effects of cholera enterotoxin on adenosine 3'5' monophosphate and neutrophil function: Comparison with other compounds which stimulate leukocyte adenyl cyclase. J. Clin. Invest. 52:698-708, 1973.
93. Wasserman, S. I., Goetzl, E. J., Kaliner, M. and Austen, K. F.: Modulation of the immunological release of the eosinophil chemotactic factor of anaphylaxis from human lung. Immunology 26:677-684, 1974.
94. Bourne, H. B., Melmon, K. L. and Lichenstein, L. M.: Histamine augments leukocyte cyclic AMP and blocks antigenic histamine release. Science 173:743-745, 1971.
95. Lichenstein, L. M. and Gillespie, E.: Inhibition of histamine release by histamine is controlled by an H_2-receptor. Nature (Lond.) 244:287-288, 1973.
96. Kaliner, M., Orange, R. P. and Austen, K. F.: Immunologic release of histamine and slow-reacting substance of anaphylaxis from human lung. IV Enhancement by cholinergic and alpha adrenergic stimulation. J. Exp. Med. 136:556-567, 1972.
97. Kaliner, M. and Austen, K. F.: Hormonal Control of the Immunologic Release of Histamine and Slow-Reacting Substance of Anaphylaxis From Human Lung. In 'Cyclic AMP,' Cell Growth, and the Immune Response. Braun, W., Lichtenstein, L. M. and Parker, C. W. (Eds.) Springer-Verlag, p. 163-175, Berlin, 1974.
98. Austen, K. F. and Orange, R. P.: Bronchial asthma: The possible role of the chemical mediators of immediate hypersensitivity in the pathogensis of subacute chronic disease. Amer. Rev. Resp. Dis. 112:423-436, 1975.
99. Kaliner, M. and Austen, K. F.: Immunologic Release of Chemical Media-

tors from Human Tissues. Annual Review of Pharmacology, Elliott, H. W., George, R. and Okun, R. (Eds.), Annual Review, Inc., p. 177-189, California, 1975.
100. Mathé, A. A., Yen, S.-S., Sohn, R. J. and Hedqvist, P.: Release of prostaglandins and histamine from sensitized and anaphylactic guinea pig lungs: Changes in cyclic AMP levels. Biochem. Pharm. (in press) 1976.
101. Lewis, R. A., Wasserman, S. I., Goetzl, E. J. and Austen, K. F.: Formation of slow-reacting substances of anaphylaxis in human lung tissue and cells before release. J. Exp. Med. 140:1133-1146, 1974.

8

Role of Cyclic Nucleotides in Gastroinestinal Diseases

ROBERT A. LEVINE

This review will attempt to update past reviews (1–6) and emphasize recent investigations in man concerning a possible regulatory role of cyclic nucleotides in gastrointestinal physiology and clinical disease. The reader is referred to my last three reviews for a discussion of the physiologic importance of adenosine 3',5'-monophosphate (cyclic AMP) in hepatic metabolism (1–3).

The adenylate and guanylate cyclase systems located in the cell membrane appear to participate in practically every event associated with ion transport. For that reason emphasis will be given to critically examine the role which cyclic nucleotides may play in health and disease in control of bile, gastric, intestinal, and pancreatic secretion. I will also review their possible mediation in motility and their interaction in the gut with prostaglandins.

INTERCELLULAR COMPARTMENTALIZATION OF CYCLIC NUCLEOTIDES

Considerable caution must be exercised in interpretation of data obtained with mixed cell populations, as in the gastrointestinal tract. With the development of fluorescent immunocytochemical procedures (7), it is evident that cyclic nucleotides and their respective cyclases can be localized histologically to various cell areas providing clues for the location of different receptor sites for cyclic AMP and guanosine 3',5'-monophosphate (cyclic GMP). For example, guanylate and adenylate cyclases are found, respectively, in the intestinal brush border and lateral membrane border locations. This technique further implies separate roles of cyclic nucleotides in cellular function and indicates that hormonal stimulation might cause a redistribution of either cyclic AMP or cyclic GMP in a particular tissue without altering the total tissue level of these nucleotides. Thus the important new concept of so-called physiologically active cyclic nucleotide pools relates to their possible compartmentalization within the cell. Since measurement of intracellular cyclic AMP or cyclic GMP represents the sum total of a heterogenous group of cells, some misleading information may result from the failure to measure small, but probably more significant changes in secretion-related cyclic AMP and cyclic GMP pools, particularly as related to bile, pancreatic and gastric secretion. It is possible that regional changes in cyclic nucleotide concentration may not be reflected in alterations of total cellular nucleotide content. Moreover, different sensitivity to hormones or agents may affect separate cyclases, providing for increases in cyclic nucleotides within particular regions of the cell and accounting for a more highly selective control of cellular function.

CYCLIC NUCLEOTIDE ACTIVITY IN HUMAN GASTROINTESTINAL TISSUES AND FLUIDS

Few reports of cyclic nucleotide levels in gastrointestinal tissues and fluids have appeared in the literature. Published values have varied depending upon extraction and purification methods and assay procedures. We measured cyclic AMP by the Gilman protein-binding assay and radioimmunoassay and cyclic GMP by radioimmunoassay alone in various tissues and fluids from rats, dogs,

Table 1
Mean Gastrointestinal Cyclic Nucleotide Concentrations

Source	Cyclic AMP	Cyclic GMP
(Fluid (pmole/ml):		
Human gastric juice	11	2
Canine gastric juice	14	5
Human pancreatic juice	24	1
Human bile	169	6
Baboon bile	5,833	-
Canine bile	10,300	-
Rodent bile	100	-
Tissue (pmole/gm):		
Human stomach	650	20
Canine stomach	400	20
Rat stomach	600	9
Human jejunum	1,000	90
Rodent jejunum	950	45
Human ileum	835	-
Rodent ileum	950	45
Rodent colon	980	17
Human pancreas	-	37
Baboon liver	959	-
Chicken liver	950	12
Rat liver	380	8

baboons and humans using four different column chromatography procedures. As shown in Table 1, cyclic AMP was present in all samples at levels 10- to 100-fold higher than cyclic GMP. The lowest limit of sensitivity for cyclic GMP was 1 pmole/50 mg in human gastric tissue. The least amount of cyclic GMP occurred in human pancreatic and gastric juice.

One of the major problems concerning published reports about gastrointestinal cyclic nucleotide activity relate to confirmation of the presence of cyclic nucleotides in the samples used by various assay procedures. Unless cyclic nucleotides are completely hydrolyzed by excess phosphodiesterase treatment, their presence in fluids or tissues cannot be verified. In our studies we found that phosphodiesterase treatment completely hydrolyzed and therefore confirmed the presence of cyclic nucleotides in tissues regardless of 4 different column procedures employed (8). However, their presence in gastric juice and bile was best established after phosphodiesterase treatment by only one of 4 column procedures (neutral alumina eluted with *tris*-formate buffer followed by a Dowex-1-formate column eluted with neutral formic acid). Thus selection of the proper column is necessary for the accurate measurement of cyclic nucleotides in bile and gastric juice, although tissues gave consistent results regardless of the column procedure utilized.

BILE SECRETION

The possible role of cyclic AMP on bile formation has been the subject of recent speculation (9). Infusion of cyclic AMP or its dibutyryl derivative (DBcAMP) in experimental animals has led to a variety of responses. Some animal studies have shown increased bile flow of the nonbile salt-dependent fraction while others failed to document any alteration in bile volume or bile salt output after exogenous DBcAMP or cyclic AMP. Glucagon, which is structurally similar to secretin, is a potent choleretic in the dog. Since glucagon mediates some of its hepatic actions through cyclic AMP, it seemed reasonable to look for cyclic AMP as a second messenger in the choleretic response of secretin. Secretin is known to produce an increased flow of bicarbonate-rich, bile-acid–poor bile in experimental animals and humans. Secretin appears to stimulate bile flow rate at a site distal to the canaliculus.

We have performed studies to determine if cyclic AMP plays a role in secretin's action on bile formation and whether cyclic AMP is elaborated at a canalicular or ductular location (10). We studied cyclic AMP content in human, baboon and canine bile before and during secretin choleresis. Furthermore, we have determined how the effect of exogenous cyclic AMP on human bile formation compares with secretin and other known choleretic agents. We measured the cyclic AMP concentration in human bile after DBcAMP, aminophylline, glucagon and triketocholanoic (dehydrocholate) acid administration.

Patients were studied with either hepatic duct fistulae or after insertion of T-tubes into the common duct following cholecystectomy utilizing an inflatable balloon so as to assure complete collection of hepatic bile and preclude the flow of any pancreatic secretion out of the T-tube.

Cyclic AMP output in the bile in response to intravenous secre-

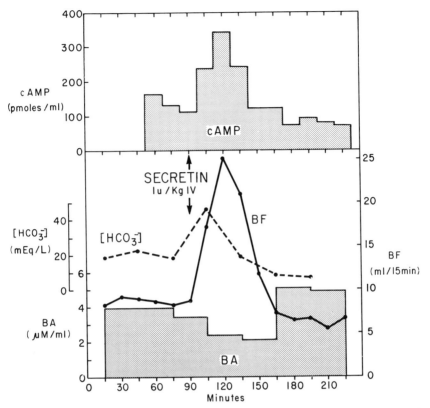

FIG. 1. Effect of secretin injection on bile flow (BF), and concentration of bile acids (BA), biliary bicarbonate [HCO^-_3] and cyclic AMP in bile from a patient with a hepatic duct fistula. Note the increased cyclic AMP and bicarbonate concentrations, increased bile flow and reduced bile acid concentration for approximately 1 hr after intravenous secretin injection. From Levine and Hall (10).

tin was measured in 11 patients, 12 baboons, and 15 dogs. Secretin was given to patients with bile drainage tubes as an intravenous bolus (1 U/kg). In baboons and dogs both secretin infusion (4 U/kgm/hr) and bolus injection (1 U/kgm) were used. In baboons cyclic AMP was also determined in liver, extrahepatic duct tissue, and in perfusate from isolated segments of extrahepatic bile ducts.

Secretin induced a marked choleresis in all 3 species. A typical example is shown in Fig. 1. In humans biliary cyclic AMP concentration increased an average ($\pm$ 1 SE) of 68% $\pm$ 12 percent and in baboons fourfold, but no increase occurred in dogs. In baboons, cyclic AMP concentration increased in both bile duct tissue and perfusate from isolated bile ducts concomitant with secretin choleresis, but not in liver. In humans, the choleretic effects of sodium dehydrocholate, aminophylline and glucagon were compared to DBcAMP. All agents increased bile flow two to threefold (Fig. 2). Cyclic AMP concentration in bile markedly increased after glucagon and DBcAMP but not after sodium dehydrocholate and aminophylline (Fig. 3). The choleretic and biliary cyclic AMP responses to aminophylline were potentiated by a concomitant injection of secretin. Figure 4 illustrates in a typical patient the augmented cyclic AMP concentration in bile induced by simultaneous secretin and aminophylline administration.

The addition of water to the bile by cyclic AMP is strikingly similar to its action in the intestine and salivary glands (see below). Our observations that dehydrocholate-induced choleresis was not associated with alterations in biliary cyclic AMP output tends to support the concept that bile salt choleresis at the canalicular level does not participate in cyclic AMP formation. In contrast, the mechanism responsible for secretin-induced choleresis is presumably located principally in the large ductal or ductular epithelium rather than at the level of the bile canaliculus, and seems likely to be mediated by cyclic AMP generation. Our study in baboons using isolated segments of common duct perfused with saline allows us to conclude more emphatically than in man that secretin stimulates the egress of cyclic AMP from the major bile ducts. The failure to find a change in cyclic AMP in liver biopsies from baboons receiving secretin infusions further implicates bile ductal epithelial cells as the sole source of cyclic AMP in the mediation of secretin choleresis in baboons. We conclude that cyclic AMP is implicated in secretin choleresis in both humans and baboons, but not in dogs. The bile duct appears to be the site of cAMP elaboration induced by secretin in baboons and probably is also in man.

While choleretic effects of glucagon have been demonstrated in dogs previously, the potent choleretic effect of this hormone has not been emphasized in man. Glucagon increased biliary flow to a similar degree as secretin and DBcAMP (Fig. 2). However, some caution against implicating cyclic AMP as a mediator of secretin choleresis is raised by the fact that despite these comparable choleretic responses in man elicited by all three agents, the magnitude

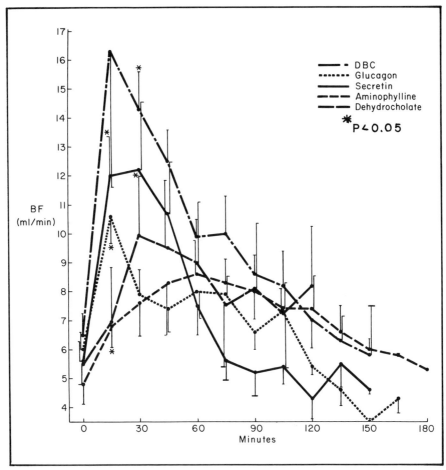

2 Effect of DBcAMP (DBC), glucagon, secretin, aminophylline and dehydrocholate on bile flow in man. Bile flow (BF) is plotted on the ordinate against time (min) on the abscissa for each group of agents administered. Each of 11 subjects acted as their own control. Control bile flow at time 0 represents an average of at least three preceding 15-min collection periods. Mean values ± 1 SE are shown for each 15-min collection period. The asterisks indicate the time when bile flow initially became significant ($p < 0.05$). The dose administered intravenously for each agent were: DBcAMP = 0.25 mgm/kgm/min × 30 min; glucagon = 1 mgm; secretion = 1 U/kgm; aminophylline = 500 mgm and dehydrocholate (Decholin) = 500 mgm. From Levine and Hall (10).

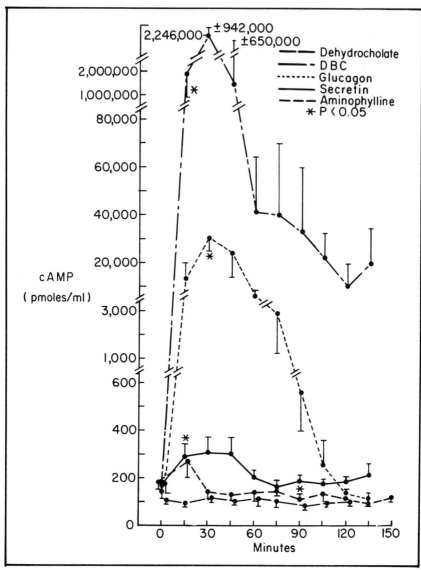

3 Effect of dehydrocholate, DBcAMP (DBC), glucagon, secretin, and aminophylline on biliary cyclic AMP concentration in man. The asterisks indicate initial significant values ($p < 0.05$). Aminophylline and dehydrocholate did not significantly increase cyclic AMP concentrations. From Levine and Hall (10).

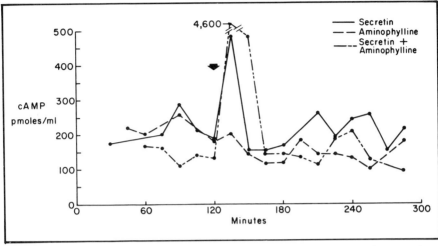

4 Effect of intravenous secretin (1 U/kgm), aminophylline (500 mgm), and secretin plus aminophylline given in randomized order on three separate days to a postcholecystectomized patient with an indwelling T-tube. Note nine-fold augmented cyclic AMP response in bile when both drugs were simultaneously administered.

of the increased cyclic AMP output in bile was less after secretin compared to that after glucagon and, of course, DBcAMP. Presumably glucagon, in contrast to secretin, stimulates the hepatic cyclic AMP generating system in man, as it does in animals, accounting for the greater output of cyclic nucleotide in the bile after glucagon. The manner by which glucagon, or possibly other peptide choleretics, might increase bile flow, whether through a mechanism of cyclic AMP formation at either a canalicular or ductular site, remains unknown. Further studies in a variety of species will be required to assess the possible regulatory role of cyclic AMP on bile formation after glucagon administration.

GASTRIC SECRETION

A variety of chemical substances stimulate normal regulation of gastric secretion and probably act directly on the parietal, or acid-secreting cell. They include acetylcholine, gastrin, histamine, digested protein and "entero-oxyntin", a postulated intestinal-phase hormone that reaches the parietal cell from intestinal endocrine cells via the blood (11). All such regulators of acid secretion act on receptors of the parietal cell wall and do not enter the cell. Whether cyclic nucleotides are the intracellular messengers that directly regulate the activity of the cell and thus gastric acid secretion is unsettled.

A mass of experimental data has accumulated concerning the possible role of cyclic AMP and cyclic GMP in control of gastric acid secretion (see Refs. 1–6, 12, 13 for reviews). A positive qualitative, quantitative and temporal correlation between the effects of chemical agents on cyclic nucleotide levels and secretory response of the stomach has been observed by some investigators, primarily in the frog, rat and rabbit. Others, however, have reported inconsistent results relating secretagogue-induced acid output with alterations in cyclic nucleotide metabolism. There are few studies concerning the role of cyclic nucleotides on human gastric secretion and I will, therefore, review this subject in more detail after briefly summarizing important experimental animal data.

The role of cyclic AMP in gastric secretion has been documented best in the frog gastric mucosa in vitro, but recent evidence indicates that gastric secretory responses in this preparation may differ from those in vivo in mammalian stomach. The existence of multiple cyclic nucleotide phosphodiesterases in frog gastric mucosa, similar to the situation in the intestine, has been documented (14). In addition there appears to be "activators" and "inhibitors" of cyclic AMP phosphodiesterase (15).

The fact that cyclic AMP appears to be involved in gastric secretion in frog gastric mucosa may be misleading since this tissue represents a heterogenous group of cells of which the acid-secreting cells comprise only a small portion. Thus, increases in intracellular cyclic AMP content may not reflect the increases in the acid-secreting cells themselves.

Increased concentration of adenylate cyclase or cyclic AMP after histamine stimulation has been reported in guinea pig, rabbit and canine gastric mucosa (16–18). Others, however, have been unable to demonstrate significant changes in adenylate cyclase or cyclic AMP in dog stomach after histamine stimulation (19, 20). The recent observation (21) that metiamide, a specific H_2-receptor blocker which inhibits histamine-stimulated gastric acid secretion, prevents histamine activation of canine gastric mucosal adenylate cyclase, supports a possible role of cAMP as a mediator of histamine-induced acid secretion in dogs.

Our studies on stimulated gastric secretion in dog and man have demonstrated that exogenous cyclic AMP inhibits gastric secretion in vivo (12). The canine data suggest that the major effects of exogenous cyclic AMP, as well as certain other adenine nucleotides, may be on the local microcirculation. Cyclic AMP reduced mucosal blood flow, as measured by aminopyrine clearance. Whether the reduction in mucosal blood flow was primary or secondary to decreases in gastric secretion could not be determined.

A study in innervated canine gastric fistula showed that histamine increased gastric mucosal cyclic AMP about twofold after 60 min of stimulation suggesting that release of cyclic AMP into the gastric juice was not a simple washout phenomenon (16). Data in our labortaory (unpublished) in both Heidenhain pouch, antral pouch and intact dogs failed to confirm these reported studies. We were unable to demonstrate any alteration of cyclic AMP or cyclic GMP levels in either antral of fundic gastric mucosa during a 1 hr infusion with either histamine or pentagastrin. Moreover, the concentration of cyclic AMP in the gastric juice was also unchanged.

Amer (13) observed that gastrin and cholecystokinin (CCK), which stimulate acid secretion, activate phosphodiesterase isolated from rabbit gastric mucosa and gallbladder and reduce intracellular cyclic AMP levels. Histamine also stimulated phosphodiesterase from rabbit fundic mucosa suggesting that it shares a common mechanism with other gastrointestinal hormones which induce gastric secretion. In support of this hypothesis are the observations that parenteral administration of such agents as prostaglandins, glucagon and secretin all inhibit pentagastrin and histamine-stimulated gastric acid secretion yet increase both adenylate cyclase and intracellular cyclic AMP activity in many cell systems.

Cholinergic stimulation has been shown to elevate cyclic GMP levels in certain tissues. Although the stomach is the recipient of constant cholinergic discharge, there is little data relating human gastric cyclic GMP activity with acid secretion. Perhaps this relates to the more recent development of the cyclic GMP radioimmunoassay and to the belated search for a biologic role for this nucleotide. There is no evidence that cholinergic stimulation of the stomach is via adenylate cyclase. Acetylcholine and other cholinergic agents are without effect on the adenylate cyclase activity of broken cell preparations (5, 13). Recently cyclic GMP was observed to be increased in canine fundic mucosa after vagal stimulation or acetylcholine infusion (22), suggesting that cyclic GMP may serve some second messenger role in cholinergic stimulation of the stomach. These observations fit Amer's concept that cyclic GMP mediates gastric acid secretion (13).

Much of the recent data in humans concerns the measurement of cyclic nucleotides in gastric juice. The interest in gastric juice cyclic nucleotide activity relates to the changes in extracellular levels that may reflect alterations in intracellular levels of cyclic nucleotides in response to certain chemical substances and hormones.

To date 3 published reports in man show increased cAMP or cGMP output in human gastric juice after betazole (Histalog) or pentagastrin secretion (16, 23, 24). In only one of these reports was intracellular cyclic AMP measured in gastric mucosa and it failed to change after pentagastrin secretion (24). In the 2 subjects who had hyperchlorhydria the gastric mucosal levels of cyclic AMP after stimulation were similar to those of control subjects (24).

What can be the significance of finding an elevated output of cyclic nucleotides in gastric juice under basal or stimulated conditions? The answer to this question probably will remain unsettled until there is evidence that the cyclic nucleotides measured are derived entirely from the parietal cells. The development of an in vitro system of pure parietal cells in which cyclic nucleotide production could be measured remains a major challenge for future research. We have attempted to resolve this problem by studying in dogs, as noted before, the differential response of antral (non-acid-secreting) and fundic (acid-secreting) mucosa to secretagogue stimulation and by investigating in man gastric cyclic nucleotide acitvity in those clinical situations of achlorhydria (i.e., pernicious anemia [PA] and excessive basal hyperchlorhydria (peptic ulcer [PU] and Zollinger-Ellison [ZE] disease).

Our initial study in man (12) showed inhibition of betazole (Histalog)-stimulated gastric acid secretion and volume by exogenous cyclic AMP injection and no alteration of basal gastric secretion. Subsequently we studied fasting normal subjects and patients with dyspepsia who were divided into 3 treatment groups. In the first group of 15 studies, intramuscular betazole was given at the concentration of 1.5 mgm/kgm. In the second group of 10 experiments histamine acid phosphate was infused intravenously at a constant rate of 0.01 mgm/kgm/hr to achieve submaximal gastric stimulation. The antihistamine, diphenydramine hydrochloride was infused concomitantly at the dose of 25 mgm/hr. In the 3rd group of experiments prostaglandin A_1 (PGA_1), an inhibitor of stimulated gastric secretion, was infused for 15 min at the dose of 1 μgm/kgm/min during secretory stimulation by histamine (see results of PGA_1 studies in section below).

The mean fasting cyclic AMP concentration in human gastric juice averaged 11 pmole/ml and was less than that of the dog (14 pmole/ml). Bile juice concentrations were much higher than that of gastric juice in both the dog and human (see Table 1) and for that reason any samples with bile reflux were discarded. Betazole and/or histamine significantly increased both volume, acid and cyclic AMP secretion into the gastric juice. The control rate of cyclic AMP, expressed as pmole/min, significantly increased at 30 min and subsequently remained significantly elevated threefold above control values during the duration of stimulatory response (25). pH also significantly decreased from basal values of 3.35 to 0.98. The fact that the secretion rate of cyclic AMP was persistently elevated during stimulation of acid and volume, and that cyclic AMP concentration did not significantly change during this period, raised the possibility that the cyclic nucleotide was released into the gastric juice by a simple "washout" phenomenon.

In order to evaluate the influence of endogenous gastrin and intraluminal pH on cAMP activity and its possible source of origin, studies were performed in 6 patients with documented PA and hypergastrinemia in 3 patients with ZE disease and hypergastrinemia and in 12 patients with PU and normal serum gastrin (26). Fasting gastric samples, collected every 15 min before and after pentagastrin (6 μgm/kgm) or maximal betazole stimulation (1.5 mgm/kgm), were analyzed for volume, pH, titratable acidity, sodium and radioimmunoassayable cyclic AMP. Sodium output was considered an index of nonparietal or intestinal fluid entering the stomach lumen. Basal cyclic AMP (pmole/ml) and sodium (mEq/liter) concentrations were, respectively, similar in PA and PU, but lower in ZE. After betazole stimulation cyclic AMP and sodium concentrations remained, respectively, unaltered in PA and ZE but decreased 1/3 and threefold in PU ($p < 0.05$), proportional to the increased volume. Concomitantly, cyclic AMP and acid outputs were, respectively, unchanged in ZE, increased 3- and 10-fold in PU ($p < 0.01$) and transiently decreased 3-fold and unchanged in PA ($p < 0.05$). Mean fasting cyclic AMP concentrations in gastric mucosal hiopsies were 820 pmole/gm in PA, 786 pmole/gm in ZE and 864 pmole/gm in PU.

The presence of cyclic AMP in unstimulated gastric juice in PA and its transient decreased output after betazole, suggest an origin and dilution derived from nonparietal secretions or transmucosal exudation. Low basal cyclic AMP concentrations in ZE and increased cyclic AMP output after betazole in PU both appear to be proportional to parietal volume flow. Intra- or extracellular gastric cyclic AMP concentration did not correlate with serum gastrin, acid output, associated disease or age. Gastrin appears not to modulate gastric cyclic AMP activity in man (26).

We have subsequently measured fasting gastric juice and correlated intra- and extracellular cyclic nucleotide activity with the acid secretory response to both betazole and pentagastrin in 20 subjects, including patients with dyspeptic symptoms and normal volunteers. The intracellular cyclic GMP content of gastric tissue was biopsied by the Woods type suction biopsy tube as previously done for cyclic AMP. Biopsy specimens of gastric mucosa were taken before and at various time intervals (5 to 60 min) during secretagogue-induced secretion. Identification of both cyclic nucleotides in the presence of phosphodiesterase treatment was confirmed in gastric juice and in gastric mucosal biopsy specimens.

The concentration of cyclic AMP in gastric juice, in contrast to its total output in the juice, diminished during stimulation by betazole and/or pentagastrin. There was a 36 percent reduction in the mean cyclic AMP concentration in gastric juice after both secretagogues and a maximal 60% reduction after both secretagogues when peak values were determined. The extracellular concentration of cyclic AMP in gastric juice did not reflect alteration in the intracellular mucosal concentration of cyclic AMP.

A greater reduction in gastric juice cyclic GMP compared to cyclic AMP concentration after betazole and pentagastrin stimulation was observed. There was a 50 percent mean decrease after both secretagogues and an 81% peak decrease. Thus, there is a much greater dilutional effect of secretagogues on cyclic GMP compared to cyclic AMP concentration in gastric juice. In contrast to cyclic AMP, the output in gastric juice of cyclic GMP, in terms of pmole/min, was less. Thus, while both nucleotides increased acid secretion, the increased output appears to have had a greater dilutional effect on cyclic GMP concentration in gastric juice. The intracellular concentration of cyclic GMP was unchanged after stimulation by either betazole or pentagastrin (mean ± SE gastric cyclic GMP basal value = 21.6 ± 1.5pm/g; poststimulation value = 19.2 ± 2.5 pmole/gm). Again, as with cyclic AMP, the intracellular levels of cyclic GMP did not appear to reflect the reduction in nucleotide concentration seen in the gastric juice.

In summary, concentrations of cyclic nucleotides in gastric juice are greater in the basal state than during stimulation by either betazole or pentagastrin. These secretagogues fail to increase cyclic nucleotide concentration in the gastric juice or in gastric mucosal biopsy specimens, yet both increased acid secretion. If there were significant changes in the cyclic nucleotide content of gastric-acid-producing (parietal) cells, they did not appear to correlate with any alterations in gastric tissue or juice cyclic nucleotide concentrations.

The increased cyclic AMP and cyclic GMP efflux in stimulated human and canine gastric juice without an increase in cyclic nucleotide concentration may be a "washout" phenomenon or may be physiologically related to acid production at the cellular level. The apparent increase of cyclic AMP and cyclic GMP in a greater volume of secretion fails to indicate whether cyclic AMP and cyclic GMP in gastric juice were actually secreted, whether they were released from dead or dying cells which may have been shed from the mucosa during stimulation, or whether they leaked through the "tight junctions" of the gastric mucosa from the circulating plasma.

Our data suggest that any acid increase induced by secretagogues appears to be due to a mechanism independent of an effect on intra- or extracellular cyclic AMP and cyclic GMP concentrations. Gastric juice cyclic nucleotide output does not appear to reflect dynamic intracellular mucosal events. The increase in gastric acid secretion after betazole or pentagastrin may be mediated by a mechanism independent of the secretagogue's possible effect on cyclic nucleotide output. The possibility exists that simple plasma clearance of cyclic nucleotides could best explain their presence in gastric juice. Our preliminary studies (unpublished) showed that plasma concentration of cyclic nucleotides are unchanged after secretagogue stimulation.

Interaction with Prostaglandins

Our studies on the inhibitory effects of PGA_1 on stimulated gastric secretion suggested that cyclic AMP output decreased transiently, and at the same time acid and volume output were inhibited. Cyclic AMP concentration did not change. In two subjects studied under basal conditions, PGA_1 failed to alter gastric secretion, cyclic AMP output or concentration.

Studies on the effect of prostaglandins on canine gastric juice cyclic AMP during stimulation and inhibition were conduced in unanesthetized dogs with vagally denervated fundic pouches. After a 16-hr fast, gastric secretion was stimulated with either pentagastrin or histamine. In four canine studies prostaglandin E_1 (PGE_1), given after chlorpromazine pretreatment to prevent nausea and vomiting, was rapidly injected at a dose of 25 μgm/kgm after a steady secretory rate was obtained. Volume significantly increased as did cyclic AMP output in the first and second 15-min collection periods after secretagogue administration. A 10-fold maximal increase in cyclic AMP production after stimulation was found and is similar to that observed previously by Bieck and colleagues in innervated pouch dogs (16). Cyclic AMP concentration failed to significantly change, however. PGE_1 inhibited cyclic AMP and volume output in three of four dogs studied, but did not alter cyclic AMP concentration in the juice.

PGA_1 and PGE_1 inhibited cyclic AMP output concomitant with inhibition of gastric secretion in the human stomach and canine denervated gastric pouch, respectively. The failure of gastric juice cyclic AMP output to be associated with a similar alteration of cyclic AMP concentration in gastric juice contrasts with our studies on bile secretion in man where secretin-induced choleresis was associated with significant elevations of both cyclic AMP concentration and output. The potential therapeutic role of oral prostaglandin E_2 and its methyl ester analogs as important inhibitors of human gastric secretion and in preventing peptic ulcer disease is one of the great challenges in medical therapeutics of this decade (2). As yet the antisecretory action of prostaglandins does not appear to be mediated by cyclic nucleotides.

PANCREATIC SECRETION

Several animal studies indicate that secretin elevates cyclic AMP concentrations in the pancreas. However, it must be remembered that the pancreas is heterogenous gland, and more than one event associated with accumulation of cyclic AMP may be taking place in different cells or in different intracellular compartments. The acinar cells comprise the greater part of the pancreas, and one would expect that the cyclase enzymes prepared from the whole gland would be more sensitive to hormones than the Islets of Langerhans, since the endocrine part of the pancreas is only a small fraction of the total pancreatic tissue. Although cyclic AMP may be involved in insulin secretion, and defective insulin release in diabetics is associated with decreased adenylate cyclase activity of the beta cell, electrolyte secretion from the panrceas appears to be responsive to hormones which simultaneously activate the cyclic AMP system, at least on membrane fragments of the pancreas.

A further confusing possibility relates to the fact that hormones may not alter tissue levels of cyclic AMP in the pancreas but involve alteration in calcium binding and transport. The relationship between calcium and cyclic AMP in electrolyte secretion by the pancreas is still uncertain but there is probably a strong interrelationship between cyclic AMP and calcium in the pancreas since an absolute requirement exists for calcium in the external medium of the isolated pancreas for secretion of electrolytes and water.

Recent experiments on adenylate cyclase activity in plasma membranes from isolated pancretaic exocrine cells stimulated by various hormones provide further support for the concept that the same polypeptide hormones which stimulate pancreatic exocrine secretion and intestinal secretion (see below), and inhibit gastric secretion, appear to increase adenylate cyclase activity (27). These important experiments in plasma membranes of isolated guinea pig pancreatic exocrine cells showed that vasoactive intestinal peptide (VIP) and secretin, but not glucagon or gastrin, increased adenylate cyclase activity. VIP was the most potent stimulator of adenylate cyclase although CCK (formerly called pancreozymin), and secretin also induced a response on the enzyme system. VIP, secretin and glucagon are similar both in their chemical structure and in their spectrum of biological activities. Since all three hormones increase adenylate cyclase activity from the plasma membranes of a variety of tissues including fat cells, hepatocytes, and, in the case of VIP and secretin, pancreatic acinar cells, they may alter pancreatic function by activating adenylate cyclase and increasing cellular cyclic AMP. The study in isolated pancreas further documented that the membrane receptor for CCK is functionally distinct from the membrane receptor with which VIP and secretin interact (27). The one criticism of the preparation used by these investigators is that they were not dealing with a homogeneous cell population, since pancreatic ductal cells probably represent 20 percent of the total preparation. Thus any changes observed in this preparation were occurring both in acinar and ductal cells, but to a much lesser degree in the ductal cells.

The aforementioned studies support the concept that there is compartmentalization of cyclic nucleotides in most tissues. The same investigators have recently found that CCK, both the octapeptide and the heptapeptide, as well as other muscarinic agents, mobilize membrane-bound calcium, activate guanylate cyclase, and increase cellular cyclic GMP during calcium mobilization (J. D. Gardner, personal communication). Using a calcium ionophore to resolve whether or not divalent cations move across membranes, it was found that the ionophore increased membrane-bound calcium and increased cellular cyclic GMP concentrations 12-fold, thus mimicking the effect of CCK. This suggests that cyclic GMP but not cyclic AMP is involved in the initial sequence of events in the CCK action on plasma membranes and calcium mobilization in pancreas. Since the ionophore mimicked the hormone's action by stimulating calcium transport through plasma membranes or by releasing calcium from intracellular stores, the data suggests that CCK also works in a similar fashion.

A great deal of conflicting data has accumulated concerning the possible in vivo effect of cyclic AMP and theophylline on pancreatic exocrine secretion (1, 4). Some studies have reported that theophylline potentiates the effects of CCK or has a secretinlike effect, while others failed to observe either an increase in enzyme concentration or output in a variety of species, and even noted suppression of secretory output following theophylline administration. It is conceivable that theophylline may exert its action through a mechanism independent of effects on secretin or CCK and without any change in intracellular levels of cyclic AMP. It is of interest that both glucagon and vasopressin exert a suppressive effect on pancreatic secretion, both of which agents are known to activate adenylate cyclase in a variety of tissues. In general, theophylline or exogenous cyclic AMP administration in animals appears to increase enzyme output, not affect secretion in the resting gland, and probably decrease secretion in the stimulated gland.

Effect of Exogenous Cyclic AMP on Human Pancreatic Secretion

The possible role of cyclic AMP in pancreatic secretion was studied in a 72-year-old woman who had a resection of an adenomatous polyp arising in the third portion of the duodenum close to the papilla of Vater with subsequent partial papillectomy and re-implantation of the common bile and pancreatic ducts into the duodenum (28). At surgery, both the common and pancreatic ducts appeared to be larger than normal, and it was possible to place a T-tube into the common duct with the long arm going into the duodenum and a catheter into the pancreatic duct so that specimens of pure bile and pancreatic juice could be obtained.

Following infusion of DBcAMP at the rate of 0.25 mgm/kgm/min for 30 min, pancreatic flow increased 85 percent, pancreatic amylase 295 percent and pancreatic lipase 330 percent (Fig. 5). In addition bile flow also increased, as previously discussed, and shown in the lower panel of Fig. 5. In view of the sharp increase in pancreatic enzyme concentration and output lasting longer than 60 min induced by exogenous DBcAMP, consideration should be given to a possible mediating role of cyclic AMP in hormonal control of human pancreatic secretion.

Interaction with Prostaglandins

Several studies have reported effects of prostaglandins on pancreatic secretion but all have yielded contradictory results (2, 5). Unfortunately, there have been no investigations concerning the effect of prostaglandins on adenylate or guanylate cyclase or on cyclic nucleotide levels, but they, too, would be difficult to interpret because of the heterogeneity of the cell types in the pancreas. Currently, it appears that cyclic nucleotides may mediate the pancreatic response to certain hormones, particularly VIP, secretin, CCK, and, perhaps, to acetylcholine and prostaglandins.

INTESTINAL SECRETION

Cyclic AMP appears to play an important role in intestinal water and electrolyte transport. As indicated earlier, adenylate cyclase is located in the lateral membrane and not in the brush border.

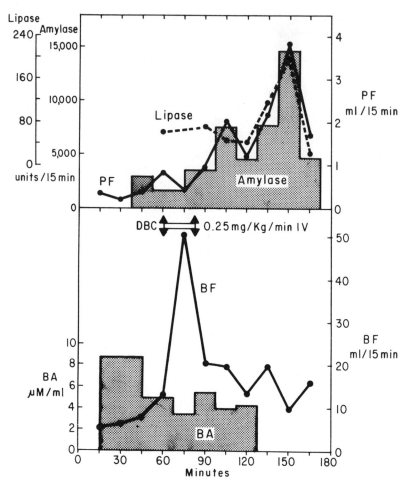

5 Effect of DBcAMP (DBC) infusion on pancreatic flow (PF), serum amylase and lipase (top panel); and on bile flow (BF) and bile acid (BA) concentration (bottom panel). From Levine (2).

Moreover, at least two distinct adenylate cyclases may exist in the intestinal cell, one responsible for sodium transport and the other for water transport.

In vitro and in vivo animal experiments have shown that the small intestine has high levels of guanylate cyclase and cyclic GMP.

Immunocytochemical localization of both cyclic nucleotides and their respective cyclases in the intestine demonstrate that the major part of cyclic GMP is located at the brush border membrane while cyclic AMP is found predominately in the basolateral membrane of the brush border cells and in the cytoplasm (7). A study in rat colonic mucosa localized adenylate cyclase also to the brush border (29). In single intestinal cell populations, Dr. R. E. Crane recently identified guanylate cyclase in rat and human intestinal brush border, while adenlyate cyclase was found in the lateral membrane (personal communication). In terms of further cell localization, guanylate cyclase in the rat intestinal epithelial cell has been found to be greater in villus than in crypt cells while adenylate cyclase was increased in the crypt portions rather than villus cells (30). This villus-to-crypt gradient of guanylate cyclase in the differentiated villus cell suggests a specialized role of the enzyme possibly related to factors regulating cellular proliferation.

One attractive theory to explain the interaction of hormones or agents with the intestinal cyclases involves the concept of membrane fluidity, wherein the hormone receptors are concentrated in areas of the cell surface, which do not correspond to their site of action or to localization of the cyclase systems. Thus, the hormone-receptor complex, once formed, could migrate rapidly by diffusion along the membrane and periphery of the cell eventually reaching its site of action (31). The cell would have discrimination with respect to locus of cyclic nucleotide production (presumably close to its site of action) and contact with the bathing fluid or medium containing the regulatory substances (i.e., hormones).

The variable time period observed in animal experiments between the onset of secretory effects, increase in cyclic AMP or adenylate cyclase activity and the exposure of an agent or hormone to the mucosa could be explained by this "mobile receptor theory" (31). Whether active agents are absorbed directly into the cytoplasm or move laterally within the membrane remains unresolved, but under active investigation in several laboratories.

Cyclic AMP has been implicated as a key regulator in cholera enterotoxin, *Escherichia coli* enterotoxin, prostaglandin, and VIP-induced small intestinal secretion. These agents not only act on intestinal cells and cause diarrhea but can also activate the adenylate cyclases of other kinds of mammalian cells. Cholera enterotoxin does not seem to influence cyclic GMP in intestinal mucosa but the effects of other agents on this nucleotide have not yet been studied. (For further discussion of cholera toxin see following chapter.)

Prostaglandins, Toxins, Bile Salts and Hormone-Induced Diarrheas

A cholereic diarrhea is a feature associated with certain hormone-secreting tumors including medullary carcinoma of the thyroid, tumors of neural-crest origin and non-beta cell tumors of the pancreatic islet cells (1–5). These watery diarrheas may exert their effect through prostaglandins, since elevations in circulating prostaglandins have been found in these conditions and because oral or parenteral administration of exogenous prostaglandins is accompanied by severe diarrhea and colicky abdominal pain (5). In man, prostaglandins administered intraluminally or by other parenteral routes of administration induce net secretion of water and electrolytes similar to that observed after exposure of the intestine to cholera enterotoxin (2, 5).

A great deal of experimental data shows that prostaglandins enhance mucosal adenylate cyclase and cyclic AMP activity. Unlike cholera, prostaglandins exert a more rapid effect on the adenylate cyclase-cyclic AMP system and quickly induce diarrhea. Recent experiments in vitro and in vivo have provided evidence against an intermediary role for prostaglandins in the pathogenesis of cholera toxin-induced intestinal secretion or activation of adenylate cyclase. Indomethacin, a potent inhibitor of prostaglandin synthesis, failed to alter cholera toxin-induced fluid secretion in perfused rabbit loops and did not affect tissue cyclic AMP levels (32).

Two new peptides isolated from the mucosa of hog small intestine have been demonstrated to stimulate canine intestinal secretion. These are VIP and gastric inhibitory peptide (GIP). Recent studies by Barbezat and Grossman (33) indicate that VIP and GIP, in addition to glucagon and pentagastrin, promptly stimulate intestinal secretory output over control values with little change in electrolyte concentrations. The effect of these stimulatory peptides on gut water secretion is similar to that of cyclic AMP and cholera toxin. Pentagastrin, glucagon, calcitonin, and secretin do not alter cellular cyclic AMP in rabbit ileal mucosa (5). Cholera toxin may also release VIP or GIP (33).

In addition to cholera toxin, prostaglandin and certain hormones, a variety of pathogenic microorganisms have been shown to increase adenylate cyclase. Various nonspecific diarrheal diseases caused by bacterial toxins that immunologically cross-react with cholera toxin are responsible for greater morbidity and mortality than cholera. For example, a significant portion of cases of traveler's diarrhea are, apparently, caused by *E. coli* enterotoxin. There is preliminary evidence that patients with tropical sprue and other diarrheas in the tropics are in the net secretory state and have an abundance of intestinal microorganisms that produce high concenrations of intraluminal ethanol and secretory enterotoxin (F. A. Klipstein, personal communication). As ethanol and cholera toxin both stimulate intestinal adenylate cyclase (2, 5), it is conceivable that patients with chronic diarrhea are in the secretory state due to alterations in the adenylate cyclase-cyclic AMP system resulting from stimulation by bacteria-induced ethanol or enterotoxins, or both.

Recently, E. coli and other species of coliform bacteria have been shown to elaborate an enterotoxin that activates adenylate cyclase (34). Since such species commonly invade the small intestine of persons with malnutrition and tropical sprue, it is conceivable that chronic exposure of the intestinal mucosa to the enterotoxins elaborated by these bacteria may be a factor in the pathogenesis of sprue. Perhaps adenylate cyclase or cyclic AMP are playing an important modulating role in the development of clinical sprue. At the present time there is no firm data to support this hypothesis, but this area is worthy of future clinical investigation.

Other bacterial species also produce enterotoxins. Shigellosis is one of the most important clinically as dysentery is a major epidemiologic problem world-wide. Recent studies by Donowitz and colleagues (35) in rabbit ileal mucosa failed to show cyclic AMP mediation of Shigella enterotoxin stimulation of intestinal fluid and electrolyte secretion. Thus *Shigella* and cholera enterotoxins' action on cyclic AMP markedly differ, yet both produce electrolyte secretion in vitro and in vivo. It would be important to similarly study intestinal cAMP levels after exposure to other enterotoxins produced by bacteria which induce diarrhea (staphylococci, clostridia, and pseudomonas). One nonenterotoxin elaborating enteropathogen causing diarrhea, *Salmonella typhimurium*, has recently been studied in rabbit ileal loops (36). This bacteria stimulated intestinal secretion and adenylate cyclase, although the mechanism remains unclear. Invasion of the mucosa per se was not found to be a sufficient stimulus for either fluid secretion or cyclase activation (36).

One could speculate that congenital chloride diarrhea, an autosomal recessive disease manifested by persistent, life-threatening, watery diarrhea with a uniquely high chloride concentration of stool water, may be related to defects in the cyclic AMP system. Although recent human studies detailing its pathogenesis have not implicated cyclic AMP (37), future attemps to assay adenylate cyclase or cyclic AMP activity in the intestinal mucosa of patients with this disease and in other forms of diarrhea may be helpful in determining a possible mechanism wherein cyclic AMP might be of pathogenic importance in diarrhea.

Cyclic AMP in colonic mucosa may also play a role in modulating water and electrolyte transport. Studies in rat colon preparations show that cyclic AMP mediates bile salt-induced secretion (38). Clinical studies on bile-salt–induced diarrhea such as in regional ileitis or small bowel resection would be required to confirm this experimental investigation.

Evidence Against Cyclic AMP-Mediation of Human Intestinal Secretion

VIP is the most potent stimulator of the intestinal cyclic AMP system. Its role in "pancreatic cholera" is now well recognized. This syndrome consists of a watery diarrhea with severe hypokalemia, often hypocalcemia and achlorhydria (39). We recently observed a 65-year-old female with a 3 mo history of hypokalemia and intractable diarrhea and elevated levels of circulating VIP (11,000 pgm/ml) and gastrin (850 pgm/ml). She was found to have islet cell hyperplasia of the pancreas but normal levels of cyclic AMP and cyclic GMP in the jejunal mucosa and in the plasma. This is the first demonstration in man of normal small intestinal cyclic nucleotide content in VIP-induced diarrhea. If confirmed, it fails to support the concept that cyclic AMP mediates secretion in clinical states of diarrhea.

The effects of a number of hormones on intestinal mucosal cyclic AMP and cyclic GMP levels in intact tissue is currently under investigation in our laboratory. Many questions remain. Is the cyclic AMP or cyclic GMP affecting ion transport representative of only a small part of the total mucosal cyclic nucleotides? Does cyclic AMP or cyclic GMP play a role in the physiologic regulation of electrolyte transport? If so, what hormones or agents control the process and are they of physiologic importance? Hopefully, answers to these questions will be forthcoming soon from the basic laboratory and from clinical studies.

GASTROINTESTINAL MOTILITY

The regulation of motility is the one aspect of gastrointestinal control mechanisms which is least well understood. The dominant regulatory mechanism in gut motility is the nervous system, triggered by luminal distention by means of food and modulated by extrinsic innervation. The most intensively studied gut motor activity is inhibition, which appears to be a property of some enteric and other hormones. The adrenergic nerves and catecholamines are key regulators of intestinal inhibition. It has been demonstrated that the intestinal smooth muscle response to acetylcholine is associated with an increase in cyclic GMP (5, 13). On the other hand, cyclic AMP mediates the inhibitory response to beta-receptor activation of the gut (1). Thus, the classic opposing influences of cyclic GMP and cyclic AMP, respectively, appear to exist with regard to stimulation and inhibition of gut motility.

Cyclic AMP or other adenine nucleotides promptly inhibited basal or prostigmine-stimulated gastric motor activity in dog and man (1). As in the case of gastric secretion, the inhibitory action was nonspecific, conceivably owing to alterations in the microcirculation. Cyclic AMP administration did not, however, change canine gastroepiploic or mesenteric arterial blood flow (1, 12).

Adrenergically induced intestinal relaxation is probably mediated through stimulation of both alpha and beta receptors (1). Theoretically, an increase in the concentration of cyclic AMP by catecholamines should be associated with smooth muscle relaxation, although cyclic AMP increases contractility of cardiac and skeletal muscles (1). Exogenous cyclic AMP, adenosine, and other adenine nucleotides inhibit motility in isolated rabbit ileum before and after alpha and beta adrenergic blockage (1, 40). It is likely that cyclic AMP mediates only the beta receptor responses and is not involved in intestinal inhibitory responses to alpha receptor activation (41). Since the inhibitory response of the gut to cyclic AMP is nonspecific, adenosine may represent the active relaxant component that acts on a common receptor in the smooth muscle cell membrane

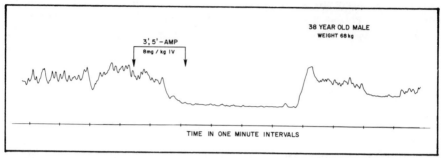

6 Inhibition of small intestinal motor activity by cyclic AMP in an unanesthetized volunteer subject recorded by balloon kymography. Acute injection abolished intestinal motility and for 5 min. From Levine et. al. (42).

(40). Intestinal relaxation to catecholamines need not necessarily be mediated by cyclic AMP simply because the cyclic nucleotide may produce the same type of response. Nevertheless, such data do not exclude the possibility that cAMP may be an intermediate in catecholamine or other hormonally induced responses of the gut.

Studies in our laboratory showed that cyclic AMP inhibits intact canine ileal motility in the presence of alpha and beta adrenergic blockage (42, 43). Studies have also been performed in volunteer subjects using balloon kymography (42). Acute intravenous injections of cyclic AMP rapidly abolished intestinal motility and tonus for 4 to 9 min, followed by a prompt return of the motility pattern to a control status in six studies in four subjects (Fig. 6). Whether or not the inhibitory effects on gastrointestinal motility were related to changes in blood flow could not be determined, however, the inhibitory responses demonstrated by cyclic AMP in isolated rabbit ileum suggests that this is a direct effect of the nucleotide of its adenosine moiety (40, 43).

Inhibition of basal or prostigmine-stimulated motor activity by cyclic AMP was also observed in eight of nine studies using chronic, canine gastric, jejunal, or ileal fistulae preparations (42). Unlike cyclic AMP, administration of 2',3'-AMP promptly increased motility and tone for a duration of 2 to 12 min in two of three small bowel and in two gastric fistula preparations. Eight studies employing 5'-AMP and ATP failed to consistently alter the motility pattern.

Interaction with Prostaglandins

Prostaglandins have been found in the gastrointestinal tract of man and animals and have been extensively studied in isolated gastrointestinal muscle, where they generally contract longitudinal muscle and relax circular muscle (2). Whether they act via cyclic nucleotide alterations is not known. Since prostaglandins are marked vasodilators in vivo, their actions on intestinal motility may be affected by influencing blood flow. Prostaglandins have been shown to increase mesenteric blood flow in anesthetized dogs, associated with inhibition of intestinal motility (2).

Evidence that prostaglandins influence human intestinal motility relates to prior reported studies using intravenous infusion of PGE_1 which caused abdominal cramps. In addition, oral administration of 0.8 to 3.2 mgm resulted, after a delay of up to several hours, in loose watery stools, due to increased intestinal propulsion (2). However, it is not known whether the effect of injected PGE_1 is caused by the parent compound or by a metabolite acting locally on the bowel wall after absorption, nor has the possibility of an action on water absorption or some other factor which might influence the gut been excluded (46). While it seems unlikely that inhibition or contraction of the gut is secondary to vascular changes per se, effects in vitro demonstrating a stimulatory (contractile) effect on gastrointestinal smooth muscle contrasts with in vivo studies. The few reported in vivo studies indicate that gut motility is affected in different ways by exogenously administered prostaglandins, depending upon the particular species involved, the preparation used, and the area of the gastrointestinal tract studied. Previous in vivo studies utilizing PGE_1 infusions have demonstrated that canine gastric and intestinal motility was inhibited.

We have observed that PGE_1 administered rapidly into the blood stream of unanesthetized dogs induced a biphasic motility response in the ileum consisting of an initial increase in the intraluminal pressure followed by prolonged inhibition (43). In isolated rabbit ileum, PGE_1 caused only contraction of the longitudinal muscle. The in vitro and in vivo responses to PGE_1 were not prevented by pretreatment with phentolamine and propranolol, although the latter adrenoreceptor antagonists prevented the characteristic relaxant action of epinephrine on isolated and intact ileum (43). Based upon the use of these adrenoreceptor blockers, our data suggest that PGE_1 does not act by local release of sympathomimetic amines or on adrenergic receptors. The effects of prostaglandins on motility are also not mediated through the cholinergic system or by serotonin release (1).

PGE_1 has a relaxant action on the lower esophageal sphincter, which may be mediated by cyclic AMP. Both PGE_1 and DBcAMP produced dose-dependent relaxation of the lower esophageal sphincter in the anesthestized oppossum (2, 3). It is of further interest that glucagon, like secretin and CCK, inhibits resting lower esophageal sphincter tone in humans (45). During pentagastrin stimulation this inhibition occurred at glucagon doses lower than those required to cause inhibition under control conditions. Thus certain polypeptides and prostaglandins may relax the esophageal sphincter via interaction with the cyclic AMP-adenylate cyclase system.

Other Hormones

The effects of glucagon on isolated rabbit ileum and on intact human esophageal, jejunal and colonic motility are inhibitory, presumably mediated by cyclic AMP (1). The possible relationship between other gastrointestinal polypeptide hormones and cyclic AMP on motility is an area of rank speculation. Contractions of isolated rabbit gallbladder strips induced by CCK were not blocked by atropine or adrenergic receptor blockage, but were produced by imidazole, a drug that activates phosphodiesterase and lowers intracellular cyclic AMP. This suggests that CCK may act by lowering intracellular cyclic AMP, a finding supported by the activation of phosphodiesterase by CCK in rabbit gallbladder (13), and the relaxation of the gallbladder by cyclic AMP, theophylline and glucagon (1).

Furthermore, in guinea pig gallbladder, cyclic AMP relaxed CCK-induced contractions and CCK lowered the content of cyclic AMP by activating phosphodiesterase (46). In contrast, CCK increased cyclic AMP content in the isolated sphincter of Oddi of the cat concomitant with a relaxant effect (47). Beta receptor stimulators, glucagon and theophylline also relax the sphincter activity while imidazole contract it, supporting the concept of a mediating role of cAMP in relaxation of the sphincter (48).

Studies by Amer (49) have confirmed that the action of CCK in gallbladder is associated with activation of cyclic GMP in that tissue. This is another classical example of the so-called "Yin-Yang" theory in that CCK as well as cholinergic agents stimulate gallbladder mucosa associated with increased cyclic GMP levels while concomitantly cyclic AMP activity in that tissue is decreased.

Table 2
Summary of Gastrointestinal Effects of Cyclic AMP and Cyclic GMP

Tissue	Process Affected	Activity Change		Change in Concentration	
		Cyclic AMP	Cyclic GMP	Cyclic AMP	Cyclic GMP
Bile Ducts	Bile secretion	↑	?	↑	?
Salivary gland	Amylase, fluid secretion	↑	?	↑	?
Pancreas	Amylase, fluid secretaion	↑	?	↑	?
Esophagus	Low sphincter	↓	?	?	?
Stomach	Motility	↓	?	?	?
Stomach (in vitro)	Acid, fluid secretion	↑	?	↑	?
Stomach (in vivo)	Acid, fluid secretion	↓ of 0	?	0	0
Intestine	Motility	↓	?	?	↑
Intestine	Chloride, bicarbonate, fluid secretaion	↑	?	↑	?
Gallbladder	Motility, Oddi sphincter	↓	↑	↓	↑

↑, Increase; ↓, decrease; ?, not tested; 0, no effect.

SUMMARY

This review emphasizes the possible regulatory role which cyclic nucleotides may play in gastrointestinal physiology and clinical disease. Caution should be given to interpretation of cyclic nucleotide data, since the gastrointestinal tract contains mixed cell populations and measurement of total tissue levels of cyclic AMP and cyclic GMP may not reflect regional intracellular alterations. Thus the concept of "physiologically active" or secretion-related cyclic nucleotide pools may better reflect their compartmentalization and regional changes within gastrointestinal cells. Furthermore, unless cyclic nucleotides are completely hydrolyzed by excess phosphodiesterase treatment, their presence in fluids or tissues cannot be verified. The actions of cyclic AMP and cyclic GMP on the gastrointestinal tract and their possible physiologic role are summarized in Table 2. Cyclic AMP appears to play a key regulatory role in intestinal electrolyte and fluid secretion but its mediating role in bile, pancreatic, salivary, and gastric secretion and in gastrointestinal motility is less well established. There is evidence for direct stimulation of acid by cyclic AMP in isolated amphibian gastric mucosa but in the in vivo mammalian stomach, gastric secretion both in health and disease appears to be inhibited or unaltered by exogenous cyclic AMP administration and to be independent of changes in intra- or extracellular cyclic AMP or cyclic GMP concentrations. While prostaglandins may interact with cyclic nucleotides in the gut and are also potent inhibitors of gastric acid secretion, their antisecretory action does not appear to be mediated by cyclic AMP.

REFERENCES

1. Levine, R. A.: The role of cyclic AMP on hepatic and gastrointestinal function. Gastroenterology 59:280-300, 1970.
2. Levine, R. A.: The role of cyclic AMP in prostaglandins in hepatic and gastrointestinal functions. In Prostaglandins and Cyclic AMP: Biological Actions and Clinical Applications, (ed.) R. H. Kahn and W. E. M. Lands, Academic Press, 1973, pp. 75-117.
3. Levine, R. A.: The role of cyclic AMP on hepatic and gastrointestinal function. Endocrinology of the Gut, (ed.) W. Y. Chey and F. B. Brooks, C. B. Slack, Inc., 1974, pp. 339-353.
4. Scratcherd, T. and Case, R. M.: The role of cyclic adenosine-3′,5′-monophosphate (AMP) in gastrointestinal secretion. Gut 10:957-961, 1969.
5. Kimberg, D. V.: Cyclic Nucleotides and their role in gastrointestinal secretion. Gastroenterology 67:1023-1064, 1974.
6. Levine, R. A.: Cyclic AMP in digestive physiology. Am. J. Clin. Nutrition 26:876-881, 1973.
7. Ong, S. H., Whitley, T. H., Stowe, N. W. and Steiner, A. L.: Immunohistochemical localization of 3′:5′-cyclic AMP and 3′:5′-cyclic GMP in rat liver, intestine, and testis. Proc. Nat. Acad. Sci., USA, 72:2022-2026, 1975.
8. Schwartzel, E. H. Jr., Bachman, S. and Levine, R. A.: Cyclic nucleotide activity in gastrointestinal tissues and fluid. Fed. Proc. 35:583, 1976 (abstract).
9. Erlinger, S., Dhumeaux, D.: Mechanisms and control of secretion of bile water and electrolytes. Gastroenterology 66:281-304, 1974.
10. Levine, R. A. and Hall, R. C.: Cyclic AMP in Secretion Choleresis—Evidence for a regulatory role in man and baboons but not in dogs. Gastroenterology 70:537-544, 1976.
11. Grossman, M. I.: The chemicals that activate the "on" switches of the oxyntic cell. Mayo Clin. Proc. 50:515-518, 1975.
12. Levine, R. A. and Wilson, D. E.: The role of cyclic AMP in gastric secretion. Ann. of N.Y. Acad. Sci. 185:363-375, 1971.

13. Amer, M. S.: Cyclic AMP and gastric secretion. Amer. J. Dig. Dis *17*: 945-953, 1972.
14. Sung, C. P., Wielbelhous, V. D., Jenkins, B. C., Adler-Creutz, P., Hirschowitz, B. I. and Sachs, G.: Heterogeneity of 3',5'-phosphodiesterase of gastric mucosa. Am. J. Physiol. *223*:648-650, 1972.
15. Ray, T. K. and Forte, J. G.: Demonstration of an "activator factor" and an "inhibitor factor" in the cyclic AMP phosphodisterase from oxyntic cells of bullfrog gastric musoca. FEBS Lett. *20*:205-2981 1972.
16. Bieck, P. R., Oates, J. A., Robison, G. A. and Adkins, R. B.: Cyclic AMP in the regulation of gastric secretion in dogs and humans. Am. J. Physiol. *224*:158-164, 1973.
17. Sung, C. P., Jenkins, B. C., Hackney, V., Racey Burns, L., Spenny, J. G., Sachs, G. and Wiebelhaus, V. D.: Adenyl and guanyl cyclase in mammalian gastric mucosa. Gastroenterology *64*:808, 1973 (abstract).
18. Perrier, C. V., Laster, L.: Adenyl cyclase activity of guinea pig gastric mucosa; stimulation by histamine and prostaglandins. J. Clin. Invest. *49*: 73a, 1970 (abstract).
19. Mao, C. C., Jacobson, E. D. and Shanbour, L. L.: Mucosal cyclic AMP and secretion in the dog stomach. Am. J. Physiol. *225*:893-896, 1973.
20. Mao, C. C., Shanbour, L. L., Hodgins, D. S. and Jacobson, E. D.: Adenosine 3'5'-monophosphate (cyclic AMP) and secretion in the canine stomach. Gastroenterology *63*:427-438, 1972.
21. Dousa, T. P., Code, C. F.: Effect of histamine and its methyl derivatives on cyclic AMP metabolism in gastric mucosa and its blockage by an H_2 receptor antagonist. J. Clin. Invest. *53*:334-337, 1974.
22. Eichhorn, J. H., Salzman, E. W. and Silen, W.: Cyclic GMP response in vivo to cholinergic stimulation of gastric mucosa. Nature *248*:238-239, 1974.
23. Bower, R. H., Sode, J. and Lipshutz, W.: Cyclic GMP and gastric acid secretion. Dig. Dis. *19*:582, 1974.
24. Domschke, W., Domschke, S., Rosch, W., Classen, M. and Demling, L.: Failure of pentagastrin to stimulate cyclic AMP accumulation in human gastric mucosa. Scand. J. Gastroent. *9*:467-471, 1974.
25. Levine, R. A. and Washington, A.: Increased cyclic AMP production in human gastric juice in response to secretagogues. Gastroenterology *64*: 863, 1973 (abstract).
26. Levine, R. A. and Schwartzel, E. H., Jr.: Gastric cyclic AMP activity in Zollinger-Ellison disease and pernicious anemia. Gastroenterology *66*:730, 1974 (abstract).
27. Klaeveman, H. L., Conlon, T. P. and Gardner, J. D.: Effects of gastrointestinal hormones on adenylate cyclase activity in pancreatic exocrine cells. In *Gastrointestinal Hormones*, J. C. Thompson, (ed.), University of Texas Press, Austin, 1975, pp. 321-344.
28. Levine, R. A. and Hall, R. C.: Role of cyclic AMP in bile and pancreatic secretion in man. Gastroenterology *62*:873, 1972 (abstract).
29. Corriveau, M. and Rojo-Ortega, J. J.: Cytochemical localization of adenyl cyclase in the rat colonic mucosa. Virchows Arch. B. Cell Path. *18*:129-134, 1975.

30. Quill, H. and Weiser, M. M.: Adenylate and guanlyte cyclase activities and cellular differentiation in rat small intestine. Gastroenterology 69:470-478, 1975.
31. Cuatrecasas, P., Hollenberg, M. D., Chang, D. K. J. and Bennett, V.: Hormone receptor complexes and their modulation of membrane function. Rec. Prog. Horm. Res. 31:37-94, 1975.
32. Wilson, D. E., El-Hindi, S., Tao, P. and Poppe, L.: Effects of indomethacin on intestinal secretion, prostaglandin E and cyclic AMP: Evidence against a role for prostaglandins in cholera toxin-induced secretion. Prostaglandins 10:581-587, 1975.
33. Barbezat, G. O. and Grossman, M. I.: Intestinal secretion: stimulation by peptides. Science 174:422-423, 1971.
34. Klipstein, F. A., Horowitz, I. R., Englert, R. F. and Schenk, F. A.: Effect of *Klebsiella pneumoniae* enterotoxin on intestinal transport in the rat. J. Clin. Invest. 56:799-807, 1975.
35. Donowitz, M., Keusch, G. T. and Binder, H. J.: Effect of shigella enterotoxin on electrolyte transport in rabbit ileum. Gastroenterology 69:1230-1237, 1975.
36. Gianella, R. A., Gots, R. E., Charvey, A. N., Greenaugh, W. B. and Formal, S. B.: Pathogensis of salmonella-mediated intestinal fluid secretion: activation of adenylate cyclase and inhibition by indomethacin. Gastroenterology 69:1238-1245, 1975.
37. Bieberdorf, F. A., Gorden, P. and Fordtran, J. S.: Pathogenesis of congenital alkalosis with diarrhea. Implications for the physiology of normal ileal electrolyte absorption and secretion. J. Clin. Invest. 51:1958-1968, 1972.
38. Binder, H. J., Filburn, C. and Volpe, B. T.: Bile salt alteration of colonic electrolyte transport: Role of cyclic adenosine monophosphate. Gastroenterology 68:503-508, 1975.
39. Bloom, S. R. and Polak, J. M.: The role of VIP in pancreatic cholera. In Gastrointestinal Hormones, J. C. Thompson, (ed.), University of Texas Press, Austin, 1975, pp. 635-642.
40. Kim, T. S., Shulman, J. and Levine, R. A.: Relaxant effect of cyclic 3',5'-AMP on the isolated rabbit ileum. J. Pharmacol. Exp. Ther. 163:36-42, 1968.
41. Wilkenfeld, B. E. and Levy, B.: The effects of theophylline, diazoxide and imidazole on isoproterenol-induced inhibition of the rabbit ileum. J. Pharmacol. Exp. Ther. 169:61-67, 1969.
42. Levine, R. A., Cafferata, E. P. and McNally, E. F.: Inhibitory effect of Adenosine 3',5'-Monophosphate on gastric secretion and gastrointestinal motility in vivo. In Proceedings of the 3rd World Congress of Gastroenterology. The Third World Congress of Gastroenterology, Tokyo, 1967, pp. 408-410.
43. Kapoor, K. and Levine, R. A.: Effect of adrenergic receptor blockade on the response to prostaglandin E_1 in canine ileum. Arch Int. Pharmacol. Therap. 203:243-250, 1973.
44. Wilson, D. E.: Prostaglandins and the gastrointestinal tract. Prostaglandins 1:281-293, 1972.

45. Hoke, S. E., Reid, D. P., Hogan, W. J., Dodds, W. J., Kalkhhoff, R. K. and Arndorfer: The effect of glucagon on esophageal motor function. Clin. Res. *20*:732, 1972.(abstract).
46. Andersson, K. E., Andersson, R., Hedner, P. and Persson, C. G. A.: Effect of cholecystokinin on the level of cyclic AMP and on mechanical activity in the isolated sphincter of Oddi. Life Sci. *11*:723-732, 1972.
47. Andersson, K. E., Andersson, R. and Hedner, P.: Cholecystokinetic effect and concentration of cyclic AMP in gall-bladder muscle *in vitro*. Acta Physiol. Scand. *85*:511-516, 1972.
48. Andersson, R. and Nilsson, K.: Cyclic AMP and calcium in relaxation in intestinal smooth muscle. Nature (New Biol.) *238*:119-120, 1972.
49. Amer, M. S.: Cyclic guanosine 3',5'-monophosphate and gallbladder contraction. Gastroenterology *67*:333-337, 1974.

9

Adenylate Cyclase and the Stimulatory Effect of Cholera Toxin in the Causation of Diarrhea

GEOFFREY W. G. SHARP

Studies over the past few years have established that the diarrhea of cholera is caused by the stimulation of adenylate cyclase in mucosal epithelial cells of the small intestine (1–6). The stimulation is caused by an exoenterotoxin of *Vibrio cholerae* (7), and only brief exposure of the intestine to the toxin is required for the development of the diarrheal response (6). Stimulation of adenylate cyclase results in increased intracellular concentrations of adenosine 3′,5′-monophosphate (cyclic AMP) (1, 3, 8). These, in turn, influence the ion transport systems in the intestinal cells so that the normal net reabsorption of ions and water is converted to a frank secretion of ions and water. This general outline of the mechanism of action of cholera toxin is shown schematically in Fig. 1.

The fecal fluid formed in cholera is isotonic, contains raised bicarbonate and potassium relative to plasma concentrations, and contains little protein. Differences in fluid composition occur at different levels of the intestine, and duodenal fluid has lower bicar-

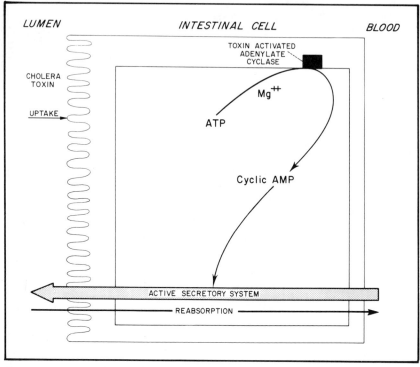

1 General outline of the action of cholera toxin, produced in the lumen of the small intestine, to cause diarrhea by changing the normal net reabsorption of ions and water to net secretion.

bonate than plasma, whereas ileal fluid has raised bicarbonate (9–11). The changes in ion and water movement are induced with no apparent damage to the mucosa as judged by light and electron microscopic examination and by studies showing little permeability change to large marker molecules (12–14). Marked closure of the intercellular spaces is one characteristic feature of the intestine during cholera toxin-induced fluid secretion, and this would be in accord with the stimulation of a blood to intestinal lumen active ion transport system (15).

Discussion of the mechanism by which cholera toxin stimulates the secretion of ions and water must necessarily involve many steps. Thus we need to know the nature and structure of the toxin, its binding and access to the cell, the manner of activation of adenylate cyclase, and the role of cyclic AMP in reversing the transport of ions. These topics will be dealt with in turn in this chapter. The discussion will cover the effects of cholera toxin on several different cell types, since the toxin appears to indiscriminately stimulate adenylate cyclase in mammalian cells (16–23). Thus it has been used as a tool to elevate intracellular cyclic AMP levels in many cell types and tissues.

Cholera toxin has been purified (24, 25) and crystallized (26) and found to be a protein of molecular weight around 84,000 (27) composed of at least two types of subunit designated A and B. The B subunits are apparently responsible for the binding of cholera toxin to cell membranes and the A subunit contains the activity responsible for stimulation of adenylate cyclase. In investigation of the structures of these components it has been suggested that the toxin might be composed of seven light units with molecular weight around 8,000 and one heavy unit of some 28,000 (28). In another report, the subunit molecular weights were estimated at 15,000 and 25,000 for the light (B) and heavy (A) subunits, respectively (29). Suport for the idea that the toxin contains four 15,000 and one 25,000 subunit has been gained from amino acid sequencing studies in which the whole toxin was sequenced by amino acid quantitation following the Edman degradation procedure (30). Amino acids were identified at the first 20 positions from the N-terminal. Where the amino acids in individual positions of the A and B chains are not similar, the molar ratio was determined. As this ratio was close to 1:4 for the A to B chains, respectively, it would appear that the toxin is composed of one active subunit and four binding subunits. The four binding subunits are identical with choleragenoid, the natural toxoid of cholera toxin. Furthermore, choleragenoid has been shown to block the action of choleragen if added to cells prior to toxin, and to prevent the binding of cholera toxin to cell membranes (31, 32). Since the intact cell appears to be much less responsive to the active subunit compared with the whole toxin, the binding subunit must play a major role in gaining access to the cell for the active subunit. The initial interaction of cholera toxin with the cell membrane is a binding to specific gangliosides in the membrane. Following the observation that mixed gangliosides could bind toxin and block the activity of the toxin (33) it was shown that the binding capacity resided in the sialidase stable monosialosyl ganglioside GGnSLC (GM_1) (34–36). Further, this treatment of intestinal cells with sialidase increased the capacity of the cells to bind toxin, presumably due to the increase in content of ganglioside GM_1 (36). Similarly, it has been shown that when fat cells are treated with GM_1 and the excess GM_1 subsequently removed by washing, the fat cells have increased toxin binding capacity and an increased responsiveness to cholera toxin (36). Thus all the available evidence suggests that ganglioside GM_1 may be the natural cell membrane receptor responsible for the uptake of cholera toxin during the initial toxin–cell interaction.

From early studies of the action of cholera toxin it was apparent that the uptake process was rapid and irreversible. Exposure of the intestine to cholera toxin for as little as 1 min, followed by washing, was sufficient to cause a subsequent full diarrheal response (37). Exposure of intestine to toxin followed by exposure to choleragenoid similarly resulted in the full response. Direct studies of the binding of cholera toxin to cells confirmed the rapid nature of the toxin–cell interaction (36, 38).

The first evidence that cholera toxin stimulated adenylate cyclase was obtained in 1971. These experiments were performed by treating intestinal cells with cholera toxin and, at suitable time intervals after the toxin, homogenizing the cells and assaying the enzyme activity (2, 3). It was shown that the increase in enzyme activity exhibited a lag phase followed by a progressive rise which was maximal after several hours. These early studies were confirmed subsequently on dog jejunum (6) and, in 1972, by reports showing that in patients with cholera, adenylate cyclase activity was elevated during the diarrheal phase of the disease (4, 5). In this study, adult patients with bacteriologically confirmed classical cholera were studied with respect to jejunal adenylate cyclase activity in the Cholera Research Laboratory in Dacca, East Pakistan. The patients received intravenous therapy to correct volume depletion, electrolyte imbalance and acidosis. Jejunal biopsies were obtained within 24 hr of the start of the disease and again 24 hr after the diarrhea had ceased completely. The average 24-hr stool volume during the initial biopsy was 10.5 liters. Biopsy material was obtained by means of the Cooke capsule localized by x-ray to the first 20 cm of jejunum beyond the ligament of Treiz. The biopsy samples were placed in oxygenated phosphate Ringer's solution kept on ice and washed in a buffer solution of $75mM$ Tris and $25mM$ $MgCl_2$ (pH 7.5). They were then homogenized in the same buffer solution, first by ten strokes of the pestle in a Ten Brock ground-glass hand homogenizer. During this treatment, connective tissue was trapped between the wall of the homogenizer and the pestle. The upper mixture of buffer and both broken and unbroken cells was transferred to a second homogenizer where cell breakage was completed by a further 10 strokes of the pestle. This homogenate was then assayed for enzyme activity. In the initial stages of this study it was necessary to define the optimal conditions for assay and, in addition, the conditions under which two biopsy samples taken several days apart could be correctly compared. Optimal assay conditions were achieved in the presence of an ATP regenerating system, $10 mM$ Mg^{++} and a pH of 7.5 with protein concentrations from the cells of 20 to 50 μgm per 50 μl incubation volume. With respect to the reproducibility of assays performed on samples obtained several days apart it was found that biopsy samples did not lose adenylate cyclase activity when maintained in Ringer solution at 4°C for as long as 3 hr. However, after homogenization the activity of adenylate cyclase declined markedly with a half-life of less than 90 min. Thus it was important to standardize the time between homogenization and assay. Thus for all paired studies on individual patients (i.e., samples assayed during the acute phase of the disease and after recovery) the time between homogenization and enzyme assay was kept constant. Similarly, the protein concentration used in the two assays was also maintained constant. Under these conditions it was found that adenylate cyclase activity was significantly increased during the acute diarrhoeal phase of cholera compared with the values obtained during convalescence. Basal activity was 7.1 pmol/mgm protein/min during cholera and 3.2 pmol/mg protein/min after recovery ($p < 0.01$, $n = 24$). All values obtained during cholera were above 3.7 pmol/mg/min and all recovery values below 4.2 pmol/mg/min except for one patient who had values of 7.0 for the acute stage and 8.7 for the convalescent stage and 8.3 two weeks after recovery. The reason for this high and anomalous adenylate cyclase activity in this patient remains unknown.

Some properties of the enzyme were then studied to see if differences might exist which could shed light on the mechanism by which the toxin stimulated the enzyme. Thus the effects of different concentrations of fluoride, which stimulates adenylate cyclase, ATP, Mg^{++}, Ca^{++}, Mn^{++}, and prostaglandin E_1 were examined. No major differences were detected in the responses of cholera affected and control tissue. Thus, in these limited tests on human jejunal adenylate cyclase, only increased enzyme activity was observed in response to challenge by cholera toxin. Results from other animal tissues however have recently exposed differences which may result in the elucidation of the mechanism of stimulation. Therefore the remainder of this account of the actions of cholera toxin will be concerned with our knowledge of its mechanism of action as derived from studies on a variety of tissues.

One feature of the action of cholera toxin which excited interest was the latent period which followed exposure to toxin and the subsequent progressive rise in adenylate cyclase activity. This response and its association with fluid secretion was shown in a study on dog jejunum (6). In this study using Thiry-Vella loops it was possible to apply cholera toxin to the jejunal lumen for only 10 minutes, to remove excess toxin by extensive washing, and then to assay the mucosal cells at different time points up to 48 hours while simultaneously measuring the water and ion movements across the epithelia. Adenylate cyclase activity was not elevated at 30 min but was significantly increased at 60 min. Activity then increased further to 3 hr, was still elevated at 24 hours but had returned to normal by 48 hr. Water movement across the jejunum similarly was not affected at 30 min and showed the normal reabsorptive flux. This was decreased, but still reabsorbing at 60 min and had changed to a secretory flux at 90 min. Again, the secretion increased up to 3 hr, was still secretory at 24 hr and was reabsorbing normally at 48 hours. A close correlation was observed between the percentage increase in adenylate cyclase activity and the net water flux. The questions arising from this study therefore are the elucidation of the latent period and the cause of gradually increasing activity of adenylate cyclase. As a first approach to the question of the latent period it seemed worthwhile to determine the localization of adenylate cyclase in the epithelial cells. That is, does adenylate cyclase reside in all the membranes of the epithelial cells or is it localized to either the brush border membranes or to the basal and lateral membranes. Accordingly brush border membranes and basal and lateral membranes were purified from mucosal epithelial cells of rabbit ileum (39). This was accomplished by conventional techniques of mild homogenization of the cells and differential centrifugation techniques (40, 41). The purified membranes were characterized by light and electron microscopic techniques and by the measurement of enzyme markers in the preparations. Thus brush border membranes were identified by microscopy and by high activity of alkaline phosphatase, low activity of succino dehydrogenase and NaK-ATPase and the presence of other membrane bound enzyme. Basal and lateral membranes in contrast were identified by the presence of large smooth surface membrane material, high NaK-ATPase and low alkaline phosphatase activity. Adenylate cyclase assays revealed that only 5 percent of the total enzyme activity was present in the brush border membrane preparations and that highest activity was in the basal and lateral membranes. When allowance is made for the contamination of the brush border preparations with fragments of basal and lateral membranes, particularly the lateral tags of membrane that can be seen adhering to the brush borders, it seems clear that adenylate cyclase is not a normal constituent of the brush border but resides in the basal and lateral membranes. It follows, therefore, that the stimulation of adenylate cyclase by cholera toxin is not the result of a direct interaction between the enzyme and the toxin. The effect of cholera toxin in the intestinal lumen must necessarily be indirect, because the site of binding of the toxin, the brush border, does not contain adenylate cyclase. Thus the mechanism of activation must involve some form of signal transmission or movement of the active subunit of cholera toxin to the enzyme it affects.

In seeking clues to the mode of action of the toxin, some unusual features are present that may lead to a greater understanding of the situation. For example, until recently the toxin has been found to stimulate adenylate cyclase only after addition to intact cells and only after a latent period of variable length. Subsequent homogenization and assay has demonstrated that the enzyme is activated in a permanent and essentially irreversible way. This is in marked contrast to the stimulatory effects of hormones which are able to stimulate the enzyme in broken cell preparations but in an easily reversible manner. Two further unusual features have arisen from studies on rat liver. In a study of the relationship of hormone activation to the stimulated state caused by cholera toxin, liver was chosen as the experimental tissue because it can be stimulated by polypeptide hormones, such as glucagon, and by catecholamines, thus presenting the opportunity for investigation of two types of hormones (42). For study of hormone effects in the liver after activation by cholera toxin, the toxin was injected into the rats intravenously and the liver removed 4 hr later, after activation of adenylate cyclase had occurred. It was found that after stimulation by cholera toxin, the stimulatory effects of both isoproterenol and epinephrine on a particulate preparation of rat liver were enhanced. Isoproterenol ($10^{-4}M$) increased adenylate cyclase activity by 10 pmoles/mg protein/10 min in control tissue and by 62 pmoles/mg protein/10 min in toxin-treated tissue. The values for epinephrine stimulation similarly were 26 pmoles and 85 pmoles respectively. Thus the responsiveness of adenylate cyclase in this tissue was markedly increased by toxin treatment. Also in this study, an attempt was made to solubilize the control and toxin-treated enzyme with detergents. It was found that the enzyme could be solubilized by homogenizing with 0.25 percent Lubrol PX and that the toxin treated enzyme was still activated after this treatment. Solubilization was defined as nonsedimentation at 200,000 g and the ability to pass through millipore filters with minimal loss of activity.

From these results it is clear that three unusual features characterize the stimulated state of adenylate cyclase after treatment with cholera toxin. The stimulated state is permanent, the enzyme activation is maintained after solubilization and the responsivity to stimulation by catecholamines is enhanced. In these respects the stimulation by cholera toxin resembles the stimulation of enzyme activity by nonhydrolysable analogs of GTP, analogues such as guanylimidodiphosphate in which the terminal phosphate is attached via a nitrogen atom rather than by oxygen (43). This aspect of the mechanism of action of cholera toxin will be referred to later.

Significant progress towards our understanding of the action of cholera toxin has been made recently. It has been shown that cholera toxin will stimulate adenylate cyclase activity in broken cell preparations if nicotinamide adenine dinucleotide is added to the membranes in the presence of a cytosol factor (44–47). Along with the increased knowledge of the process of activation which these findings bring comes an increased ability to further explore the process. Thus it is clearly easier to explore the action of cholera toxin in purified membrane systems than in whole cell situations. In following up these observations, it was found that, as in the pigeon erythrocyte preparation (45), in rat liver cholera toxin also activates adenylate cyclase in cell homogenates (46, 47). This was achieved when sufficient NAD was added to the homogenate along with the toxin. It was also apparent that a component present in the 200,000 g supernatant from the homogenate was also needed for activation. The activation process in broken-cell preparations was extremely rapid and maximal activation could be achieved after 15 min incubation at room temperature with toxin and NAD. When incubation was performed at 37°C an even more rapid stimulation was achieved.

Experiments designed to test for cytosol requirements and the time sequence in which the toxin and NAD are required were of interest because of the apparently essential role of a soluble component of the cytosol. For instance the lack of response of washed membranes to cholera toxin and NAD can be restored by the addition of 200,000 g supernatant. Incubation of cholera toxin alone or NAD with the homogenates followed by washing and addition of the complementary agent (NAD or cholera toxin, respectively) in a second incubation did not evoke a stimulatory effect suggesting a necessity for the simultaneous presence of the three agents (cholera toxin, NAD, and supernatant).

Experiments in which cholera toxin alone was incubated with homogenates, washed, and followed by incubation with supernatant and NAD suggest that cholera toxin bound during the first incubation can interact with the supernatant and NAD during the second incubation period to produce stimulation of the adenylate cyclase. However, a lack of effect of large amounts of choleragenoid to prevent the action of the toxin, suggested that binding of cholera toxin to the membrane is not an essential requirement in the activation process.

The characteristics of the stimulated adenylate cyclase, after treatment with cholera toxin and NAD were determined to see if the stimulation achieved in the broken cell state was the same as the stimulation achieved by toxin treatment of intact cells. Thus studies were performed routinely to determine the persistence of the stimulation. As found for the intact cell stimulation, the effect of cholera toxin is persistent and cannot be reversed by washing or solubilization. A comparison was also made of the effects of cholera toxin and NAD on homogenates from control livers and on homogenates from livers of rats pretreated with cholera toxin 4 hr before obtaining the livers. No evidence for a difference in the nature of the stimulation was obtained in that there was no additivity, and the effect of the toxin and NAD in vitro was overlapping with that obtained in vivo. Similarly, the effects of fluoride and toxin were partially overlapping. Final points of similarity were obtained by a comparison of the effects of hormones on control and toxin plus NAD stimulated liver preparations. As reported previously, isoproterenol and glucagon both stimulate adenylate cyclase under control and toxin-treated conditions, and isoproterenol had enhanced activity in toxin stimulated liver (42). Similar observations have been reported for turkey erythrocytes (48) and fat cells (49). In the broken-cell conditions isoproterenol again exhibited enhanced activity in the preparation activated by cholera toxin and NAD in vitro. Glucagon stimulated adenylate cyclase in both control and in vitro toxin-treated preparations as it does in the in vivo toxin-stimulated liver (42). Thus, in all comparative tests made, the stimulation of adenylate cyclase by cholera toxin and NAD in the broken cell preparation appears to be similar to that achieved in vivo or in the isolated but intact cells.

The advantages of an in vitro method for the stimulation of adenylate cyclase by cholera toxin in purified membrane preparation are obvious. The system can be simplified, cofactors identified and the scope for experimentation widened.

Since our first report that cholera toxin caused an enhancement of the effects of catecholamines on adenylate cyclase and that the activated enzyme could be solubilized in the activated state (42) we have been impressed by the similarities of the effects of cholera toxin and certain guanyl nucleotides on the enzyme. After the early report of the effect of GTP on the enzyme by Rodbell et al. (50) a number of laboratories have found that GTP enhances the effect of hormones on adenylate cyclase, and can stimulate basal activity of the enzyme. Nonhydrolyzable analogs of GTP such as guanylylimidodiphosphate (GppNHp) and guanylylmethylenediphosphate (GppCH$_2$p) have a similar but greater effect than GTP on both basal and hormone-stimulated adenylate cyclase activity, presumably because of the stability of the terminal phosphate group (51). These stimulatory effects are irreversible, resisting dilution, washing or solubilization. Interference of the stimulatory effect of sodium fluoride by GTP or GppCH$_2$p has also been reported.

Thus the effects of cholera toxin and the analogues of GTP are similar in that they both stimulate adenylate cyclase, they both enhance certain hormone effects and produce an essentially irreversible stimulation of the enzyme which can then be solubilized in the activated state. Because of these similarities, the interrelationship of the effects of cholera toxin and GppNHp was examined. This was facilitated by the ability to stimulate adenylate cyclase in broken cell preparations and allowed the possibility of sequential incubations with the two agents to study their interaction in detail.

Test rats were injected with cholera toxin, 1 μgm/gm body weight dissolved in saline and control rats were injected with saline.

After 4 hr the rats were killed and 1200 g pellets from the liver homogenates prepared. Adenylate cyclase activity was assayed in test and control preparations in the presence and absence of different concentrations of GppNHp. It was found that cholera toxin caused its expected large increase in enzyme activity. It was also found that GppNHp stimulated the enzyme in the control liver preparation over the concentration range tested, $10^{-6}M$ to $10^{-3}M$. In contrast to these effects on the control tissue GppNHp failed to stimulate the toxin-treated preparation. Furthermore, GppNHp at $10^{-6}M$ caused a significant inhibition of adenylate cyclase activity. Inhibitory, rather than stimulatory effects of GppNHp were found when tested on preparations stimulated in vitro by cholera toxin. Following these experiments, sequential additions of toxin and NAD, and GppNHp were studied. Four sets of conditions were examined, viz. no additions (controls); cholera toxin and NAD added at 10 min; $10^{-3}M$ GppNHp at zero time; and $10^{-3}M$ GppNHp at zero time followed by cholera toxin and NAD at 10 minutes. In these experiments GppNHp stimulated adenylate cyclase to a greater extent than did cholera toxin and NAD. More importantly, prior treatment with GppNHp completely blocked the stimulatory effect of cholera toxin.

From these results it was concluded that stimulation of adenylate cyclase by the nonhydrolyzable analog of GTP, GppNHp, blocks the stimulatory effect of cholera toxin and NAD. Furthermore, the stimulation of adenylate cyclase by cholera toxin can be inhibited by the subsequent addition of GppNHp. This mutual interference in the stimulatory effect of cholera toxin and GppNHp suggests a common site of action or at least a common site at some stage of their mechanisms of stimulation of adenylate cyclase. These observations strengthen the possibility that both agents act on the same control mechanism of the adenylate cyclase complex to stimulate the enzyme. As mentioned earlier, adenylate cyclase after stimulation by cholera toxin resembles in several respects the enzyme after stimulation by GppNHp. Thus both agents cause a marked stimulation of adenylate cyclase. The stimulation of the enzyme is essentially irreversible and is not reduced by dilution, washing, or solubilization by detergent treatment and also the effects of catecholamines are enhanced. These similarities, when combined with the ability of both cholera toxin and GppNHp to block the action of the other, suggest strongly that they share a common target site on the adenylate cyclase complex—presumably the nucleotide regulatory site. This site appears to have a major role in setting the rate of adenylate cyclase activity and the extent of stimulation by hormones. Pfeuffer and Helmreich (43) have detected a guanyl nucleotide-binding protein, which they suggested might regulate the activity of adenylate cyclase; thus interaction of guanylnucleotides with this membrane bound component of the adenylate cyclase complex would essentially change the activity of the enzyme. Binding of GppNHp to turkey erythrocyte membranes (52) and GTP to a solubilized preparations of dog myocardium (53) has been reported and related to changes in adenylate cyclase activity. A series of papers from Rodbell and his associates (54–57) has developed a model for the effect of guanyl nucleotides and glucagon on adenylate cyclase in hepatic adenylate cyclase. In this model, GppNHp induces the formation of an intermediate transition state, with no increase in enzyme activity until the transition state isomerizes to a high activity state. Glucagon accelerates the isomerization rate. An action of cholera toxin and NAD to cause a similar state transition to one of high activity would be consistent with the known characteristics of its action.

In summarizing this work, a theory for the activation of adenylate cyclase by cholera toxin is illustrated in Fig. 2 using the small intestine as the target tissue.

It is apparent that the action of cholera toxin encompasses the binding to cell membranes and activation of adenylate cyclase. The consequence of this, an increase in intracellular cyclic AMP, results in the ion transport effects which reverse the net reabsorption of ions and water to secretion. Little evidence is available with regard to the mechanism of this action of cyclic AMP. By analogy with other systems it might be assumed that activation of protein kinase (62) and phosphorylation of a key component of the ion transport system(s) might be involved. While increased phosphorylation of protein in the intestine has been observed in response to cholera toxin treatment (63) this in no way relates necessarily to the diarrhoeal response. Whatever the mechanism it seems likely that the action of cyclic AMP is to inhibit a coupled NaCl influx process across the brush border of the mucosal epithelial cells and to unmask or stimulate a neutral NaCl efflux process resulting in net secretion (64–66). Certainly the mechanisms of glucose and amino acid absorption coupled to sodium are intact and functional so that they form the basis for the success of oral fluid and electrolyte therapy in the treatment of cholera (67, 68).

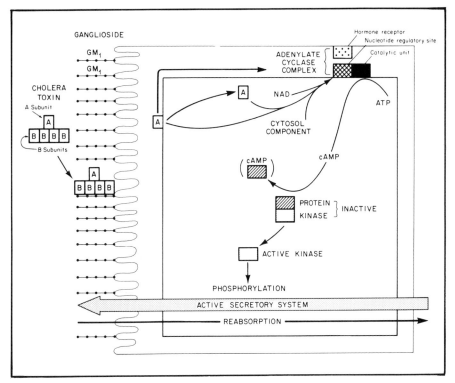

2 Schematic representation of a hypothesis for the action of cholera toxin in stimulating adenylate cyclase and ion secretion in an epithelial cell of the small intestine. Cholera toxin, composed of binding subunits and an active subunit, interacts in a rapid and irreversible manner with GM_1 ganglosides on the external surface of the brush border membrane of the cell. Following the initial interaction the active subunit gains access to the interior of the cell. Here, in combination with an as yet unknown cytosol component and NAD, it affects the function of the nucleotide regulatory site of the adenylate cyclase complex. The result of this change is a permanent stimulation of adenylate cyclase located in the basal and lateral membranes of the cell. The stimulation of adenylate cyclase results in increased concentrations of cyclic AMP, which, in turn, affects the activity of protein kinases in the cell. One consequence of the increased activity of protein kinase is increased phosphorylation of a critical component of the ion transport systems of the epithelial cells so that the normal reabsorptive flux of ions is converted to secretion. Water follows the secretory ion flux into the intestinal lumen and diarrhea results.

SUMMARY

The diarrhea of cholera is caused by an exoenterotoxin of *Vibrio cholerae*, a protein of molecular weight estimated at 84,000. The toxin is composed of subunits responsible for binding to cells and an active subunit responsible ultimately for the accumulation of diarrheal fluid in the lumen of the small intestine. In producing its effect the toxin is thought to bind rapidly and irreversibly with the sialidase resistant mono-sialosyl ganglioside designated GGnSLC or GM_1. Following this initial interaction there is a latent period followed by a progressive rise in adenylate cyclase activity and increase in the intracellular concentration of adenosine 3',5'-monophosphate (cyclic AMP). Studies on the localization of adenylate cyclase in the intestinal cells suggest that the enzyme is not present in the brush border membranes, the initial site of toxin-cell interaction, but instead is present in the basal and lateral cell membranes. Thus the stimulation of adenylate cyclase must necessarily be an indirect process. From enzyme assays performed on jejunal mucosa obtained by biopsy from patients with cholera and after recovery, it appears that adenylate cyclase activity is stimulated to double its normal level during the acute phase of the disease. It has been shown that even after a brief application of the toxin to the intestine, followed by extensive washing, that the enzyme is progressively stimulated over the next several hours.

The stimulated enzyme has several characteristic features. The raised activity is permanent and cannot be restored to normal by washing the membranes or by "solubilizing" with nonionic detergents. Furthermore the affected enzyme demonstrates enhanced responses to stimulation by catecholamines. In these respects, stimulation of adenylate cyclase by cholera toxin resembles the stimulation caused by certain non-hydrolysable analogues of GTP, analogs such as guanylylimidodiphosphate and guanylylmethlenediphosphate.

Current evidence suggests that to stimulate adenylate cyclase, cholera toxin requires an unknown component of cytosol and adequate NAD. From studies on the interactions of guanylylimidodiphosphate and cholera toxin on adenlyate cyclase it appears possible that the action of the toxin is exerted on the nucleotide regulatory site of the membrane bound adenylate cyclase complex.

REFERENCES

1. Schafer, D. E., Lust, W. D., Sircar, B. and Goldberg, N. O.: Elevated concentration of adenosine 3',5' monophosphate in intestinal mucosa after treatment with cholera toxin. Proc. Nat. Acad. Sci. 67:851-856, 1970.
2. Sharp, G. W. G. and Hynie, S.: Stimulation of intestinal adenyl cyclase by cholera toxin. Nature 229:226-269, 1971.
3. Kimberg, D. V., Field, M., Johnson, J., Henderson, A. and Gershon, E.: Stimulation of intestinal mucosal adenyl cyclase by cholera enterotoxin and prostaglandins. J. Clin. Invest. 50:1218-1230, 1971.
4. Chen, L. C., Rohde, J. E. and Sharp, G. W. G.: Intestinal adenyl cyclase activity in human cholera. Lancet 1:939-941, 1971.
5. Chen, L. C., Rohde, J. E. and Sharp, G. W. G.: Properties of adenyl cyclase from human jejunal mucosa during naturally acquired cholera and convalescence. J. Clin. Invest. 51:731-740, 1972.
6. Guerrant, R. L., Chen, L. C. and Sharp, G. W. G.: Intestinal adenyl cyclase activity in canine cholera: correlation with fluid accumulation. J. Inf. Dis. 125:377-381, 1972.
7. De, S. N.: Enterotoxicity of bacteria-free culture filtrate of Vibrio cholerae. Nature 183:1533-1534, 1959.
8. Sharp, G. W. G., Hynie, S., Lipson, L. C. and Parkinson, D.: Action of cholera toxin to stimulate adenyl cyclase. Trans. Assoc. Amer. Phys. 84: 200-211, 1971.
9. Banwell, J. G., Pierce, N. F., Mitra, R. C., Brigham, G. J., Cavanasos, R. I., Keimowitz, D. S., Fedson, J., Thomas, J., Gorbach, S. L., Sack, R. B. and Mondal, A.: Intestinal fluid and electrolyte transport in human cholera. J. Clin. Invest. 49:183-195, 1970.
10. Carpenter, C. C. J., Sack, R. B., Feeley, J. C. and Steenberg, R. W.: Site and characteristics of electrolyte loss and effect of intraluminal glucose in experimental canine cholera. J. Clin. Invest. 47:1210-1220, 1968.
11. Sack, R. B., Carpenter, C. C. J., Steenberg, R. W. and Pierce, N. F.: Experimental cholera: a canine model. Lancet 2:206-207, 1966.
12. Gangerosa, E. J., Beisel, W. R., Benyajati, C., Sprinz, H. and Piyaratn, P.: The nature of gastrointestinal lesion in Asiatic cholera and its relation to pathogenesis: a biopsy study. Am. J. Trop. Med. Hyg. 9:125-135, 1960.
13. Elliott, H. L., Carpenter, C. C. J., Sack, R. B. and Yardley, J. H.: Small bowel morphology in experimental canine cholera. A light and electron microscopic study. Lab. Invest. 22:112-120, 1970.
14. Norris, H. T. and Majno, G.: On the role of the ileal epithelium in the pathogenesis of experimental cholera. Am. J. Pathol. 53:263-279, 1968.
15. DiBona, D. R., Chen, L. C. and Sharp, G. W. G.: A study of intracellular spaces in the rabbit jejunum during acute volume expansion and after treatment with cholera toxin. J. Clin. Invest. 53:1300-1307, 1970.
16. Vaughan, M., Pierce, N. F. and Greenough, W. B.: Stimulation of glycerol production in fat cells by cholera toxin. Nature 226:658-659, 1970.

17. Zieve, P. D., Pierce, N. F. and Greenough, W. B.: Stimulation of glycogenolysis by purified cholera exotoxin in disrupted cells. Johns Hopkins Med. J. *129*:299-303, 1971.
18. Bourne, H. R., Lehrer, R. I., Lichtenstein, L. M., Weissman, G. and Zurier, R.: Effects of cholera enterotoxin on adenosine 3'-5' monophosphate and neutrophil function. J. Clin. Invest. *52*:698-708, 1973.
19. Wolff, J., Temple, R. and Hope-Cook, G.: Stimulation of steroid secretion in adrenal tumor cells by choleragen. Proc. Nat. Acad. Sci. *70*:2741-2744, 1973.
20. Wollheim, C. B., Blondel, B. and Sharp, G. W. G.: Effect of cholera toxin on insulin release in monolayer cultures of the endocrine pancreas. Diabetologia *10*:783-787, 1974.
21. O'Keefe, E. and Cuatrecasas, P.: Cholera toxin mimics melanocyte stimulating hormone in inducing differentiation in melanoma cells. Proc. Nat. Acad. Sci. *71*:2500-2504, 1974.
22. Boyle, J. M. and Gardner, J. D.: Sequence of events mediating the effects of cholera toxin on rat thymocytes. J. Clin. Invest. *53*:1149-1158, 1974.
23. Mashiter, K., Mashiter, G. D., Hauger, R. L. and Field, J. B.: Effects of cholera and E. coli enterotoxins on cyclic adenosine 3',5'-monophosphate levels and intermediary metabolism in the thyroid. Endocrinology *92*:541-549, 1973.
24. Finkelstein, R. A. and LoSpolluto, J. J.: Production of highly purified choleragen and choleragenoid. J. Inf. Dis. *121*:563-572, 1970.
25. Finkelstein, R. A. and LoSpolluto, J. J.: Pathogenesis of experimental cholera: Preparation and isolation of choleragen and choleragenoid. J. Exp. Med. *130*:185-202, 1969.
26. Finkelstein, R. A. and LoSpolluto, J. J.: Crystalline cholera toxin and toxoid. Science *175*:529-530, 1972.
27. LoSpolluto, J. J. and Finkelstein, R. A.: Chemical and physical properties of cholera exo-enterotoxin (choleragen) and its spontaneously formed toxoid (choleragenoid). Biochem. Biophys. Acta. *257*:158-166, 1972.
28. Lonnroht, I. and Holmgren, J.: Subunit structure of cholera toxin. J. Gen. Microbiol. *76*:417-424, 1973.
29. Van Heyningen, S.: Cholera toxin: Interaction of subunits with ganglioside GM_1. Science *183*:656-657, 1974.
30. Jacobs, J. W., Niall, H. D. and Sharp, G. W. G.: The amino terminal sequence of cholera toxin subunits. Biochem. Biophys. Res. Comm. *61*:341-345, 1974.
31. Pierce, N. F.: Differential inhibitory effects of cholera toxoids and ganglioside on the enterotoxins of vibrio cholerae and escherichia coli. J. Exp. Med. *137*:1009-1023, 1973.
32. Lichtenstein, L. M., Henney, C. S., Bourne, H. R. and Greenough, W. B.: Effects of cholera toxin on in vitro models of immediate and delayed hypersensitivity. Further evidence for the role of adenosine 3',5'-monophosphate. J. Clin. Invest. *52*:691-697, 1973.
33. Van Heyningen, W. E., Carpenter, C. C. J., Pierce, N. F. and Greenough, W. B.: Deactivation of cholera toxin by ganglioside. J. Inf. Dis. *124*:415-418, 1971.

34. Van Heyningen, W. E.: On the similarity of tetanus and cholera toxins. Naunyn-Schmiedebergs Arch. Exp. Path. Pharmak. *276*:289-295, 1973.
35. King, C. A. and Van Heyningen, W. E.: Deactivation of cholera toxin by a sialidase resistant monosialosylganglioside. J. Inf. Dis. *127*:639-647, 1973.
36. Cuatrecasas, P.: Gangliosides and membrane receptors for cholera toxin. Biochemistry *12*:3558-3566, 1973.
37. Pierce, N. F., Greenough, W. G. and Carpenter, C. C. J.: Vibrio cholerae enterotoxin and its mode of action. Bacteriol. Rev. *35*:1-13, 1971.
38. Walker, W. A., Field, M. and Isselbacher, K. J.: Specific binding of cholera toxin to isolated intestinal microvillous membranes. Proc. Nat. Acad. Sci. *71*:320-324, 1974.
39. Parkinson, D. K., Ebel, H., DiBona, D. R. and Sharp, G. W. G.: Localization of the action of cholera toxin on adenyl cyclase in mucosal epithelial cells of rabbit intestine. J. Clin. Invest. *51*:2292-2298, 1972.
40. Forstner, G. G., Sabesin, K. J. and Isselbacher, K. J.: Rat intestinal microvillous membranes. Purification and biochemical characterization. Biochem. J. *106*:381-390, 1968.
41. Quigley, J. P. and Gotterer, G. S.: Distribution of ($Na^+ + K^+$) stimulated ATPase activity in rat intestinal mucosa. Biochem. Biophys. Acta. *173*:456-468, 1969.
42. Beckman, B., Flores, J., Witkum, P. A. and Sharp, G. W. G.: Studies on the mode of action of cholera toxin. Effects on solubilized adenyl cyclase. J. Clin. Invest. *53*:1202-1205, 1974.
43. Pfeuffer, T. and Helmreich, E. J. M.: Activation of pigeon erythrocyte membrane adenylate cyclase by guanylnucleotide analogues and separation of a nucleotide binding protein. J. Biol. Chem. *250*:867-876, 1975.
44. Gill, D. M. and King, C. A.: The mechanism of action of cholera toxin in pigeon erythrocyte lysates. J. Biol. Chem. *250*:6424-6432, 1975.
45. Gill, D. M.: The involvement of nicotinamide adenine dinucleotide in the action of cholera toxin "in vitro". Proc. Nat. Acad. Sci. *72*:2064-2068, 1975.
46. Flores, J. and Sharp, G. W. G.: Effects of cholera toxin on adenylate cyclase. Studies with guanylylimidodiphosphate. J. Clin. Invest. *56*:1345-1349, 1975.
47. Flores, J., Witkum, P. A. and Sharp, G. W. G.: Activation of adenylate cyclase by cholera toxin in rat liver homogenates. J. Clin. Invest. *57*:450-458, 1976.
48. Field, M.: Mode of action of cholera toxin: Stabilization of catecholamine-sensitive adenylate cyclase in turkey erythrocytes. Proc. Nat. Acad. Sci. *71*:3299-3303, 1974.
49. Hewlett, E. L., Guerrant, R. L., Evans, D. J. and Greenough, W. B.: Toxins of Vibrio cholerae and Escherichia coli stimulate adenylate cyclase in rat fat cells. Nature *249*:371-373, 1974.
50. Rodbell, M., Birnbaumer, L., Pohl, S. L. and Kraus, H. M. J.: The glucagon sensitive adenyl cyclase system in plasma membranes of rat liver. An obligatory role of guanyl nucleotides in glucagon action. J. Biol. Chem. *246*:1877-1882, 1971.
51. Londos, C., Salomon, Y., Lin, M. C., Harwood, J. P., Schramm, M., Wolff, J. and Rodbell, M.: 5'guanylylimidodiphosphate, a potent activator of

adenylate cyclase systems in eukaryotic cells. Proc. Nat. Acad. Sci. *71*: 3087-3090, 1974.
52. Spiegel, A. M. and Aurbach, G. D.: Binding of 5'-guanylyl-imidodiphosphate to turkey erythrocyte membranes and effects of β adrenergic-activated adenylate cyclase. J. Biol. Chem. *249*:7630-7636, 1974.
53. Lefkowitz, R. J.: Guanosine triphosphate binding sites in solubilized myocardium. Relation to adenylate cyclase activity. J. Biol. Chem. *250*:1006-1011, 1975.
54. Schramm, M. and Rodbell, M.: A persistent active state of the adenylate cyclase system produced by the combined actions of isoproterenol and guanylylimidodiphosphate in frog erythrocyte membranes. J. Biol. Chem. *250*:2232-2237, 1975.
55. Salomon, Y., Lin, M. C., Londos, C., Rendell, M. and Rodbell, M.: The hepatic adenylate cyclase system. I. Evidence for transition states and structural requirements for guanine nucleotide activation. J. Biol. Chem. *250*:4239-4245, 1975.
56. Lin, M. C., Salomon, Y., Rendell, M., and Rodbell, M.: The hepatic adenylate cyclase system. II. Substrate binding and utilization and the effects of magnesium ion and pH. J. Biol. Chem. *250*:4246-4252, 1975.
57. Rendell, M., Salomon, Y., Lin, M. C., Rodbell, M. and Berman, M.: The hepatic adenylate cyclase system. III. A mathematical model for the steady state kinetics of catalysis and nucleotide regulation. J. Biol. Chem. *250*: 4253-4260, 1975.
58. Pierce, N. F., Carpenter, C. C. J., Elliott, H. L. and Greenough, W. G.: Effects of prostaglandins, theophylline, and cholera exotoxin upon transmucosal water and electrolyte movement in the canine jejunum. Gastroenterology *60*:22-32, 1971.
59. Matuchansky, C. and Bernier, J. J.: Effect of prostaglandin E_1 on glucose, water and electrolyte absorption in the human jejunum. Gastroenterology *64*:1111-1118, 1973.
60. Cummings, J. H., Newman, A., Misiewicz, J. J.: Effect of intravenous prostaglandin E_2 on small intestinal function in man. Nature *243*:169-171, 1973.
61. Field, M.: Intestinal secretion. Gastroenterology *66*:1063-1084, 1974.
62. Walsh, D. A. and Ashby, C. D.: Protein kinases: aspects of their regulation and diversity. Rec. Progr. Horm. Res. *29*:329-359, 1973.
63. Lucid, S. W. and Cox, A. C.: The effect of cholera toxin on the phosphorylation of protein in epithelial cells and their brush borders. Biochem. Biophys. Res. Commun. *49*:1183-1186, 1972.
64. Field, M.: Ion transport in rabbit ioeal mucosa. II. Effects of cyclic 3',5'-AMP. Am. J. Physiol. *221*:992-997, 1971.
65. Field, M., Fromm, D., Al-Awqati, Q. and Greenough, W. G.: III. Effect of cholera enterotoxin on ion transport across isolated ileal mucosa. J. Clin. Invest. *51*:796-804, 1972.
66. Al-Awqati, Q., Cameron, J. L. and Greenough, W. B.: Electrolyte transport in human ileum; effect of purified cholera exotoxin. Am. J. Physiol. *224*:818-823, 1973.

67. Pierce, N. F., Sack, R. B., Mitra, R. C., Banwell, J. G., Brigham, K. L., Fedson, D. S. and Mondal, A.: Replacement of water and electrolyte losses in cholera by an oral glucose-electrolyte solution. Ann. Intern. Med. 70: 1173-1181, 1969.
68. Cash, R. A., Nalin, D. R., Forrest, J. N. and Abrutym, E.: Rapid correction of acidosis and dehydration of cholera with oral electrolyte and glucose solution. Lancet 2:549-550, 1970.

Copyright 1977, Spectrum Publications, Inc.
Clinical Aspects of Cyclic Nucleotides

10

Cyclic AMP in Saliva, Salivary Glands and Gingival Tissues

L. DANIEL SCHAEFFER

Within the past decade, considerable evidence has been obtained for various aspects of cyclic nucleotide metabolism in salivary glands (1–8), and the relationship of this metabolism to cyclic nucleotide content in saliva (9–16). Many lines of evidence have indicated that salivary amylase secretion is mediated through adrenergic beta receptors (1, 2, 4, 6) and is enhanced by exogenously administered adenosine 3',5'-monophosphate (cyclic AMP) and butyryl derivatives of cyclic AMP as well as by methylxanthine derivatives. However, it is not known if the cyclic AMP content in saliva is changed by pathological conditions. Therefore, we measured cyclic AMP in saliva of healthy children, children with Down's syndrome and in saliva from children with salivary gland cancer.

In addition to this, we measured cyclic AMP content and the enzymatic properties of cyclic AMP phosphodiesterase in normal and moderately inflamed gingival tissues. Initial data suggests a metabolic abnormality in the cyclic nucleotide pathway in diseased periodontal tissues.

CYCLIC AMP CONTENT IN HUMAN SALIVA

Whole Saliva and Parotid Saliva

Parotid saliva samples were collected using a modification of the "Carlson-Crittenden" parotid cup (17). Stimulation parotid saliva samples were obtained either by applying a 10 percent solution of citric acid to the tongue or by placing "sour-grape" candy lozenges on the subjects tongue. Cyclic AMP concentrations were assayed with modifications of the protein bindery method described by Gilman (18) and alpha amylase activity was assayed by the amylopectic azure method (19). Units were expressed as 37°C. Protein was measured by the method of Lowry et al. (20).

Our initial studies (21) confirmed the findings of Stefanovich and Wells (22) that cyclic AMP was present in human saliva. The values reported by Stefanovich and Wells were 10^3 fold higher than the values we calculated and this discrepancy was undoubtedly due to interfering agents in their assay procedure (22). Recently, Kanamori et al. (11, 12) have published values for salivary cyclic AMP which were similar to those reported by our laboratory (see Table 1). Also, of interest are the reports by Grower et al. (23, 24) that gingival fluid contains approximately 100-fold higher cyclic AMP levels than those measured in serum and saliva.

The data in Table 1 were obtained from healthy normal adult male and female subjects. The mean values for cyclic AMP concentrations calculated on either a per unit volume or per mg protein basis from stimulated parotid saliva samples, were found to differ significantly from unstimulated samples ($p < .01$). Moreover, when amounts of cyclic AMP were normalized for various rates of flow, greater significant differences between stimulated and unstimulated salivas were noted. For example, within a one minute time interval an average total of 32.5 pmoles/ml of cyclic AMP was found in stimulated parotid saliva while only 2.8 pmoles/ml/min were found in unstimulated saliva.

Down's Syndrome

Down's syndrome subjects (trisomy of autosomal chromosome 21) have been reported to have severe periodontal disease (25, 26),

Table 1. Cyclic AMP Content in Saliva and Alpha Amylase
Secretion of Normal Human Subjects

Samples	Cyclic AMP		Flow Rate	Alpha Amylase
	pmole/ml[a]	pmole/mg[b]	ml/min	Units/ml[c]
Whole Saliva (5)	22.6 ± 42[d]	7.4 ± 0.92	ND[e]	ND
Parotid Saliva				
Stimulated (24)[f]	37.4 ± 1.80	12.1 ± 0.78	0.87 ± 0.08	128.3
Unstimulated (20)	23.6 ± 2.80	6.3 ± 0.82	0.12 ± 0.10	97.4

[a]Pincomoles cyclic AMP/ml saliva.
[b]Picomoles/mgm of protein.
[c]Micromoles reducing sugar/min/mgm protein at 37°C.
[d]All values are mean ± SEM.
[e]Not determined.
[f]Stimulated saliva obtained by placing citric acid candies on tongue.

minimal caries activity (22–29) and various other dental anomalies (30) including abnormalities of certain constituents of saliva (10, 31, 32). Therefore, our laboratory undertook an investigation to determine whether cyclic AMP concentrations were altered in Down's syndrome subjects at comparable age groups of normal subjects. Preliminary data of the range, mean, and standard error of the mean of ages, sex, decayed, and filled teeth (DF), and whole saliva concentrations of cyclic AMP is shown in Table 2. Assessments of cyclic AMP in saliva from normal and Down's syndrome subjects exhibited a significant difference between the two groups ($p < .01$), the values obtained were 10.2 and 29.6 pmoles cyclic AMP/ml, respectively. The mean age of the normal age groups was 11 years while that of the Down's syndrome was 13 years.

The data in this study indicates that genetic lesions associated with nondisjunction of autosomal chromosome predisposed the Down's syndrome children to low caries incidence, and high salivary levels of cyclic AMP, whereas normal children had a higher incidence of caries, but a much lower saliva concentration of cyclic AMP. Results obtained from patients with Down's syndrome could possibly be related to either abnormal cyclic AMP metabolism in salivary glands or perhaps to greater concentrations of this cyclic nucleotide in the blood, or both. Further studies are needed to de-

Table 2. Cyclic AMP Concentration in Whole Saliva in Normal and Down's Syndrome Subjects

Subject	Age Years	Sex	DF[1]	Cyclic AMP pm/ml[2]
Normal				
1	12	F	6	17.20
2	10	M	8	8.46
3	12	M	11	3.44
4	9	F	17	11.58
	11[3]		10.5[3]	10.2 ± 1.52[4]
Down's syndrome				
1	14	M	2	46.62
2	16	M	10	37.80
3	13	M	0	25.46
4	13	M	2	31.9
5	13	F	5	28.1
6	12	M	4	30.5
7	12	F	3	36.3
8	13	F	1	26.7
9	12	M	0	23.6
10	12	M	5	17.3
11	12	M	1	27.9
12	13	F	2	33.5
13	11	M	14	40.8
	13[3]		3.5[3]	20.6 ± 1.40[4]

[1] represents number of teeth either decayed or filled
[2] picomoles cyclic AMP/ml saliva
[3] mean value
[4] mean ± standard error of mean

termine the exact origin of cyclic AMP in saliva from Down's syndrome subjects.

Saliva from Children with Salivary Gland Cancers

The objective of this study was to determine the cyclic AMP content in saliva obtained from children undergoing radiation treatment for malignant cancers of the parotid salivary glands. Clinical changes in the oral environment after radiation treatment to the

Table 3. Cyclic AMP Concentrations in Whole and Parotid Saliva from Normal and Salivary Gland Cancer Subjects

	Cyclic AMP pm/ml	
	Salivary Gland Cancer Subjects	Control Subjects
Whole Saliva	55.2 (8) ± 7.9	11.6 (5) ± 1.2
Parotid Saliva		
Stimulated	38.4 (8) ± 1.9	20.5 (5) ± 3.2
Unstimulated	55.2 (8) ± 4.7	N.D.

Legend: is as described in Table 1

salivary glands are well documented (33–37), and these changes may include modification of the cyclic AMP content. The experimental group consisted of 8 children who had undergone approximately 5000 rads of radiation at a level of 1000 rad/wk of partial or total coverage of the major salivary glands. As shown in Table 3 cyclic AMP concentration in whole saliva of these children is significantly higher (55.2 ± 7.9 pmoles/ml) than in healthy controls (11.6 ± 1.2 pmoles/ml $p < .05$). However, the mean stimulated parotid flow rates were significantly lower in the radiation therapy patients when compared with normals, 0.15 and 0.68 ml/min, respectively. These results indicate that radiation-induced changes in saliva flow in patients with salivary gland cancers did not change the total amount of cyclic AMP being released.

CYCLIC AMP CONTENT AND CYCLIC NUCLEOTIDE PHOSPHODIESTERASE IMBALANCE IN HUMAN INFLAMED GINGIVAL BIOPSIES

Cyclic AMP and its respective degradative enzyme, phosphodiesterase (PDE), have been implicated in various disease processes (38–41). For example, numerous investigators have shown cyclic AMP effects on cell growth and on morphologic changes in cultured tumorigenic and nontumorigenic cell lines (42–46).

To date, no one has reported whether similar types of alterations may be involved in inflamed periodontal tissues. An alteration in the PDE activity in periodontal tissue may affect the normal physiologic responses of this tissue and, therefore, lead to an abnormal or diseased state. In an attempt to test this possibility, we have compared cyclic AMP concentrations and PDE activities in normal human periodontal tissues with those from "inflamed" periodontal tissue.

Gingival tissue used in this study was obtained from patients undergoing surgical periodontal treatment. Samples were taken primarily from the lingual and labial areas of maxillary and mandibular regions. Normal gingival tissue was obtained from patients undergoing corrective surgery not complicated by periodontitis. All tissue specimens used in this study were quick-frozen on dry ice and weighed immediately after surgical removal from the patients. The periodontal index, age, sex and detailed medical history were recorded for each patient at the time of gingival biopsy and specimens were homogenized one minute at full speed with a Polytron tissue homogenizer. To an aliquot of crude homogenate TCA, 10 percent final concentration, was added for cyclic AMP determination. Another aliquot of this crude homogenate was used for phosphodiesterase analyses. The remaining aliquots were then centrifuged 30,000 g for 30 minutes, and the supernatant and pellet was analyzed for cyclic AMP content and phosphodiesterase activities. Cyclic AMP concentrations were assayed with modifications of the protein binding method described by Gilman (47). The assay for cyclic AMP phosphodiesterase is as that described by Russell (48) and Thompson (49). The concentration of 3 H-cyclic AMP in this assay was approximately 1 to 2 pmoles/assay. Phosphodiesterase activity is expressed as pmoles cyclic AMP hydrolyzed/min/mgm protein at 30°C. All measurements were standardized by protein determinations (50).

These results (Table 4) show that the homogenate and the 30,000 xg supernatant of normal tissue specimens contain 46 and 64 pmoles cyclic AMP/mg of protein, respectively, whereas the diseased specimens of protein exhibited 26 pmoles cyclic AMP/mgm of proteins for the homogenate and 47 pmoles of cyclic AMP/mgm proteins for the supernatant. No significant differences were observed for cyclic AMP in the 30,000 g pellets.

Diseased tissues contained approximately two fold more phosphodiesterase activity in the homogenates and supernatants than normal tissues. Phosphodiesterase activities in the 30,000 g pellet of inflamed gingival were higher than normal gingival, 0.63 and 0.46 units, respectively. Furthermore, preliminary evidence was obtained from the kinetic properties of cyclic AMP phosphodiesterases. Ap-

Table 4. Cyclic AMP Concentration and Cyclic AMP
Phosphodiesterase Activity
in Normal and Inflamed
Gingival Biopsies

A. Cyclic AMP: pm camp/mg protein[1]

	Normal Tissue (9)	Inflamed Tissue (7)
Homogenate	45.85 ± 6.17	26.50 ± 5.80
Supernatant	64.12 ± 9.84	45.15 ± 4.49
Pellet	17.32 ± 0.96	21.94 ± 1.67

B. Phosphodiesterase: Units PDE/mg protein[2]

	Normal Tissue (9)	Inflamed Tissue (7)
Homogenate	0.37 ± 0.07	0.73 ± 0.15
Supernatant	0.32 ± 0.12	0.65 ± 0.16
Pellet	0.46 ± 0.05	0.63 ± 0.05

[1] picomoles cyclic AMP/mg Lowry protein
[2] picomoles cyclic AMP hydrolyze/min/mg Lowry protein

parently both inflamed and normal gingival tissues exhibited the presence of two enzymes since the Lineweaver-Burk plots showed two apparent affinity constants on visual inspection. The Michaelis-Menten constants for inflamed gingivae were 110 and 1.7 μM of cyclic AMP for the respective apparent low and high affinity forms of the enzymes. Normal tissue exhibited apparent KM values of 200 and 11.8 μM of cyclic AMP. The half maximal velocities (V 1/2 max) for inflamed and normal gingival were 11.9 and 97.2 pmoles cyclic AMP/mgm protein/min, respectively.

Our experiments demonstrate a metabolic abnormality in the cyclic nucleotide pathway in diseased periodontal tissues. No attempt was made in this study to correlate the varying degrees of severity of inflamed gingiva to either cyclic AMP concentrations or to cyclic AMP phosphodiesterase activities.

Some recent experimental findings obtained in other types of tissues may help in the interpretation of our observations. Voorhees (40) has shown that psoriatic skin lesions are deficient in overall cyclic AMP concentration. Furthermore, he has shown that these cells have potentially abnormal phosphodiesterase and adenylate cyclase (41). It has been shown that hepatoma cells exhibit significantly higher phosphodiesterase activities which may be due to changes in genetic expression (38). Also, some forms of epithelium derived from human mammary tumors have exhibited significantly higher cyclic AMP phosphodiesterase values (51).

The mechanisms that underlie such deviations in cyclic AMP concentrations and cyclic nucleotide phosphodiesterase activity in inflamed periodontal tissue remain unknown. The higher levels of phosphodiesterase activity with subsequent lower overall tissue concentrations of cyclic AMP may be of paramount importance in the development of both hyperplasia and hypertrophy of this disease. The results obtained in this study corroborate those obtained for other tissue types and may be related to the overall rate of cell division. Studies are currently underway to determine whether or not adenylate cyclase activity levels exhibit similar changes to that of phosphodiesterase or if an additional parameter in the second messenger concept may be involved in periodontal disease.

SUMMARY

In this report I have estimated cyclic AMP concentrations in human oral fluids, namely whole saliva and parotid saliva from normal individuals as well as from Down's syndrome patients and from patients undergoing irradiation therapy for parotid salivary gland tumors.

Parotid saliva cyclic AMP levels were significantly higher in stimulated parotid saliva than those found in either unstimulated parotid saliva or whole saliva and is approximately two fold higher than those of blood serum. Saliva cyclic AMP concentrations were found to be approximately three fold higher in Down's syndrome individuals and five fold higher in patients undergoing radiation therapy for salivary gland cancers.

Preliminary data, involving cyclic AMP metabolism in normal and inflamed gingival tissues biopsied from human subjects showed that overall cyclic AMP phosphodiesterase (low Km) activities in diseased human gingival tissues were found to be two fold higher than in normal gingival tissue. Also, the concentration of cyclic AMP was significantly lower in the inflamed gingival tissue.

ACKNOWLEDGMENTS

The author's appreciation is extended to (1) Toshka Alster for expert technical assistance, (2) Dr. Roger Sanger, Dr. Jim Stenger, and Dr. Hugh Kopel of the Pedodontics Departments at Childrens Hospital, Los Angeles, and the USC School of Dentistry for assistance in obtaining saliva samples, (3) Dr. Alan Sproles and Dr. Tom Watson, Pacific State Hospital, for saliva samples from Down's syndrome patients, and (4) Dr. Roger Stambaugh, Department of Periodontics and Biochemistry for supplying gingival biopsy specimens. Support for this research was in part from the California Dental Association Grant, the Interdepartmental Cancer Grant from the American Cancer Society, and the General Research Support Grant provided by the USPHS Division of the National Institute of Dental Research.

REFERENCES

1. Schaeffer, L. D.: Detection and significance of cyclic AMP in human parotid saliva. In: Mechanisms of exocrine secretions, S. S. Han, L. Sreebny, R. Suddick (eds.) pp. 61-76, University of Michigan Press, Ann Arbor, 1974.
2. Schramm, M., and Selinger, Z.: The functions of cyclic AMP and calcium as alternative second messengers in parotid gland and pancreas. J. Cyclic Nucleotide Res. 1:181-192, 1975.
3. Bdolah, A., and Schramm, M.: The function of 3',5'-Cyclic AMP in enzyme secretion. Biochem. Biophys. Res. Commun. 18:452-545, 1965.
4. Badbad, H., Benzi, R., Bdolah, A., and Schramm, M.: The mechanism of enzyme secretion by the Cell. IV. Effects of inducers, substrates, and inhibitors on amylase secretion by rat parotid slices. Europ. J. Biochem. 1:96-101, 1967.
5. Schramm, M.: Amylase secretion in rat parotid slices by apparent activation of endogenous catecholamine. Biochem. Biophys. Acta. 165:546-549, 1968.
6. Monnard, P., and Schorderet, M.: Cyclic Adenosine 3',5'-monophosphate concentration in rabbit parotid slices following stimulation by secretaagogues. Eur. J. Pharmacol. 23:306-310, 1973.
7. Kanamori, T., Hayakawa, T., and Nagatsu, T.: Adenosine 3',5'-monophosphate-dependent protein kinase and amylase secretion from rat parotid gland. Biochem. Biophys. Res. Commun. 57:394-398, 1974.
8. Wells, H.:Functional and pharmacological studies on the regulation of salivary gland growth, In: Secretory mechanisms of salivary glands, L. H. Schneyer and C. A. Schneyer (eds.), pp. 178-207, Academic Press, New York, 1967.
9. Schaeffer, L. D., Sproles, A., and Krakowski, A.: Detection of cyclic AMP in parotid saliva of normal individuals. J. Dent. Res. 52:629, 1973.

10. Sproles, A. C.: Cyclic AMP concentration in saliva of normal children and children with Down's Syndrome. J. Dent. Res. 52:915-917, 1973.
11. Kanamori, T., Nagatsu, T., and Matsumeto, S.: Origin of cyclic adenosine monophosphate in saliva. J. Dent. Res. 54:535-539, 1975.
12. Kanamori, T., Kuzuya, H., and Nagatus, T.: Excretion of cyclic GMP and cyclic AMP into human parotid saliva. J. Dent. Res. 53:760, 1974.
13. Lemon, M. J. C., and Bhoola, K. D.: Excitation-secretion coupling in exocrine gland properties of cyclic AMP phosphodiesterase and adenylate cyclase from the submaxillary gland and pancreas. Biochem. Biophys. Acta 385:101-113, 1975.
14. Lillie, J. H. and Han, S. S.: Secretory protein synthesis in the stimulated rat parotid gland. J. Cell. Biol. 59:708-721, 1973.
15. Ekfors, T., Chang, W. W. L., Bressler, R. S., and Barka, T.: Isoproterenol accelerates the postnatal differentiation of rat submandibular gland. Develop. Biol. 29:38-47, 1972.
16. Herman, G., and Rossignol, B.: Regulation of protein secretion and metabolism in rat salivary glands. Eur. J. Biochem. 55:105-110, 1975.
17. Sproles, A., and Schaeffer, L. D.: An advanced Design of the Carlsen-Crittenden Cup for the Collection of Human Parotid Fluid. Biomat. Med. Dev. Art. Organs. 2:95-100, 1974.
18. Gilman, A. G.: A protein binding assay for adenosine 3',5' cyclic monophosphates. Proc. U.S. Natl. Acad. Sci. 67:305-311, 1970.
19. Rinderknecht, H., Marbach, E. P., Carmack, C. R., Conteas, C., and Geokos, M. C.: Clinical evaluation of an alpha amylase with insoluble starch label with remazolbrilliant blue Camylopectin azure. Clin. Biochem. 4:162-174, 1971.
20. Lowry, O., Rosebrough, N. J., Fan, A. L., and Randall, R. J.: Protein measurement with the Folin phenol reagent. J. Biol. Chem. 193:265, 1951.
21. Schaeffer, L. D., Sproles, A. C., and Karkowski, T.: Detection of cyclic AMP in parotid saliva of normal individuals. J. Dent. Res. 52:629, 1973.
22. Stefanovich, V., and Wslls, H.: Cyclic AMP in human saliva. Fed. Proc. 30:565, 1971.
23. Grower, M. F., Ficara, A. J., Chandler, D. W., and Kramer, G. D.: Differences in AMP levels in the gingival fluid of diabetics and non-diabetics. J. Periodontal, 46:669-672, 1975.
24. Grower, M. F., Lyon, D. R., Levin, M. P., and Chandler, D. W.: Cyclic AMP content of gingival fluid in women taking oral contraceptives. J. Oral Path. 4:291-296, 1975.
25. Cohen, M. M., and Winer, R. A.: Dental and facial characteristics in Down's syndrome. J. Dent. Res. 44:197-200, 1965.
26. Cohen, M. M., Winer, R. A., Schwarta, S., and Shklar, G.: Oral aspects of Mongolism. Oral Surg. 14:92-107, 1961.
27. Moellinger, C. E.: Down's syndrome, a review of the recent literature. J. Missouri Dent. Assoc. 46:8-13, 1966.
28. Wolf, W. C.: Caries incidence in Down's syndrome (Mongolism). J. Wisc. State Dent. Soc. 43:37, 1947.
29. Winer, R. A., and Cohen, M. M.: Dental caries in Mongolism. Dent. Prog. 2:217-219, 1962.

30. Brown, R. J., and Cunningham, W. M.: Some dental manifestations of mongolism. Oral Surg. 14:664-676, 1961.
31. Winer, R. A., and Chauncey, H. H.: Parotid saliva enzymes in Down's syndrome. J. Dent. Res. 54:62-64, 1975.
32. Winer, R. A., and Feller, R. P.: Composition of parotid and submandibular saliva and serum in Down's syndrome. J. Dent. Res. 51:449-454, 1972.
33. Blotzis, G. G., and Robinson, J. E.: Oral tissue changes caused by radiation therapy and their management. Dent. Clin. N. Amer. 643-656, 1968.
34. English, J. A.: Morphological effects of irradiation of the salivary glands in rats. J. Dent. Res. 34:4, 1955.
35. Phillips, R. M.: X-ray induced changes in function and structure of the rat parotid gland. J. Oral Surg. 28:422-427, 1970.
36. Van den Brank, H. A.: Radiation damage to the salivary glands. Brit. J. Radiol. 42:700, 1969.
37. Brown, L. R., Dreizen, S., Handler, S., and Johnston, D. A.: Effect of radiation-induced xerostomia on human oral microflora. J. Dent. Res. 54: 740-750, 1975.
38. Clark, J.: Cyclic adenosine 3',5'-monophosphate phosphodiesterase activity in normal, differentiating, regenerating, and neoplastic liver. Cancer Res. 33:356-361, 1973.
39. Thompson, W. J., Little, S. A., and William, R. H.: Effect of insulin and growth hormone on rat liver cyclic nucleotide phosphodiesterase. Biochemistry 12:1889-1894, 1973.
40. Voorhees, J. J., and Duell, E. A.: Psoriasis as a possible defect of the adenylate-cyclic AMP cascade. A defective chalone mechanism. Arch. Dermatol. 104:352-358, 1971.
41. Voorhees, J. J., Duell, E. A., Stawiski, M., and Harrell, E. R.: Cyclic nucleotide metabolism in normal and proliferating epidermis, In Adv. Cyclic Nucleotide Metabolism, Vol. 4, (Eds.) P. Greengard and G. A. Robison, Raven Press, New York, pp. 117-162, 1974.
42. Smith, E. E., and Handler, A. H.: Apparent suppression of the tumorigenicity of human cancer cells by cyclic AMP. Res. Comm. Chem. Path. Pharmacol. 5:863-866, 1973.
43. Johnson, G. S., Friedman, R. M., and Pastan, I.: Restoration of several morphological characteristics of normal fibroblasts in sarcoma cells treated with adenosine 3',5'-cyclic monophosphate and its derivatives. Proc. U.S. Natl. Acad. Sci. 68:425-429, 1971.
44. Johnson, G. S., and Pastan, I.: Role of 3',5'-adenosine monophosphate in regulation of morphology and growth of transformed and normal fibroblasts. J. Natl. Cancer Inst. 38:1377-1387, 1972.
45. Sheppard, J. R.: Restoration of contact inhibited growth to transformed cells by dibutyryl adenosine 3',5'-cyclic monophosphate. Proc. U.S. Natl. Acad. Sci. 68:1316-1320, 1971.
45. Sheppard, J. R.: Restoration of contact inhibited growth to transformed cleotide phosphodiesterases in neoplastic human mammary tissues. Can. Res. 36:60-66, 1976.

11

Cyclic Nucleotides in Hemostasis and Thrombosis

MICHAEL L. STEER
EDWIN W. SALZMAN

The arrest of bleeding in man is achieved through interaction of blood platelets, soluble plasma protein clotting factors, and the blood vessel wall. When a blood vessel is divided, it constricts, narrowing the effective bleeding lumen. Platelets in blood issuing from the wound adhere to exposed subendothelial connective tissue, and an aggregate of platelets forms, effectively closing off the bleeding orifice. A series of enzymatic reactions between protein clotting factors interacting in the plasma in the vicinity of the platelet plug lead to formation of a fibrin capsule that reinforces the mass of platelets and provides it with rigidity and strength required to resist arteriolar pressure, when the constricted vessel later relaxes. The process of the disease thrombosis is similar to that of hemostasis, except that the initiating event is apt to be exposure of the blood to subendothelial components of an ulcerated atherosclerotic plaque or an artificial surface or entrance into the bloodstream of activated coagulation factors derived from damaged tissues.

The discussion that follows will confine itself to the role of cyclic nucleotides in the function of blood platelets. There is no evidence linking these compounds to interactions of the plasma clotting factors. Their participation in the function of vascular smooth muscle is considered elsewhere in this volume.

MECHANISM OF PLATELET FUNCTION

Blood platelets are disc-shaped cytoplasmic fragments derived from fragmentation of a precursor bone marrow cell, the megakaryocyte. They circulate in the blood for 8 to 10 days and are then removed, chiefly by the spleen. They measure 2 to 3 μm in diameter and about 0.5 μm in thickness. Normal whole blood contains 150 to 250,000 platelets/mm^3. The platelet contains no nucleus, no DNA, and little RNA or rough endoplasmic reticulum, and it therefore has little ability to synthesize proteins. The platelet contains mitochondria and gains energy from either aerobic or anaerobic glycolysis. Its discoid shape appears to be maintained by a marginal bundle of microtubules, which lie along its greatest circumference beneath the plasma membrane. Other intracellular inclusions include several varieties of storage granules, including glycogen granules, so-called α-granules containing acid hydrolases, and dense osmophilic granules containing biologically active substances such as serotonin (5-HT), epinephrine, and adenine nucleotides, as well as calcium and potassium. The origin of the ATP and ADP within these granules has not been established, but these compounds appear to be sequestered from the metabolic adenine nucleotide pool and are not labelled by incubating platelets with radioactive adenine or adenosine for short periods. In contrast, the 5-HT and catecholamines within the dense granules are taken up by the platelet from extracellular fluid and concentrated in these granules.

Platelets do not adhere to normal endothelial cells, but at the cut end of a blood vessel or at areas of trauma where endothelial denudation exists they will adhere to subendothelial connective tissue, especially to collagen fibrils. They spread out on the underlying surface and undergo internal rearrangements, including centralization of intracellular granules, whose membranes apparently fuse with the walls of an intracellular canalicular system communicating with the external plasma medium. The contents of the granules are discharged into the plasma. This secretory process, known as the "release reaction," is energy dependent and utilizes the metabolic pool of ATP for energy. There is evidence that the release reaction is triggered by an increase in cytoplasmic Ca^{2+}, which may come from outside the platelet (via increased Ca^{2+} influx) or from Ca^{2+} sequestered in intracellular sites (see below). Release may be triggered by a variety of stimuli besides adhesion of platelets to collagen fibrils and other surfaces, including thrombin produced by simultaneous activation of plasma coagulation. Substances secreted by the platelet include serotonin, catecholamines, adenine nucleotides, calcium, potassium, fibrinogen, acid hydrolases, and a cationic protein with antiheparin activity ("platelet factor 4"). Among the products released particular note is due adenosine diphosphate (ADP), which has the capacity to induce platelet aggregation. ADP is thought to be responsible for subsequent development of the mass of aggregated platelets that composes the bulk of a hemostatic plug or thrombus.

Many of these events can be duplicated in vitro in model systems. For example, addition of ADP to a suspension of platelets in citrated plasma (platelet-rich plasma, PRP) results in a prompt shape change from disc to spiculated sphere (possibly analogous to spreading on a surface) and to formation of loose aggregates which may subsequently disperse. At higher concentrations, ADP causes release of platelet constituents and mobilization of additional platelets into large irreversibly joined clumps. Epinephrine and thrombin have similar effects in vitro. The platelet release reaction appears to play a pivotal role in the process of platelet aggregation in vitro by linking the first or reversible phase of aggregation to the second or irreversible phase of aggregation. Other agents, such as collagen fibrils and arachidonic acid do not stimulate a reversible phase of platelet clumping but directly induce release of platelet constituents and platelet aggregation. Although it has been suggested that the release reaction induced by ADP itself is an in vitro artifact requiring unphysiologically low concentrations of divalent cations (1), it is generally believed that irreversible aggregation following the release reaction induced by collagen, thrombin, or other substances is triggered by substances secreted by the platelet, particularly ADP. The process of secretion itself appears to be intimately related to prostaglandin metabolism (see below). The details of these phenomena have been extensively reviewed (2–6).

OVERVIEW OF CYCLIC NUCLEOTIDES AND RELATED PATHWAYS IN PLATELETS

Inhibition of platelet aggregation in vitro by adenosine 3',5'-monophosphate (cyclic AMP) was reported in 1965 (7), and the first demonstration of changes in endogenous cyclic AMP with platelet activity followed four years later (8–11). It was subsequently shown that many pharmacological agents known to affect platelet function influence the activity of platelet adenylate cyclase or the several platelet phosphodiesterases, and a large literature has developed. It is now accepted that an increase in platelet cyclic AMP content inhibits the release reaction and platelet aggregation. There is evidence that a reduction in platelet cyclic AMP accompanies aggregation and release, but the data are conflicting, and the significance of the relationship is not settled. The more recent demonstration of a rise in platelet guanosine 3',5'-monophosphate (cyclic GMP) associated with activation of platelets has provided additional material for consideration. Detailed references to these observations are given below.

The interrelationships of cyclic nucleotide metabolism and the prostaglandin system have been investigated. PGE_1 was found in 1966 to be a potent inhibitor of platelet activity (12) and was subsequently discovered to stimulate platelet adenylate cyclase. PGE_2 has similar actions at high concentrations, but at lower concentrations, similar to its endogenous levels, the compound has the reverse effect. Platelet production of endogenous PGE_2 and $PGF_{2\alpha}$ in response to thrombin has been described, and inhibition of these reactions by aspirin was subsequently reported. It was found that aspirin blocked the fall in basal levels of cyclic AMP and inhibited the rise of cyclic GMP otherwise seen in response to platelet aggregating agents, although there is controversy concerning the latter point.

Since aspirin is known to produce a bleeding disorder characterized by defective platelet clumping and inhibition of the platelet release reaction, an association of prostaglandin metabolism with cyclic nucleotides in platelet function seems very likely. More recent experiments have suggested the following pathways: Thrombin, ADP, collagen fibrils and probably other agents activate phospholipase A_2 at the platelet surface, releasing arachidonic acid from platelet membrane phospholipids and thereby setting off a series of reactions that ultimately lead to formation of biologically active prostaglandin endoperoxides and thromboxanes. Formation of the endoperoxides is blocked by aspirin. The possible relationships of endoperoxides and thromboxanes to cyclic nucleotide metabolism are at present the subject of active investigation. The details of these reactions are considered below.

AGENTS THAT AFFECT PLATELET AGGREGATION

Catecholamines

Catecholamines are present in platelet storage granules and are secreted in the course of the release reaction. Their importance in hemostasis and thrombosis is by no means clear but their effects on platelet function in vitro are striking, and their mechanism of action has been extensively studied. The ability of catecholamines to affect human platelets has been well documented (13–17).

Platelets respond to both α and β-adrenergic stimulation. Early observations showed that α-adrenergic stimulation leads to platelet aggregation and β-adrenergic stimulation to disaggregation or inhibition of aggregation, and it has been suggested that these responses are mediated by the adenylate cyclase system. The evidence to be summarized below strongly suggests that the β-adrenergic effect is mediated through stimulation of adenylate cyclase and elevated platelet levels of cyclic AMP. On the other hand, although most subsequent investigations have focused on the α-adrenergic response, because the principal physiological effect of catecholamines on platelets appears to be promotion of aggregation, the mechanism for this effect of α-stimulation remains unclear. Although α-stimulation may decrease platelet cyclic AMP levels, other factors not totally elucidated probably also play a role in α-stimulated platelet aggregation.

Platelets contain measurable levels of cyclic AMP. Many investigators have attempted to measure the change in platelet cyclic AMP that accompanies adrenergic stimulation. In general, these studies have been performed in one of three ways: (1) the platelet is prelabelled with radioactive adenine, and formation of labelled cyclic AMP in the intact platelets is measured; (2) the platelet is prelabelled with radioactive adenosine, and formation of labelled cyclic AMP in the intact platelet is similarly measured; (3) a change in total platelet cyclic AMP following stimulation is observed by assaying total cyclic AMP content in platelet-rich plasma or in platelets separated from plasma by centrifugation or other means. Unfortunately, these three approaches do not yield the same results. For example, when radioactive adenine has been used to label the nucleotide pool, no decrease in basal cyclic AMP is noted after α-adrenergic stimulation, although α-stimulation is noted to decrease the accelerated rate of cyclic AMP formation found in the presence of PGE_1 (18–26). In one notable exception (23, 28), Haslam noted a fall in the rate of cyclic AMP formation produced by epinephrine in the presence of a β-blocker but, in the absence of specific blockers, the overall effect of epinephrine was to increase the rate of labelled cyclic AMP formation from adenine-labelled ATP. When platelet ATP is prelabelled by incubation with labelled adenosine, however, α-stimulation has uniformly been noted to decrease basal rates of cyclic AMP formation (22, 29–31) as well as to decrease the accelerated rate of cyclic AMP formation induced by PGE_1 (22, 29, 30). When total cyclic AMP content of PRP is measured, most studies show that α-adrenergic stimulation decreases the degree of elevation noted with PGE_1 (22, 32–34), but there is disagreement concerning the effect of α-stimulation on basal cyclic AMP formation rates. Thus, α-stimulated reduction of the basal level of cyclic AMP has been noted (22, 31, 34), whereas others found no change in the basal levels of cyclic AMP. The results that show differing changes in cyclic AMP using adenosine as opposed to adenine to label the platelet suggest that the ATP pool is not homogenous and that the cyclic AMP which is affected by α-adrenergic agents may be derived primarily from the ATP formed from adenosine and not adenine. Alternatively, since adenylate cyclase is stimulated by adenosine (see below), it is possible that labelled cyclic AMP detected after labelling with adenosine is not truly "basal."

Adenylate cyclase activity has also been measured directly in disrupted platelets. In general, β-stimulation results in increased adenylate cyclase activity (13), while α-stimulation has been noted to inhibit basal adenylate cyclase activity and to decrease the degree of activation by PGE_1 (21, 30, 35, 36). It has been claimed that α-adrenergic inhibition of basal adenylate cyclase activity is seen only in the absence of a phosphodiesterase inhibitor (30), which might suggest that the α-adrenergic "inhibition of basal adenylate cyclase" may actually reflect an α-adrenergic stimulation of phosphodiesterase. This suggestion is supported by other studies (37) which show that epinephrine shifts the equilibrium distribution of phosphodiesterases from the high to the low K_m form—an effect that would result in an increased conversion rate of cyclic AMP into 5' AMP at low cyclic AMP concentrations. However, other studies (35, 36) have shown that the epinephrine induced decrease in cyclic AMP and adenylate cyclase activity can occur in the presence of phosphodiesterase inhibitors, and the epinephrine effect on phosphodiesterase is not blocked by α-adrenergic inhibitors (38). Further studies are, therefore, needed to clarify the direct effect, if any, of α-adrenergic agents on platelet adenylate cyclase.

Other circumstantial evidence suggests that α-adrenergic stimulated platelet aggregation may result from or be related to a decrease in platelet cyclic AMP. For example, epinephrine induced aggregation is inhibited by agents that elevate cyclic AMP levels, such as PGE_1, dibutyryl cyclic AMP, and phosphodiesterase inhibitors (22, 35, 39, 40). Epinephrine itself counteracts the inhibition of aggregation induced by these agents (41). However, epinephrine can induce aggregation even when cyclic AMP levels remain above basal values if platelets are preincubated with low concentrations of PGE_1 or other inhibitors of aggregation (19, 23, 26). Thus, it is clear that aggregation can occur without a decrease in the total platelet content of cyclic AMP, although it is possible that alterations in specific intracellular compartments of the nucleotide could still be important. Epinephrine may have other, cyclic AMP-independent effects that promote aggregation. Epinephrine has been reported to increase platelet cyclic GMP content, and this effect is said to precede platelet aggregation (42, 43). Epinephrine induced aggregation is enhanced by the addition of dibutyryl cyclic GMP or cyclic GMP in low concentrations, although higher concentrations are inhibitory (44).

The relationship between α-adrenergic stimulation and prostaglandin biosynthesis has yet to be elucidated. Epinephrine stimulation results in the formation of PGE_2 and $PGF_{2\alpha}$ coincident with release and the second or irreversible phase of epinephrine induced aggregation (45), implying that epinephrine stimulation also leads to endoperoxide formation, and more direct evidence for this effect has been reported by Smith et al. (46). This presumption is supported by the finding that aspirin prevents the second phase of epinephrine induced aggregation (135). The aspirin effect can be overcome by preincubating platelets with arachidonic acid (47), and low concentrations of arachidonic acid enhance epinephrine induced aggregation (48). The converse, that is, epinephrine enhancement of arachidonic acid induced aggregation, is also noted (49).

Although the platelet is known to take up and store catecholamines, uptake of catecholamines is not required for α-adrenergic stimulation of aggregation (50).

ADP

The platelet release reaction and aggregation can be induced by the addition to PRP of adenosine 5'-diphosphate (ADP) or many other stimuli. During the release reaction, ADP, serotonin, and other substances are secreted by the platelet, regardless of whether the release reaction was induced by ADP or by other aggregating agents such as epinephrine or collagen. It has been suggested that the endogenous ADP which is released acts as a common mediator of the subsequent wave of aggregation initiated by other stimuli (51). This suggestion is supported by the observation that ADP traps such as the combination of phosphoenolpyruvate and pyruvate kinase, which converts ADP to ATP, and other enzyme systems that remove ADP can inhibit aggregation induced by epinephrine, thrombin, collagen and serotonin (52).

The mechanism(s) by which ADP induces release and aggregation remain incompletely understood. There is much circumstantial evidence to suggest that the ADP effect may involve alterations in platelet cyclic nucleotide levels. Exogenous cyclic AMP or dibutyryl cyclic AMP inhibits ADP-induced aggregation (10, 35). In addition, agents such as PGE_1 and phosphodiesterase inhibitors which elevate platelet cyclic AMP levels also inhibit ADP induced aggregation (8, 10, 53–58). The dual effects of increasing cyclic AMP and inhibition of aggregation with these agents parallel each other (8, 53), and the order of potency of phosphodiesterase inhibitors as inhibitors of aggregation and inhibitors of phosphodiesterase is similar (53). Besides PGE_1, other prostaglandins also inhibit ADP-induced aggregation, and the order of potency for inhibition of aggregation and elevation of cyclic AMP is the same (10). Adenosine is a potent inhibitor of the actions of ADP on platelet function, an effect that was at one time interpreted as the result of competition of ADP and adenosine for common "receptors" (59, 60). It is now known that adenosine stimulates platelet adenylate cyclase (61) and elevates platelet cyclic AMP levels (62). Further evidence that cyclic AMP plays a role in ADP-induced aggregation comes from studies describing the effect of ADP itself on platelet cyclic AMP levels. Some studies have shown that ADP added to PRP causes a decrease in total platelet cyclic AMP (basal level) (11, 31, 34) although other reports claim that ADP has no effect on total basal platelet levels of cyclic AMP (32, 33, 63). There is general agreement, however, that ADP decreases the elevation in total platelet cyclic AMP induced by PGE_1 (32, 33, 63). In addition, several studies have reported the effect of ADP on the rate of cyclic AMP formation in intact platelets. In these studies, the platelet was prelabelled with either radioactive adenine or adenosine. If adenine was used to label the ATP pool, ADP was found not to decrease the basal level of cyclic AMP (19, 23, 26) but was found to decrease the degree of elevation of cyclic AMP with PGE_1 (19, 20, 23, 26). Using adenosine to prelabel the ATP pool, all studies have shown that ADP decreases the basal as well as the PGE_1 stimulated rate of cyclic AMP formation (8, 22, 25, 31). These findings suggest, as noted in reference to the effect of epinephrine, that adenosine and adenine may not be incorporated into the same ATP pool and that ADP inhibits basal adenylate cyclase conversion of ATP to cyclic AMP only for that portion of ATP which originates from adenosine but not from adenine. This effect of decreasing basal cyclic AMP levels is blocked by aspirin (64). In addition to its effects on platelet cyclic AMP, ADP has been noted to elevate platelet levels of cyclic GMP (31, 42) suggesting that this nucleotide may also contribute to ADP-induced aggregation.

Aggregation induced by ADP is associated with a rise in platelet protein bound phosphate. Cyclic AMP itself does not increase the amount of protein bound phosphate but accelerates its turnover. Because the phosphate in each case comes from ATP, it has been suggested that the opposing effects of ADP and cyclic AMP could be due to competition between these two agents for the same pool of ATP (65, 66).

The mechanism by which ADP alters platelet cyclic AMP levels is, at present unknown. No direct effect of ADP on adenylate cyclase or phosphodiesterase in broken cell preparations has been reported. The inhibition of basal cyclic AMP formation in intact platelets is blocked by aspirin (64), but the inhibition of PGE_1-stimulated cyclic AMP formation by ADP is not blocked by aspirin (34; Mills, D. C. B., personal communication). It is known, however, that ADP-induced aggregation can proceed without ADP uptake by platelets (67, 68) suggesting that ADP acts at some point on the surface membrane. Both ADP and epinephrine have been found to activate platelet phospholipase, which liberates arachidonic acid and leads to the formation of endoperoxides, thromboxane, and prostaglandins (69, 70). Low concentrations of arachidonic acid enhance ADP induced aggregation (48). It would appear, therefore, that ADP can affect cyclic AMP formation in two ways: (1), via prostaglandins and endoperoxides (the decrease in basal cyclic AMP which is blocked by aspirin) and (2), direct inhibition of cyclic AMP formation (if cells are pre-activated with PGE_1).

Thrombin

A proteolytic enzyme formed from prothrombin during coagulation of plasma, thrombin acts on fibrinogen to begin its conversion into fibrin. Thrombin is also a powerful stimulant of the platelet release reaction and platelet aggregation at concentrations too low to cause blood clotting. These effects are inhibited by hirudin, a specific antithrombin derived from the leech, and by heparin, which activates a natural plasma inhibitor of thrombin. It would appear that the thrombin-induced platelet effects results from hydrolysis of a surface membrane protein. Trypsin and papain, which, like thrombin, are proteolytic enzymes specific for arginine residues of polypeptides can also induce platelet release and aggregation (71). The importance of thrombin-induced proteolysis is supported by the observation that active-site–inhibited thrombin is unable to induce release or aggregation and by the finding that competitive inhibitors of thrombin induced release and aggregation also act as competitive inhibitors of hydrolysis of synthetic substrates by thrombin (71). Incubation of platelets with thrombin leads to release of a specific platelet membrane protein (thrombin sensitive protein, TSP) (72, 73). In addition, and with a very similar time course, thrombin inhibits adenylate cyclase in intact platelets (36, 72–74) and in broken cell preparations (35, 36). These events appear to precede the thrombin induced release reaction (72). All three of these thrombin effects (i.e., TSP release, adenylate cyclase inhibition, and platelet release and aggregation) can be inhibited by dibutyryl cyclic AMP or agents such as PGE_1 or phosphodiesterase inhibitors which elevate platelet cyclic AMP levels (35, 72, 75–77). Thrombin itself decreases platelet cyclic AMP (22, 25). It should be mentioned that this observation has been challenged by the finding that, in the presence of EDTA, thrombin increases cyclic AMP levels in platelets (78). This latter observation may be related to the presence of EDTA. Thrombin also stimulates phosphorylation of a specific plasma membrane protein, which occurs simultaneously with thrombin induced release reaction (79). Phosphorylation of this protein in intact platelets is inhibited by agents that elevate platelet cyclic AMP levels (79). This protein is, therefore, to be distinguished from substrates for cyclic AMP-dependent protein kinases, several of which have been described in platelets (80–83). There is evidence that several such enzymes exist in the platelet membrane and in the cell sap where they differ from one another in their kinetics and in their sensitivity to calcium inhibition. Their natural substrates have not been identified.

Although release of TSP and phosphorylation of membrane protein subsequent to thrombin exposure of platelets are well documented, there are no data that explain how, if at all, these events are linked to thrombin induced release and aggregation. One might hypothesize that the phosphorylated protein in some way regulates calcium fluxes, such that after phosphorylation, calcium influx might be accelerated and release and aggregation triggered. Similarly, one might hypothesize that the TSP represents some essential component of adenylate cyclase. In its absence, adenylate cyclase would be inhibited and cyclic AMP levels fall. Alternatively, TSP release could reflect other membrane events such as phospholipase activation and conversion of arachidonic acid into active endoperoxides. This possibility is supported by the finding that thrombin stimulation leads to the formation of PGE_2 and $PGF_{2\alpha}$ (84), especially in the presence of arachidonic acid (85), and that aspirin and indomethacin inhibit PGE_2 formation after thrombin addition (86). However, aspirin does not prevent the thrombin induced release reaction (87), at least at high thrombin concentrations, making this latter hypothesis unlikely to be the entire explanation. It is also conceivable that membrane phosphorylation secondary to thrombin results from stimulation of guanylate cyclase and elevation of platelet cyclic GMP (43). There are of course not even speculations to explain how this nucleotide might mediate release and aggregation. No cyclic GMP-dependent protein kinase has thus far been identified in platelets. Further efforts are needed to clarify the mechanisms by which thrombin induces the platelet release reaction and aggregation.

Collagen

When platelets are exposed to collagen, either at the site of blood vessel injury or in vitro, the release reaction is initiated and platelets aggregate. It has been generally believed that collagen must be in a particulate form and possess the native triple helical conformation in order to stimulate aggregation. Recently, however, platelet aggregation and the release reaction have been shown to be stimulated by the denatured α (I) chain and by cyanogen bromide peptides derived from collagen (31). These results question the requirement for particulate native collagen. Little is understood concerning the steps that couple platelet-collagen interactions with the aggregation that follows. According to a prominent although as yet unproven hypothesis, collagen is bound to the active site of the membrane enzyme, collagen:glucosyl transferase, which normally transfers glucose from UDP-glucose to the incomplete prosthetic groups of collagen (88). Evidence for such a phenomenon includes the observation that platelet aggregation with collagen can be blocked by agents that block the active site of this enzyme or that complete the reaction and destroy the enzyme-substrate (platelet-collagen) complex (88). There are many characteristics of collagen induced aggregation that resemble those described for other aggregating agents. These include observations which suggest that a decreased platelet level of cyclic AMP may, at least in part, contribute to aggregation. Thus, collagen decreases total platelet cyclic AMP (22 disputed by 33) and decreases the rate of cyclic AMP formed from ATP prelabelled with radioactive adenosine (but not adenine) (22, 25, 31). Collagen added to broken platelet preparations inhibits adenylate cyclase (35). Cyclic AMP and, to a greater extent, dibutyryl cyclic AMP inhibit collagen induced aggregation (35). Preincubation with PGE_1, which elevates platelet cyclic AMP, inhibits collagen stimulated platelet release of ATP, ADP and serotonin (58), and collagen decreases total platelet cyclic AMP levels after they have been elevated by PGE_1 (33).

A role for cyclic GMP in mediating collagen-induced platelet processes has also been proposed. Collagen and the active collagen peptides elevate platelet cyclic GMP, and this elevation precedes aggregation (31, 42, 89, 90). Collagen activates guanylate cyclase in intact platelets (31, 42, 90). In addition, collagen or collagen peptide-induced aggregation is enhanced by cyclic GMP or dibutyryl cyclic GMP (44). However, these is also evidence that suggests that collagen-induced aggregation is not mediated by cyclic GMP. This includes the finding that, although collagen-induced release and aggregation can be blocked by aspirin, the elevation in cyclic GMP levels after collagen stimulation is not blocked by aspirin (90). Unlike the inhibition of platelet aggregation and release by dibutyryl cyclic AMP which is enhanced by a period of preincubation during which platelet uptake of the cyclic nucleotide occurs (35), promotion of platelet activities by cyclic GMP or dibutyryl cyclic GMP is not increased by preincubation (44).

It would seen likely that the collagen-induced effects are, in some way, linked to alterations in prostaglandin metabolism. This may involve activation of phospholipase and liberation of arachidonic acid (40). Collagen stimulated formation of PGE_2, which is prevented by aspirin, has been reported (91). Collagen-induced aggregation in enhanced by arachidonic acid (48). One presumes that the more active endoperoxides and thromboxanes are also generated and that these may play a role in collagen induced aggregation. Formation of endoperoxides has been reported to result from addition of arachidonic acid to a suspension of washed platelets (46, 92, 93). It should be noted that an increase in platelet cyclic GMP is induced by arachidonic acid, but this is said to follow aggregation rather than to precede it (94), in contrast to the sequence reported with collagen (31, 90) in which the cyclic GMP rise precedes aggregation. These discordant data confuse the suggested explanation whereby collagen-induced liberation of arachidonic acid and generation of endoperoxides leads, through a rise in cyclic GMP, to platelet aggregation.

Prostaglandins

The prostaglandins are a group of 20 carbon fatty acids structurally related to prostanoic acid and chemically distinguished by substituent groups that differ from this compound. Much information has recently appeared to define the major metabolic pathways for prostaglandins.

Unstimulated platelets contain little or no pre-formed prostaglandins or, for that matter, even the free fatty acid precursors of prostaglandins (84). One of the first steps in prostaglandin biosynthesis appears to be mobilization of membrane phospholipids and hydrolysis of these esters to yield free fatty acids. This step may involve activation of phospholipase A, which has been shown to be an action of thrombin, epinephrine, and other aggregating agents (66, 67). The released fatty acids include 8,11,14 eicosatrienoic acid, which contains three double bonds and is the precursor of the "1" series of prostaglandins, (i.e., those containing one double bond) and 5,8,11,14 eicosatetranoic acid (arachidonic acid) which contains four double bonds and is the precursor of the "2" series of prostaglandins (i.e., those containing two double bonds) (95). In platelet phospholipid hydrolysates, the latter outweighs the former by a factor of about 30:1, which probably accounts for the fact that little or none of the "1 series" of prostaglandins occur naturally in platelets, as contrasted to the well documented levels of the "2 series" detectable after platelet stimulation (96). Arachidonic acid is converted, by the action of a cyclo-oxygenase, into cyclic endoperoxides which are potent stimulants of the platelet release reaction and aggregation. These endoperoxides (PGG_2 and PGH_2; also known as LASS or labile aggregation stimulatory substance) are short lived and rapidly converted into a series of other compounds. Among these compounds are the prostaglandins PGE_2 and $PGF_{2\alpha}$, which are not stimulants of platelet aggregation, but which augment or inhibit aggregation induced by other stimuli, depending on their concentration. Recently, it has been reported that the endoperoxides are converted into a compound called thromboxane A_2, which appears to be identical to the previously identified but not isolated "rabbit aorta constricting substance". Thromboxane A_2 is a powerful stimulant of the platelet release reaction and aggregation. It has been suggested that the physiological effects attributed to prostaglandins and endoperoxides are actually mediated by the transient appearance of thromboxane A_2 (96a). Thromboxane A_2 is rapidly inactivated by conversion to thromboxane B_2, a stable and biologically inert product. That conversion to a thromboxane may not be an absolute requirement for biological activity of endoperoxides is suggested by the report that a synthetic azo-derivative of PGH_2, incapable of such transformation, has potent platelet aggregating ability (97).

The enzymatic steps involved in conversion of PGG_2 and PGH_2 to thromboxane A_2 and conversion of thromboxane A_2 into thromboxane B_2 have not been studied in detail, and no information describing these processes is available. It would appear that these reactions are rapid, since the products appear only transiently. Therefore, one might suspect a great potential for regulatory mechanisms to modulate platelet aggregation by affecting these steps.

More information is available to characterize the conversion of arachidonic acid into the endoperoxides PGG_2 and PGH_2. This step is catalyzed by a fatty acid cyclo-oxygenase and may be accelerated by trace levels of PGG_2 and PGH_2 (98). Much of our understanding of this process results from the fact that many nonsteroidal anti-inflammatory agents (aspirin, phenylbutazone, indomethacin) can inhibit the cyclo-oxygenase by covalently modifying it. For aspirin, this appears to involve acetylation of the enzyme (99). Aspirin also inhibits the decrease in platelet cyclic AMP induced by ADP, collagen, epinephrine, or thrombin (64).

Administration of arachidonic acid intravenously to experimental animals results in sudden death. Platelet aggregates are found in the lungs (100). Addition of arachidonic acid to platelet suspensions induces the platelet release reaction and aggregation, which can be blocked by aspirin. There is formation and release of PGE_2 and $PGF_{2\alpha}$ from the platelets (48, 91). Preincubation of platelets with arachidonic acid but not with 8,11,14-eicosatrienoic acid inhibits the aspirin induced block of epinephrine stimulated platelet release reaction and aggregation (47). The aspirin effect cannot be overcome by addition of PGE_2 or $PGF_{2\alpha}$ (40). The prostaglandins PGE_2 and $PGF_{2\alpha}$ are also formed during epinephrine and collagen induced aggregation (45).

It is currently believed that the prostaglandins, endoperoxides, and thromboxanes are primarily involved as mediators of the platelet release reaction and, therefore, only indirectly contribute to the aggregation which follows. Recent evidence suggests that prostaglandin endoperoxides, at low concentrations directly induce platelet clumping without release (Salzman, unpublished). Thus, the thesis that endoperoxides act as mediators of platelet activation by thrombin and other agents by direct initiation of release and thus of irreversible aggregation may be an over simplification.

Our knowledge of the effects of the "1 series" of prostaglandins is based primarily on data dealing with exogenous PGE_1. Since platelet phospholipid hydrolysates contain very little 8,11,14-eicosatrienoic acid, it is not surprising that little or no PGE_1 is normally formed in the platelet membrane. The pharmacological effects of PGE_1 on platelet function are profound, however, and may help elucidate some of the physiologically important biochemical events associated with aggregation. The effect of PGE_1, which is now well documented, is to inhibit aggregation induced by ADP, epinephrine, thrombin and many other agents, and this inhibition of aggregation is associated with PGE_1 stimulation of platelet adenylate cyclase and an increased level of platelet cyclic AMP (9, 10, 23, 34–36, 39, 53, 54, 56, 69, 72, 74, 101). These parallel effects have suggested that cyclic AMP acts to inhibit aggregation and this suggestion is supported by the finding that PGE_1-induced inhibition of aggregation is enhanced by other agents (such as phosphodiesterase inhibitors or exogenous dibutyryl cyclic AMP) that also elevate platelet cyclic AMP levels (53).

The effects of PGE_2 on platelet cyclic AMP are more complicated than those of PGE_1. PGE_2 enhances adenylate cyclase activity, elevates platelet cyclic AMP, and inhibits platelet aggregation and release at concentrations greater than $10^{-6}M$, although its effect is not so great as that of PGE_1 (22, 58, 65, 101). At lower concentrations PGE_2 has been observed to have effects on platelet function opposite to those of PGE_1, including augmentation of aggregation (54, 58, 101, 103), inhibition of adenylate cyclase, and reduction of cyclic AMP levels (22, 54, 65, 101). The last of these points has been disputed (58, 104), but the concentration of ADP and collagen employed in one of the latter reports (104) is so high that it does not appear that enhancement of its effect would have been possible. However, in view of this controversy, attribution of the action of PGE_2 to its effect on cyclic AMP levels must be regarded as tentative.

PGD_2, which elevates the level of platelet cyclic AMP by stimulating adenylate cyclase (105; Salzman, E. W., unpublished), inhibits platelet aggregation with over twice the potency of PGE_1 (106, 107). The drug is of particular interest as a possible antithrombotic agent, since it is very stable and survives oral ingestion, absorption from the gut, and circulation through the lungs (107).

Calcium

External Ca^{2+} appears to be required for platelet aggregation and release (8, 51, 108–110) although the change in shape that precedes aggregation and at least part of the release reaction can occur in the absence of external calcium, conceivably by mobilizing internal Ca^{2+} (108, 110, 111). Further evidence that internal Ca^{2+} can contribute to release is the observation that, in the absence of external Ca^{2+}, the divalent cation ionophore A23187 can induce the release reaction (112–114). These findings have implicated Ca^{2+} in the mediation of the platelet release reaction and suggested that this divalent cation may function, in the platelet, much as it does in other cells to couple a stimulus to the secretory response (115). Inhibition of divalent cation ionophore induced release and aggregation by PGE_1, and dibutyryl cyclic AMP have been reported (114, 116). Freedman and Detwiler (117), using higher ionophore concentrations, however, did not observe this effect. Aspirin and other inhibitors of prostaglandins synthesis fail to inhibit ionophore induced platelet activity (114, 116) suggesting that ionophore, and presumably calcium, related effects are not mediated by prostaglandins or related compounds.

The intracellular events which relate to Ca^{2+} mediated stimulus-secretion coupling have not been documented. It has been suggested that Ca^{2+} and cyclic AMP affect this process in opposing ways, and several models have been proposed to explain this phenomenon (116–118). Booyse and associates have proposed an attractive model that involves protein phosphorylation and dephosphorylation (118). They have suggested that cyclic AMP, through activation of a cyclic AMP dependent protein kinase, promotes phosphorylation of a membrane bound protein which, in its phosphorylated state, binds Ca^{2+}. In addition, they suggest that the phosphorylated Ca^{2+}-binding protein augments Ca^{2+} efflux and/or inhibits Ca^{2+} influx, which results in decreased intracellular Ca^{2+} concentrations. This model would suggest that inhibitors of aggregation such as dibutyryl cyclic AMP, phosphodiesterase inhibitors, PGE_1, and β-adrenerigc agents, which elevate platelet cyclic AMP, would also decrease free platelet Ca^{2+} concentrations. Conversely, aggregating agents such as ADP, thrombin, collagen, and α-adrenergic drugs that decrease platelet cyclic AMP would also elevate platelet free Ca^{2+} concentrations, perhaps by dephosphorylation of the Ca^{2+}-binding protein, which, according to the model, would decrease the efficiency of the Ca^{2+} extrusion pump. Elevation of platelet Ca^{2+} levels might lead to release and aggregation by mechanisms involving Ca^{2+} stimulation of the contractile proteins within the platelet (115). It should be noted that there is evidence that cyclic AMP-stimulated protein kinases isolated from platelet subcellular fractions are inhibited by calcium (80). Massini and Luscher (119) have reported that the influx of extracellular calcium after exposure of platelets to thrombin is too slow to be incriminated as the cause of the thrombin-induced release reaction and have suggested that an association with subsequent clot retraction is more likely. Elevation of platelet cytoplasmic calcium levels by transfer of the cation from a bound intracellular form, perhaps in mitochondria or endoplasmic reticulum, would still be possible as an explanation for the action of ionophores in induction of release and as a possible link of various aggregating agents with the activation of contractile proteins (120).

CLINICAL FEATURES

Disease States

In 1967 several laboratories almost simultaneously described a bleeding disorder due to abnormal platelet function, characterized by a deficiency in release of platelet constituents, in platelet aggregation by collagen, and in the second wave of platelet aggregation by ADP or epinephrine (121–123). The disorder has been termed "thrombopathia" or "thrombocytopathia" and has subsequently been shown to include two probably distinct syndromes. In one, termed "storage pool disease", platelet granules apparently lack the ability to concentrate serotonin and adenine nucleotides which are present in much lower concentration than in normal platelets (124–126). The disorder, which was initially described in patients with albinism (127, 128) but has subsequently been observed in non-albinoes as well, appears to be an inherited state transmitted as an autosomal dominant. The second syndrome, which is less common, is characterized by a defective release reaction but a normal content of adenine nucleotides and serotonin in storage granules. The platelets in this disorder behave as if affected by aspirin (see below) (125, 126, 129).

More recent studies have demonstrated that prostaglandin/cyclic AMP pathways are abnormal in both these states. Platelets with the aspirin-like disorder fail to synthesize prostaglandins upon exposure to thrombin and collagen (96; Czapek, E., Deykin, D. and Salzman, E. W., manuscript in preparation), and a defect in platelet cyclo-oxygenase activity in such patients has subsequently been described (130). The result is failure to form prostaglandin endoperoxides. The state thus resembles the case with aspirin (see below). The usual decline in cyclic AMP produced by aggregating agents is absent in patients with this disorder (Czapek, E., Deykin, D. and Salzman, E. W., manuscript in preparation). Somewhat unexpectedly, platelets from patients with "storage pool disease" were also found to be defective in formation of prostaglandins when exposed to aggregating agents (131), and they lack the usual cyclic AMP response as well (Czypek, E., Deykin, D. and Salzman, E. W., manuscript in preparation). Weiss and associates suggest the abnormality results from a flaw in the platelet response to endoperoxides (132).

A mixture of platelets from patients with storage pool disease and those from a subject given aspirin reveals correction of the defects and apparently normal platelet function (133). Presumably, endoperoxides generated in normal fashion by platelets with the storage pool defect induce the release reaction in platelets affected by aspirin, bypassing the blocked cyclo-oxygenase step and producing release of contents of storage granules from the latter platelets and leading to the usual response to released substances by all platelets in the mixture.

A defect in platelet aggregation has also been described in infants with deficiencies of essential fatty acids, specifically arachidonic and linoleic acids. The platelets failed to display a second phase of aggregation with ADP. After recovery induced by dietary treatment, the platelet behavior returned to normal (134).

Pharmacologic Agents

Aspirin and other nonsteroidal antiinflammatory drugs including indomethacin and phenylbutazone produce a defect in platelet function manifested by defective release of platelet constituents, absent or reduced second phase aggregation to ADP or epinephrine, defective aggregation to collagen, and varying responses to thrombin (135, 136). Prolongation of the bleeding time can be demonstrated in patients receiving these drugs and, in particularly susceptible individuals, a clinically significant bleeding tendency may be observed. These agents all inhibit platelet prostaglandin synthesis, presumably by blocking the action of the prostaglandin cyclo-oxygenase, and they inhibit the formation of prostaglandin endoperoxides (87, 137). A "receptor" protein that is acetylated by aspirin has been identified in platelets and tentatively related to the cyclo-oxygenase (99). The effect of aspirin on a platelet persists for the life of that platelet in the circulation (86). The effect of the other nonsteroidal anti-inflammatory agents, in contrast, is more transient.

As as result of its effect on platelet function, aspirin has been studied for its value as an antithrombotic drug. Evidence for its effectiveness has been reported in patients with predisposition to venous thromboembolism (138, 139), in patients suffering from transient cerebral ischemic attacks (140, 141), and in patients at risk from recurrent myocardial infarction (142).

Sulfinpyrazone was initially introduced as a uricosuric agent but may be even more valuable because of its ability to alter platelet function. This drug blocks platelet prostaglandin synthetic pathways (143) and in high concentrations in vitro can be shown to induce a defect in release and second phase aggregation similar to that produced by aspirin. It appears to be effective as an antithrombotic agent, for it has been shown to reduce the mortality from cardiovascular causes in elderly men (144), to correct abnormally shortened platelet survival in patients with prosthetic heart valves (144a), and to reduce the frequency of thrombotic occlusion of Scribner arterio-venous shunts (146). However, the dosage employed in patients, despite its clinical efficacy, is insufficient to produce a measurable defect in platelet function in vitro, so it is not certain that an effect on prostaglandin synthesis accounts for the effectiveness of sulfinpyrazone as an antithrombotic agent.

Several drugs in common use are potent inhibitors of platelet cyclic nucleotide phosphodiesterase, including dipyridamole (Persantine) (53, 147–149), papaverine (53, 147, 148, 150), methylxanthines, including caffeine and theophylline (13, 147, 148), and colchicine (116). In vitro they produce a rise in platelet cyclic AMP content (65) and can be shown to inhibit platelet aggregation and the release reaction and, at very high concentrations, even primary platelet aggregation induced by ADP (151). Methyl xanthines have also been shown to inhibit a cyclic GMP phosphodiesterase of platelets (152).

Dipyridamole has been extensively evaluated as an anti-thrombotic drug. There is evidence that, in conjunction with oral anticoagulants, dipyridamole reduces the frequency of thromboembolic complications of prosthetic heart valves (153). Patients with artificial heart valves, arterial grafts, or atherosclerosis have shortened platelet survival, presumably due to accelerated consumption in thrombotic events, and dipyridamole restores their platelet lifespan to normal (145). At clinically tolerated dosage, dipyridamole does not produce the effect on platelet function that can be seen in vitro when the drug is studied at higher concentrations in the test tube. Thus, as with sulfinpyrazone, dipyridamole's antithrombotic effectiveness has not been conclusively shown to result from inhibition of platelet activity.

CONCLUSIONS

Our present understanding of platelet biochemistry does not allow a complete description in molecular terms of the events that lead to platelet aggregation, but much of the phenomenology of this process has now been documented. The platelet response to a variety of stimuli has been reviewed in this communication. Thrombin, ADP, and α-adrenergic agents cause an initial reversible form of aggregation followed by the release reaction and irreversible clumping. Other agents, including collagen, appear to stimulate the release reaction directly, which then causes irreversible aggregation.

The data suggest that platelet aggregation involves more than one intra-cellular mediator. Compounds implicated include cyclic AMP and cyclic GMP, prostaglandin endoperoxides and thromboxanes, and ionized calcium. Prostaglandins of platelet origin may have a modulating role as well. The available information does not, unfortunately, allow one to assign a specific role to each of these agents, since their effects appear to be interrelated and interdependent.

The cyclic nucleotides, cyclic AMP and cyclic GMP, appear to have reciprocal effects on platelet activity, although it is not possible to assign definite cause and effect relationships. Cyclic GMP added to platelet rich plasma promotes platelet aggregation and a rise in platelet cyclic GMP accompanies aggregation. A decrease in cyclic GMP (by aspirin) is associated with inhibition of platelet aggregation. On the other hand, a rise in platelet cyclic AMP is noted after exposure to agents that inhibit platelet aggregation, cyclic AMP and dibutyryl cyclic AMP inhibit aggregation, and a decrease in platelet cyclic AMP accompanies exposure to many aggregating agents and may be one of the primary signals leading to aggregation. It appears that total platelet cyclic AMP need not be depressed below basal levels to allow for aggregation; a relative decrease in the level or a decrease in cyclic AMP in specific nucleotide compartments may suffice. Although platelet cyclic-AMP–dependent protein kinases have been identified, their natural substrates are unknown, and the way they affect platelet function has not been elucidated. Several attractive hypotheses have suggested that cyclic AMP, through activation of cyclic-AMP–dependent protein kinases, may regulate calcium homeostatis.

One of the earliest responses to aggregating agents appears to be activation of phospholipase A_2 in the platelet plasma membrane and generation of arachidonic acid. This 20-carbon fatty acid is converted into endoperoxides PGH_2 and PGG_2 and thromboxane A_2, which are potent stimulants of the platelet release reaction and aggregation. Little is known of their mode of action. Aspirin and other nonsteroidal anti-inflammatory agents inhibit platelet aggregation and release, apparently by selective inhibition of the conversion of arachidonic acid into endoperoxides and thromboxanes.

Current evidence suggests that ionized calcium, which plays an important role in regulating the secretory activity of other cells, is a critical element in the platelet release reaction and, therefore, aggregation as well. Suggested mechanisms include a direct effect on platelet contractile proteins and indirect effects from inhibition of cyclic AMP formation or of activation of cyclic AMP-dependent protein kinases. The interrelationships of platelet production of prostaglandins and their endoperoxide and thromboxane precursors and byproducts, metabolism of cyclic nucleotides, and calcium homeostasis require further clarification.

Many of the pharmacological agents which are clinically useful as inhibitors of platelet aggregation appear to act by increasing the platelet content of cyclic AMP (phosphodiesterase inhibitors) or inhibiting the platelet cyclo-oxygenase and thus blocking prostaglandin endoperoxide and thromboxane formation from arachidonic acid (aspirin and other non-steroidal anti-inflammatory drugs). A naturally occurring defect in platelet function has now been identified in which the cyclo-oxygenase is inactive or absent and, in these patients, the platelets behave as if they had been exposed to aspirin. It is to be expected that, in the future, other clinically important disorders of platelet function will be identified which result from disorders of platelet cyclic nucleotide metabolism.

SUMMARY

Hemostasis and thrombosis are complex phenomena involving both plasma and platelet components. Cyclic nucleotides appear to be important regulators of platelet function but have not been linked to plasma coagulation. In this communication, the processes of platelet adhesion, the release reaction, and platelet aggregation are reviewed, and evidence is presented linking these events to alterations in platelet cyclic nucleotides, prostaglandins, prostaglandin endoperoxides and thromboxanes, and ionized calcium. Several disease states are described in which abnormalities in these systems are associated with altered platelet function. Endogenous synthesis of PGD_2 by platelets has recently been reported (Oelz, O., Oelz, R., Knapp, H. R., Jr., Sweetman, B. J., Wilcox, H. G., and Oates, J. A.: Prostaglandin D_2 is formed by human platelets. Fed. Proc. 35:297 abs.).

ACKNOWLEDGMENTS

This work was supported by Program Project in Thrombosis and Atherosclerosis Grant HL 11414 and Biomaterials Program Project Grant HL 14322 and American Heart Association Grant 1301. MLS is a fellow of the Charles R. King Foundation.

REFERENCES

1. Mustard, J. F., Perry, D. W., Kinlough-Rathbone, R. L. and Packham, M. A.: Factors responsible for ADP-induced release reaction of human platelets. Am. J. Physiol. 228:1757-1765, 1975.
2. Weiss, H. J.: Platelet physiology and abnormalities of platelet function. New Eng. J. Med. 293:531-541, 580-588, 1975.
3. Marcus, A. J.: Platelet function. New Eng. J. Med. 280:1213-1220, 1278-1284, and 1330-1334, 1969.
4. Mustard, J. F. and Packham, M. A.: Factors influencing platelet function: Adhesion, release, and aggregation. Pharmacol. Rev. 22:97-187, 1970.
5. Michal, F. and Firkin, B. G.: Physiological and pharmacological aspects of the platelet. Ann. Rev. Pharmacol. 9:95-118, 1969.
6. Salzman, E. W.: Role of platelets in blood-surface interactions. Fed. Proc. 30:1503-1509, 1971.
7. Marcus, A. J. and Zucker, M. B.: *The Physiology of Blood Platelets*. Grune and Stratton, Inc., New York, 1965.
8. Vigdahl, R. L., Marquis, N. R. and Tavormina, P. A.: Platelet aggregation II. Adenyl cyclase, prostaglandin E_1 and calcium. Biochem. Biophys. Res. Comm. 37:409-415, 1969.
9. Robison, G. A., Arnold, A. and Hartmann, R. C.: Divergent effects of epinephrine and prostaglandin E_1 on the level of cyclic AMP in human blood platelets. Pharmacol. Res. Comm. 1:325-332, 1969.

10. Marquis, N. R., Vigdahl, R. L. and Tavormina, P. A.: Platelet aggregation I. Regulation by cyclic AMP and prostaglandin E_1. Biochem. Biophys. Res. Comm. 36:965-972, 1969.
11. Salzman, E. W. and Neri, L. L.: Cyclic 3′,5′-adenosine monophosphate in human blood platelets. Nature 224:609-610, 1969.
12. Kloeze, J.: Influence of prostaglandins on platelet adhesiveness and platelet aggregation. In: Prostaglandins, Proceedings of II Nobel Symposium, S. Bergstrom and B. Samuelsson (eds.) pp. 241-252, Interscience Publishers, London, 1966.
13. Abdulla, Y. H.: β-adrenergic receptors in human platelets. J. Atherosclerosis Res. 9:171-177, 1969.
14. Clayton, S. and Cross, M. J.: The aggregation of blood platelets by catecholamines and by thrombin. J. Physiol. 169:82-83, 1963.
15. O'Brien, J. R.: Some effects of adrenaline and anti-adrenaline compounds on platelets in vitro and in vivo. Nature 200:763-764, 1963.
16. Mills, D. C. B. and Roberts, G. C. K.: Effects of adrenaline on human blood platelets. J. Physiol. 193:443-453, 1967.
17. Schwartz, C. J. and Ardlie, N. G.: Catecholamines and platelet aggregation. Circulation Res. 21, Supplement 3:187-201, 1967.
18. Harwood, J. P., Moskowitz, J. and Krishna, G.: Dynamic aspects of hormonal control of the formation of cyclic AMP (cAMP) in human and rabbit platelets. Fed. Proc. 30:285, 1971.
19. Haslam, R. J. and Taylor, A.: Role of cyclic 3′,5′-adenosine monophosphate in platelet aggregation. In: Platelet Aggregation, J. P. Caen (ed.) pp. 85-93, Massen et Cie, Paris, 1971.
20. Mills, D. C. B.: Mechanisms affecting cyclic AMP formation in human blood platelets. Circulation 43 (suppl 2) 2:81, 1971.
21. Moskowitz, J., Harwood, J. P., Reid, W. D. and Krishna, G.: The interaction of norepinephrine and prostaglandin E_1 on the adenyl cyclase system of human and rabbit blood platelets. Biochim. Biophys. Acta 230:279-285, 1971.
22. Salzman, E. W., Kensler, P. and Levine, L.: Cyclic 3′,5′-adenosine monophosphate in human blood platelets. IV. Regulatory role of cyclic AMP in platelet function. Ann. N.Y. Acad. Sci. 201:61-71, 1972.
23. Haslam, R. J.: Interactions of the pharmacological receptors of blood platelets with adenylate cyclase. Ser. Haemat. 6:334-350, 1973.
24. Wang, Y. C., Pandey, G. N., Mendels, J. and Frazer, A.: Platelet adenylate cyclase responses in depression: Implications for a receptor defect. Psychopharmacology (Berl.) 36:291-300, 1974.
25. Salzman, E. W.: Platelet aggregation and cyclic AMP. In: *Platelets, Drugs and Thrombosis*, J. Hirsh (ed.) pp. 35-42, S. Karger Co., Basel, 1975.
26. Haslam, R. J. and Taylor, A.: Effects of aggregating agents on platelet cyclic AMP levels. In: Proceedings of the International Society on Thrombosis and Haemostasis, II Congress, p. 210, Oslo, Norway, 1971.
28. Haslam, R. J. and Taylor, A.: Effects of catecholamines on the formation of adenosine 3′:5′-cyclic monophosphate in human blood platelets. Biochem, J. 125:377-379, 1971.
29. Marquis, N. R., Vigdahl, R. L. and Tavormina, P. A.: Biochemical role

of prostaglandin E_1 in the inhibition of platelet aggregation. In: Atherosclerosis: Proceedings of the II International Symposium. R. J. Jones (ed.), Springer-Verlag, New York, 1970.
30. Marquis, N. R., Becker, J. A. and Vigdahl, R. L.: Platelet aggregation. III. An epinephrine induced decrease in cyclic AMP synthesis. Biochem. Biophys. Res. Comm. 39:783-789, 1970.
31. Chiang, T. M., Beachey, E. H. and Kang, A. H.: Interaction of a chick skin collagen fragment (α 1-CB5) with human platelets: Biochemical studies during the aggregation and release reaction. J. Biol. Chem. 250: 6916-6922, 1975.
32. Cole, B., Robison, G. A. and Hartmann, R. C.: Studies on the role of cyclic AMP in platelet function. Ann. N.Y. Acad. Sci. 185:477-493, 1971.
33 McDonald, J. W. D. and Stuart, R. K.: Regulation of cyclic AMP levels and aggregation in human platelets by prostaglandin E-1. J. Lab. Clin. Med. 81:838-849, 1973.
34. Schnetzer, G. W. and Bull, F. E.: Failure of aspirin to influence cyclic 3'5' adenosine monophosphate levels in human platelets. Clin. Res. 20: 743, 1972.
35. Salzman, E. W. and Levine, L.: Cyclic 3',5'-adenosine monophosphate in human blood platelets. II. Effect of N^6-2'-0-dibutyryl cyclic 3',5'-adenosine monophosphate on platelet function. J. Clin. Invest. 50:131-141, 1971.
36. Zieve, P. D. and Greenough, W. B. III: Adenyl cyclase in human platelets: Activity and responsiveness. Biochem. Biophys. Res. Comm. 35:462-466, 1969.
37. Amer, M. S. and Marquis, N. R.: The effect of prostaglandins, epinephrine and aspirin on cyclic AMP phosphodiesterase activity of human blood platelets and their aggregation. In: *Prostaglandins in Cellular Biology*, P. W. Ramwell and B. B. Pharriss (eds.) pp. 93-110, Plenum Press, New York, 1972.
38. Amer, M. S.: Effects of epinephrine and α-adrenergic agonists on adenosine 3'-5' cyclic monophosphate phosphodiesterase from rabbit tissues. Fed. Proc. 30:220, 1970.
39. Cole, B., Robison, G. A. and Hartmann, R. C.: Effects of prostaglandin E_1 and theophylline on aggregation and cyclic AMP levels of human blood platelets. Fed. Proc. 29:316, 1970.
40. Willis, A. L.: An enzymatic mechanism for the anti-thrombotic and antihemostatic actions of aspirin. Science 183:325-327, 1974.
41. Brinson, K.: Effect of aminophylline on blood platelet reactions. Atherosclerosis 16:233-239, 1972.
42. Chiang, T. M. H. and Kang, A. H.: The role of cyclic 3',5'-guanosine monophosphate in human platelet aggregation. Fed. Proc. 34:695, 1975.
43. White, J. G., Goldberg, N. D. Estensen, R. D., Haddox, M. K. and Rao, G. H. R.: Rapid increase in platelet cyclic 3',5'-guanosine monophosphate (cGMP) levels in association with irreversible aggregation, degranulation, and secretion. J. Clin. Invest. 52:89a, 1973.
44. Chiang, T. M., Dixit, S. N. and Kang, A. H.: Effect of cyclic 3',5'-guanosine monophosphate on human platelet function. J. Lab Clin. Med., in press.
45. Smith, J. B., Ingerman, C., Kocsis, J. J. and Silver, M. J.: Formation of

prostaglandins during the aggregation of human blood platelets. J. Clin. Invest. 52:965-969, 1973.
46. Smith, J. B., Ingerman, C., Kocsis, J. J. and Silver, M. J.: Formation of an intermediate in prostaglandin biosynthesis and its association with the platelet release reaction. J. Clin. Invest. 53:1468-1472, 1974.
47. Leonardi, R. G., Alexander, B. and White, F.: Prevention of aspirin inhibition of platelet release reaction by the fatty acid precursor of platelet prostaglandins. Thromb. Res. 3:327-338, 1973.
48. Silver, M. J., Smith, J. B., Ingerman, C. and Kocsis, J. J.: Arachidonic acid-induced human platelet aggregation and prostaglandin formation. Prostaglandins 4:863-875, 1973.
49. Vargaftig, B. B. and Chignard, M.: Substances that increase cyclic AMP content prevent platelet aggregation and the concurrent release of pharmacologically active substances evoked by arachidonic acid. Agents and Actions 5:137-144, 1975.
50. Bygdeman, S. and Johnsen, O.: Effect of tryptamine on platelet aggregation and uptake of noradrenaline and 5-hydroxytryptamine. In: *Proceedings of the International Society on Thrombosis and Haemostasis*, II Congress, p. 182, Oslo, Norway, 1971.
51. Haslam, R. J.: Role of adenosine diphosphate in the aggregation of human blood platelets by thrombin and by fatty acids. Nature 202:765-768, 1964.
52. Haslam, R. J.: Mechanisms of blood platelet aggregation. In: *Physiology of Hemostasis and Thrombosis*, S. A. Johnson and W. H. Seegers (eds.) pp. 88-112, Charles C. Thomas, Springfield, Illinois, 1967.
53. Smith, J. B. and Mills, D. C. B.: Inhibition of adenosine 3':5'-cyclic monophosphate phosphodiesterase. Biochem. J. 120:20p, 1970.
54. Kloeze, J.: Relationship between chemical structure and platelet-aggregation activity of prostaglandins. Biochim. Biophys. Acta 187:285-292, 1969.
55. Kloeze, J.: Influence of prostaglandins on ADP-induced platelet aggregation. Acta Physiol. Pharmacol. Neerl. 15:50-51, 1969.
56. Irion, E. and Blomback, M.: Prostaglandins in platelet aggregation. Scand. J. Clin. Lab Invest. 24:141-144, 1969.
57. Kinlough-Rathbone, R. L., Packham, M. A. and Mustard, J. F.: The effect of prostaglandin E_1 on platelet function in vitro and in vivo. Brit. J. Hematol. 19:559-571, 1970.
58. McDonald, J. W. D. and Stuart, R. K.: Interaction of prostaglandins E_1 and E_2 in regulation of cyclic-AMP and aggregation in human platelets: Evidence for a common prostaglandin receptor. J. Lab. Clin. Med. 84: 111-121, 1974.
59. Born, G. V. R. and Cross, M. J.: The aggregation of blood platelets. J. Physiol. 168:178-195, 1963.
60. Born, G. V. R.: Uptake of adenosine and of adenosine diphosphate by human blood platelets. Nature 206:1121-1122, 1965.
61. Haslam, R. J. and Lynham, J. A.: Activation and inhibition of blood platelet adenylate cyclase by adenosine or by 2-chloroadenosine. Life Sci. 11:1143-1154, 1972.
62. Mills, D. C. B. and Smith, J. B.: The influence on platelet aggregation

of drugs that affect the accumulation of adenosine 3':5'-cyclic monophosphate in platelets. Biochem. J. 121:185, 1970.
63. Robison, G., Arnold, A., Cole, B. and Hartmann, R.: Effects of prostaglandins on function and cyclic AMP levels of human blood platelets. Ann. N.Y. Acad. Sci. 180:324-331, 1971.
64. Salzman, E. W.: Prostaglandins, cyclic AMP, and platelet function. Thromb. et Diath. Haemorr. Supp. 60:311-319, 1974.
65. Salzman, E. W.: Cyclic AMP and platelet function. New Eng. J. Med. 286:358-363, 1972.
66. Born, G. V. R.: Platelet aggregation and cyclic AMP. In: Effects of Drugs on Cellular Control Mechanisms, B. R. Rabin and R. B. Freedman (eds.) pp. 237-257, Macmillan Press, Ltd., New York, 1972.
67. Salzman, E. W., Chambers, D. A. and Neri, L. L.: Possible mechanism of aggregation of blood platelets by adenosine diphosphate. Nature 210: 167-169, 1966.
68. Salzman, E. W., Ashford, T. P., Chambers, D. A. and Neri, L. L.: Platelet incorporation of labelled adenosine and adenosine diphosphate. Thromb. et Diathes. Haemorr. 22:304-315, 1969.
69. Schoene, N. W. and Iacono, J. M.: Stimulation of platelet phospholipase A_2 activity by aggregating agents. Fed. Proc. 34:257, 1975.
70. Smith, J. B., Silver, M. J.: Phospholipase A_1 of human blood platelets. Biochem. J. 131:615, 1973.
71. Martin, B. M., Feinman, R. D. and Detwiler, T. C.: Platelet stimulation by thrombin and other proteases. Biochemistry 14:1308-1314, 1975.
72. Brodie, G. N., Baeziger, N. L., Chase, L. R. and Majerus, P. W.: The effects of thrombin on adenyl cyclase activity and a membrane protein from human platelets. J. Clin. Invest. 51:81-88, 1972.
73. Baenziger, R. L., Brodie, G. M. and Majerus, P. W.: A thrombin sensitive protein of human platelet membranes. Proc. Nat. Acad. Sci. USA 68:240-243, 1971.
74. Baenziger, N., Brodie, G., Chase, L. R. and Majerus, P. W.: The role of thrombin in platelet aggregation. J. Clin. Invest. 50:4a-5a, 1971.
75. Wolfe, S. M. and Shulman, N. R.: Inhibition of platelet energy production and release by PGE_1, theophylline and cAMP. Biochem. Biophys. Res. Comm. 41:128-134, 1970.
76. Murer, E. H.: Compounds known to affect the cyclic adenosine monophosphate level in blood platelets: Effect on thrombin-induced clot retraction and platelet release. Biochim. Biophys. Acta 237:310-315, 1971.
77. Murer, E. H.: Some aspects of the control of platelet functions. In: *Proceedings of the International Society on Thrombosis and Haemostasis*, II Congress, p. 236, Oslo, Norway, 1971.
78. Droller, M. J. and Wolfe, S. M.: Thrombin-induced increase in intracellular cyclic 3',5'-adenosine monophosphate in human platelets. J. Clin. Invest. 51:3094-3103, 1972.
79. Lyons, R. M., Stanford, N. and Majerus, P.: Thrombin-induced protein phosphorylation in human platelets. J. Clin. Invest. 56:924-936, 1975.
80. Bishop, G. A. and Rozenberg, M. C.: The effect of ADP, calcium, and some inhibitors of platelet aggregation on protein phosphokinases from human platelets. Biochim. Biophys. Acta 384:112-119, 1975.

81. Marquis, N. R., Vigdahl, R. L. and Tavormina, P. A.: Cyclic AMP-dependent platelet protein kinase. Fed. Proc. 30:423abs., 1971.
82. Salzman, E. W. and Weisenberger, H.: Cyclic AMP-dependent protein kinase from human blood platelets. In: Proceedings of the International Society on Thrombosis and Haemostasis, II Congress, p. 254, Oslo, Norway, 1971.
83. Kaulen, H. D. and Gross, R.: Demonstration of adenosine 3',5'-monophosphate dependent protein-phosphokinase in human platelets. Klin. Wschr. 50:1115, 1972.
84. Smith, J. B. and Willis, A. L.: Formation and release of prostaglandins by platelets in response to thrombin. Brit. J. Pharmacol. 40:545p, 1970.
85. Hamberg, M., Svensson, J. and Samuelsson, B.: Mechanism of the antiaggregating effect of aspirin on human platelets. Lancet 2:223-224, 1974.
86. Kocsis, J. J., Hernandovich, J., Silver, M. J., Smith, J. B. and Ingerman, C.: Duration of inhibition of platelet prostaglandin formation and aggregation by ingested aspirin or indomethacin. Prostaglandins 3:141-144, 1973.
87. Smith, J. B. and Willis, A. L.: Aspirin selectively inhibits prostaglandin production in human platelets. Nature New Biol. 231:235-237, 1971.
88. Barber, A. J. and Jamieson, G. A.: Platelet collagen adhesion characterization of collagen glucosyltransferase of plasma membranes of human blood platelets. Biochim. Biophys. Acta 252:533-545, 1971.
89. Haslam, R. J., Davidson, M. M. L. and McClenaghan, M. D.: Cytochalasin B, the blood platelet release reaction and cyclic GMP. Nature 253: 455-457, 1975.
90. Haslam, R. J. and McClenaghan, M. D.: Effects of collagen and of aspirin on the concentration of guanosine 3':5'-cyclic monophosphate in human blood platelets: Measurement by a prelabelling technique. Biochem. J. 138:317-320, 1974.
91. Silver, M. J., Hernandowich, J., Ingerman, C., Kocsis, J. J. and Smith, J. B.: Persistent inhibition by aspirin of collagen-induced platelet prostaglandin formation. In: Platelets and Thrombosis, S. Sherry and A. Scriabine (eds.) pp. 91-98, University Park Press, Baltimore, 1974.
92. Vargaftig, B. B. and Zirinis, P.: Platelet aggregation induced by arachidonic acid is accompanied by release of potential inflammatory mediators distinct from PGE_2 and PGF_2 alpha. Nature 244:114-116, 1973.
93. Willis, A. L., Vane, F. M., Kuhn, D. C., Scott, C. G. and Petrin, M.: An endoperoxide aggregator (LASS), formed in platelets in response to thrombotic stimuli-purification, identification, and unique biological significance. Prostaglandins 8:453-507, 1974.
94. Glass, D. B., Gerrard, J. M., White, J. G. and Goldberg, N. D.: Cyclic GMP formation in human platelets aggregated by arachidonic acid. Proceedings of the American Society of Hematology Abstract 98, December, 1975.
95. Silver, M. J., Smith, J. B., Ingerman, C. C. and Kocsis, J. J.: Prostaglandins in blood: Measurement, sources, and effects. Progress in Hematol. 8:235-257, 1973.
96. Smith, J. B. and MacFarlane, D. E.: Platelets. In: *The Prostaglandins.* vol. 2. P. W. Ramwell (ed.) pp. 293-343, Plenum Press, New York, 1974.

96a. Hamberg, M., Svensson, J. and Samuelsson, B.: Thromboxanes: A new group of biologically active compounds derived from prostaglandin endoperoxides. Proc. Nat. Acad. Sci. USA 72:2994-2998, 1975.
97. Corey, E. J., Nicolaou, K. C., Machida, Y., Malmsten, C. L. and Samuelsson, B.: Synthesis and biological properties of a 9,11-azo-prostanoid: highly active biochemical mimic of prostaglandin endoperoxides. Proc. Nat. Acad. Sci. USA 72:3355-3358, 1975.
98. Kolata, G. B.: Thromboxanes: The power behind the prostaglandins? Science 190:770-771, and 812, 1975.
99. Roth, G. and Majerus, P. W.: The mechanism of the effect of aspirin on human platelets. I. Acetylation of a particulate fraction protein. J. Clin. Invest. 56:624-632, 1975.
100. Silver, M. J., Kocsis, J. J., Hoch, W., Ingerman, C. M. and Smith, J. B.: Arachidonic acid causes sudden death in rabbits. Science 183:1085-1087, 1974.
101. Taylor, R. E., Stitt, E. S., Robison, G. A. and Hartmann, R. C.: Characterization of prostaglandin-cyclic AMP interactions in human blood platelets. Blood 42:994, 1973.
103. Shio, H. and Ramwell, P.: Effect of prostaglandin E_2 and aspirin on the secondary aggregation of human platelets. Nature 236:45-46, 1972.
104. Bruno, J. J., Taylor, L. A. and Droller, M. J.: Effects of prostaglandin E_2 on human platelet adenyl cyclase and aggregation. Nature 251:721-723, 1974.
105. Mills, D. C. B. and Macfarlane, D. E.: Stimulation of human platelet adenylate cyclase by prostaglandin D_2. Thromb. Res. 5:401-412, 1974.
106. Smith, J. B., Silver, M. J., Ingerman, C. M. and Kocsis, J. J.: Prostaglandin D_2 inhibits the aggregation of human platelets. Thromb. Res. 5: 291-299, 1974.
107. Nishizawa, E. E., Miller, W. L., Gorman, R. R., Bundy, G. L., Svensson, J. and Hamberg, M.: Prostaglandin D_2 as a potential antithrombotic agent. Prostaglandins 9:109, 1975.
108. Mustard, J. F., Kinlough-Rathbone, R. L., Jenkins, C. S. P. and Packham, M. A.: Modification of platelet function. Ann. N.Y. Acad. Sci. 201: 343-349, 1972.
109. Kinlough-Rathbone, R. L., Chahil, A. and Mustard, J. F.: Effect of external calcium and magnesium on thrombin-induced changes in calcium and magnesium of pig platelets. Am. J. Physiol. 224:941-945, 1973.
110. Haslam, R. J. and Rosson, G. M.: Aggregation of human blood platelets by vasopressin. Am. J. Physiol. 223:958-967, 1972.
111. Sneddon, J. M. and Williams, K. I.: Effect of cations on the blood platelet release reaction. J. Physiol. 235:625-637, 1973.
112. Feinman, R. D. and Detwiler, T. C.: Platelet secretion induced by divalent cation ionophores. Nature 249:172-173, 1974.
113. Massini, P. and Luscher, E. F.: Some effects of ionophores for divalent cations on blood platelets: Comparison with the effects of thrombin. Biochim. Biophys. Acta 372:109-121, 1974.
114. White, J. G., Rao, G. H. R. and Gerrard, J. M.: Effects of the ionophore A23187 on blood platelets. I. Influence on aggregation and secretion. Am. J. Path. 77:135-149, 1974.

115. Berridge, M. J.: The interaction of cyclic nucleotides and calcium in the control of cellular activity. Adv. Cy. Nuc. Res. 6:1-98, 1975.
116. Salzman, E. W.: Prostaglandins and platelet function. In: Advances in Prostaglandin and Thromboxane Research, vol. II, B. Samuelsson and R. Paoletti (eds.), Raven Press, New York, 1976, pp. 767.
117. Friedman, F. and Detwiler, T. C.: Stimulus-secretion coupling in platelets. Effect of drugs on secretion of adenosine 5' triphosphate. Biochemistry 14:1315-1320, 1975.
118. Booyse, F. M., Marr, J., Yang, D. C., Guiliani, D. and Rafelson, M. E.: Development of a phosphorylation-dephosphorylation model for the regulation and mechanism of platelet aggregation. In: Frontiers in Matrix Biology, S. Karger, Basel, 1976, in press.
119. Massini, P., and Luscher, E. F.: On the significance of the influx of calcium ions in stimulated human blood platelets. In press.
120. Robblee, L. S., Shepro, D. and Belamarich, F. A.: The effect of thrombin and trypsin on calcium uptake by calf platelet membranes. Microvasc. Res. 6:99-107, 1973.
121. Weiss, H. J.: Platelet aggregation, adhesion, and adenosine diphosphate release in thrombopathia (platelet factor 3 deficiency): a comparison with Glanzmann's thrombasthenia and von Willebrand's disease. Am. J. Med. 43:570-578, 1967.
122. Hardisty, R. M., and Hutton, R. A.: Bleeding tendency associated with "new" abnormality of platelet behavior. Lancet 1:983-985, 1967.
123. O'Brien, J. R.: Platelets: a Portsmouth syndrome? Lancet 2:258, 1967.
124. Holmsen, H., Weiss, H. J.: Further evidence for a deficient storage pool of adenine nucleotides in platelets from some patients with thrombocytopathia — "storage pool disease." Blood 39:197-209, 1972.
125. Weiss, H. J.: Abnormalities in platelet function due to defects in the release reaction. Ann. N.Y. Acad. Sci. 201:161-173, 1972.
126. Pareti, F. I., Day, H. J. and Mills, D. C. B.: Nucleotide and serotonin metabolism in platelets with defective secondary aggregation. Blood 44:789-800, 1974.
127. Logan, L. J., Rapaport, S. I. and Maher, I.: Albinism and abnormal platelet function. New Eng. J. Med. 284:1340-1345, 1971.
128. Hardisty, R. M., Mills, D. C. B. and Ketsa-Ard, K.: The platelet defect associated with albinism. Brit. J. Hematol. 23:679-692, 1972.
129. Czapek, E., Deykin, D., Salzman, E. W. et al: An intermediate syndrome of platelet dysfunction. Abstracts of the Congress of the International Society on Thrombosis and Haemostasis. Washington, D.C. 1972, p. 182.
130. Malmsten, C., Hamberg, M., Svensson, J. et al.: Physiological role of an endoperoxide in human platelets: hemostatic defect due to platelet cyclooxygenase deficiency. Proc. Nat. Acad. Sci. USA 72:1446-1450, 1975.
131. Willis, A. L. and Weiss, H. J.: A congenital defect in platelet prostaglandin production associated with impaired hemostasis in storage pool disease. Prostaglandins 4:783-794, 1973.
132. Weiss, H. J., Willis, A. L., Brand, H. et al.: Prostaglandin E_2 potentiation of platelet aggregation by LASS endoperoxide: absent in storage-pool disease, normal after aspirin ingestion. Brit. J. Haematol, 1976, in press.
133. Gerrard, J. M., White, J. G., Rao, G. H. R. et al.: Labile aggregation

stimulating substances (LASS): the factor from storage pool deficient platelets correcting defective aggregation and release of aspirin treated normal patients. Brit. J. Haematol. 29:657-665, 1975.
134. Friedman, Z., Lamberth, E. L., Stahlman, M. T. and Oates, J. A.: Platelet aggregation in infants with essential fatty acid deficiency. Proceedings of the International Conference on Prostaglandins, Florence, Italy, p. 47, 1975.
135. Weiss, H. J., Aledort, L. and Kochwa, S.: Effect of salicylates on hemostatic properties of platelets in man. J. Clin. Invest. 47:2169-2180, 1968.
136. Zucker, M. B. and Peterson, J.: Inhibition of adenosine diphosphate induced secondary aggregation and other platelet functions by acetyl salicylic acid. Proc. Soc. Exp. Biol. Med. 127:547, 1968.
137. Hamberg, M., Svensson, J. and Samuelsson, B.: Prostaglandin endoperoxides, a new concept concerning the mode of action and release of prostaglandins. Proc. Nat. Acad. Sci. USA 71:3824-3828, 1974.
138. Salzman, E. W., Harris, W. H. and DeSanctis, R. W.: Reduction in venous thromboembolism by agents affecting platelet function. New Eng. J. Med. 284:1287-1292, 1971.
139. Harris, W. H., Salzman, E. W., Athanasoulis, C., Waltman, A. C., Baum, S. and DeSanctis, R. W.: Comparison of warfarin, low-molecular weight dextran, aspirin, and subcutaneous heparin prevention of venous thromboembolism following total hip replacement. J. Bone Joint Surg. 56A: 1552-1562, 1974.
140. Mundall, J., Quintero, P., von Kauzza, K. N., Harmon, R. and Austin, J.: Transient monocular blindness and increased platelet aggregability treated with aspirin: a case report. Neurology 22:280-285, 1972.
141. Harrison, M. J. G., Marshall, J., Meadows, J. C. and Russell, R. W. R.: Effect of aspirin in amaurosis fugax. Lancet 2:743-744, 1971.
142. Elwood, P. C., Cochrane, A. L., Burr, M. L., Sweetname, P. M., Williams, G., Welsby, E., Hughes, S. J. and Renton, R.: A randomized controlled trial of acetylsalicylic acid in the secondary prevention of mortality from myocardial infarction. Brit. Med. J. 1:436-440, 1974.
143. McDonald, J. W. D., Ali, M., Nagai, G. R., Barnett, W. H. and Barnett, H. J. M.: Inhibition of platelet protaglandin synthetase and platelet release reaction by sulfinpyrazone. Proceedings of the American Society of Hematology Abstract #97, December, 1975.
144. Blakely, J. A. and Gent, M.: Platelets, drugs and longevity in a geriatric population. In: *Platelets, Drugs and Thrombosis*, J. Hirsh, J. F. Cade, A. S. Gallus, and E. Schonbaum (eds.) pp. 284-291, S. Karger Co., Basal, Switzerland, 1975.
144a. Weily, H. S. and Genton, E.: Altered platelet function in patients with prosthetic mitral valves. Effects of sulfinpyrazone therapy. Circulation 42:967-972, 1970.
145. Harker, L. A. and Slichter, S. J.: Studies of platelet and fibrinogen kinetics in patients with prosthetic heart valves. New Eng. J. Med. 283: 1302-1305, 1970.
146. Kaegi, A., Pineo, G. F., Shimizu, A., Trivedi, H., Hirsh, J. and Gent, M.: Arteriovenous-shunt thrombosis: prevention by sulfinpyrazone. New Eng. J. Med. 290:304-306, 1974.
147. Mills, D. C. B. and Smith, J. B.: The influence on platelet aggregation

of drugs that affect the accumulation of adenosine 3':5'-cyclic monophosphate in platelets. Biochem. J. 121:185-196, 1970.
148. Vigdahl, R. L., Mongin, J. and Marquis, N. R.: Platelet aggregation. IV. Platelet phosphodiesterase and its inhibition by vasodilators. Biochem. Biophys. Res. Comm. 42:1088-1094, 1971.
149. Pichard, A. L., Hanoune, J. and Kaplan, J. C.: Human brain and platelet cyclic adenosine 3',5'-monophosphate phosphodiesterases: Different responses to drugs. Biochim. Biophys. Acta 279:217-220, 1972.
150. Markwardt, F. and Hoffmann, A.: Effects of papavarine derivatives on cyclic AMP phosphodiesterase of human platelets. Biochem. Phamacol. 19:2519-2520, 1970.
151. Ardlie, N. G., Glew, G., Schultz, B. G. and Schwartz, C. J.: Inhibition and reversal of platelet aggregation by methylxanthenes. Thromb. Diath. Haemorrhag. 18:670, 1967.
152. Hidaka, H., Asano, T., Shibuya, M. and Shimamoto, T.: Cyclic GMP phosphodiesterase of human blood platelet and its inhibitors. Thromb. Diath. Haemorrag., 1976, in press.
153. Sullivan, J. M., Harkin, D. E. and Gorlin, R.: Pharmacologic control of thromboembolic complications of cardiac valve replacement. New Eng. J. Med. 284:1391-1394, 1971.

12

The Role of Cyclic AMP in Neurologic and Affective Disorders

K. M. A. WELCH
JANET NELL
EVA CHABI

Cyclic AMP has been implicated as an intracellular mediator of a variety of extracellular neurohumoral influences, both in nonmammalian and mammalian organisms (1, 2). Mammalian brain has a great capacity for the formation of cyclic AMP, the activity of the synthetic enzyme, adenylate cyclase, and the catabolizing enzyme, cyclic AMP phosphodiesterase being higher in the central nervous system (CNS) than in most other body tissues (3, 4). Extensive in vitro studies in animal models, supported by more limited studies in vivo, strongly suggest that cyclic AMP is involved in central and peripheral neurotransmission and may play a role in the control of cerebral energy metabolism (4). The precise role of cyclic AMP in brain function nevertheless still requires definition.

In the ensuing pages of this chapter those neurological diseases have been reviewed in which disorder of cyclic AMP metabolism has been either implicated or directly demonstrated. However, it seems pertinent to review first the evidence from human studies that permits extrapolation of hypotheses derived from studies in animal models to the human and perhaps clinical situation.

STUDIES IN HUMAN BRAIN AND CEREBROSPINAL FLUID (CSF)

The extensive publications that have accentuated the biological importance of cyclic AMP in the CNS have been conducted largely in the laboratory animal and as yet little is known about the nucleotide in human brain. By using broken-cell preparations obtained from frozen human brain obtained at autopsy Williams et al. (5) observed activity of both synthesizing and degrading enzymes for cyclic AMP in various regions of human brain. The highest levels of these enzymes were found in the gray matter of the cerebrum and cerebellum, hypothalamus, pineal, substantia nigra and caudate nucleus. Working with intact cell preparations of human brain obtained during neurosurgical procedures, Kodama et al. (6) found that there was a sharp contrast between levels of cyclic AMP in subcortical white and gray matter (3.3 to 21 versus 42 to 143 pmoles/mgm protein). A comparative contrast in the differential activation of adenyl cyclase between white and gray matter was also observed.

Kodama et al., explaining the second messenger role of cyclic AMP, demonstrated that cyclic AMP levels in incubated slices of gray matter were elevated 20- to 50-fold by norepinephrine (0.5-1 mM), 4- to 7-fold by histamine (1mM) or adenosine (0.2mM) and 1.5- to 2-fold by serotonin (6). However, compared to epinephrine or isoproterenol, norepinephrine was inferior in its stimulatory effect on cyclic AMP levels in human brain slices and its effect could be completely antagonized by the beta-adrenergic blocking agent propranolol, albeit minimally, by alpha-blocking agents such as phentolamine and dibenamine. In this same study phosphodiesterase activity exhibited a single Km value in the region of $0.9 \times 10^{-4}M$.

From the aforementioned studies it can be concluded that the general characteristics of the cyclic AMP system are to be found in human brain and are similar to those of other animals. Therefore, the hypotheses concerning cyclic AMP function that have been substantiated in experimental animals can presumably be extrapolated to human brain.

Cyclic AMP has been measured in human CSF although values show some variability between individual studies. In our laboratory CSF values in control patients without evidence of neurological disease ranged from 10-19 picomoles/ml. These values fall in much the same range as those reported by Robison et al. of 14.2 to 16.2 pmoles/ml in selected psychiatric and neurologic patients (7). Myllylä et al. (8) reported a larger mean value at 24.3 ± 1.4 pmoles/ml, (range 18 to 67 pmoles/ml) in CSF obtained from patients with unspecified neurological disease. Higher cyclic AMP levels in the female, compared to male, population of patients were reported from the same laboratory (9). Earlier measurements, reported by the same workers, of CSF cyclic AMP values ranging from 168 to 667 pmoles/ml were probably artifactual (9).

Cyclic AMP is apparently removed from the CSF via an active transport process which is sensitive to blockade by probenecid (10). It has been suggested that the accumulation rates of cyclic AMP in CSF after administration of probenecid to clinical patients might qualitatively reflect central turnover of cyclic AMP and be useful in the assessment of disease states that may involve disorder of cyclic AMP metabolism (10). Sebens and Korf (11) have studied cisternal CSF samples in the rabbit and showed that cyclic AMP was increased by intraperitoneal probenecid, but that this increase was not influenced by tricyclic antidepressants, haloperidol, isoprenaline or L-dopa. They did, however, show that intracisternally injected norepinephrine, isoprenaline, dopamine and histamine, as well as intravenously injected isoprenaline, increased CSF cyclic AMP levels. These results substantiate the theory that cyclic AMP in the CSF is of central origin and that a close relationship exists between cyclic AMP in brain tissue and in CSF.

Sebens and Korf found, however, that the time course of accumulation of cyclic AMP after probenecid was markedly different from that obtained for the accumulation of 5-hydroxyindoleacetic acid (5-HIAA) and homovanillic acid (HVA), the respective metabolites of serotonin and dopamine (11). Thirty minutes after probenecid injection cyclic AMP levels had reached new steady state levels in CSF, while HVA and 5-HIAA continued to accumulate over at least 4 hours. These workers, therefore, question that changes in cyclic AMP levels after probenecid can be related to those of monoamine acid metabolites. A further interesting point of their study was that intravenously administered isoprenaline induced increase of CSF cyclic AMP related to stimulation of central beta adrenergic receptors thus raising the possibility that such a procedure could be used as a test for the competency of such receptors in clinical states.

Cyclic AMP phosphodiesterase has also been demonstrated in the CSF, the activity being relatively low, amounting to a twentieth of that in plasma (12). This is so despite the fact that CSF cyclic AMP levels correlate quite closely with plasma cyclic AMP levels. Cyclic AMP phosphodiesterase present in CSF apparently originates from the CNS rather than plasma.

CYCLIC AMP DISORDER IN NEUROLOGIC DISEASE

Diseases of the Cerebral Vessels

Cerebral Ischemia and Infarction

Several authors have previously reported evidence of alteration of neuroendocrinal function after acute cerebral infarction in man (13, 14). Systemic plasma cyclic AMP levels have been found elevated in patients with cerebral infarction, higher levels being found in patients presenting with stroke of more recent onset (Fig. 1). Numerous organs and tissues, e.g., liver, kidney and muscle, contribute to the plasma cyclic AMP pool, presumably as a result of cellular release, leakage, or transport into extracellular fluid, and since stimulation of intracellular cyclic AMP formation is a property common to a wide variety of neurohumoral agents, the cause of plasma cyclic AMP elevation after recent stroke is likely to be complex. For example, increased circulation of catecholamines has been observed after cerebral ischemia in both experimental animals and man (15–17). Serotonin has also been reported as elevated in systemic blood of patients with acute cerebral infarction (13). Stress response of increased corticosteroid, as well as catecholamine, output may be a further important factor that could provoke secondary elevation of circulating plasma cyclic AMP activity.

Urinary excretion of cyclic AMP in patients with stroke was found unaltered compared to controls (18), apparently confirming that cyclic AMP excretion in urine is not altogether a reliable index of altered CNS or systemic cyclic AMP metabolism.

Cyclic AMP levels in CSF of patients with recent cerebral infarction have also been found elevated (18) (Fig. 2), suggesting that the nucleotide is released from infarcted brain into CSF. Cere-

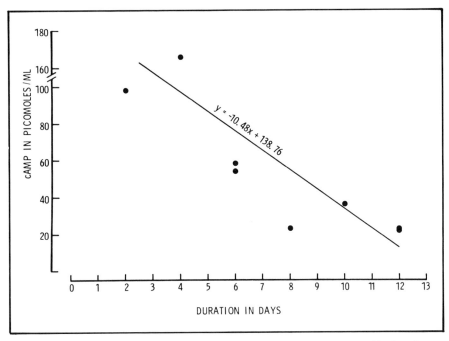

1 Correlation of cyclic-AMP in systemic venous blood with duration after onset of cerebral infarction, n = 8; r = −0.769; p <0.05. (Reproduced from Welch et al.: European Neurology, volume 13, pp. 144-154, 1975).

bral arteriovenous differences for cyclic AMP measured in the same patients showed a similar release into cerebral venous blood, perhaps facilitated by damage to the blood-brain barrier. This finding also suggested that the cyclic AMP either leaked from infarcted brain into CSF and/or cerebral venous blood or else that increased amounts of cyclic AMP in CSF were actively removed into cerebral venous blood (10).

Steiner et al. were the first to show, in animal models, that levels of cyclic AMP were elevated in the acute stage after induced cerebral ischemia (19). This has been confirmed in our laboratory in the gerbil stroke model (Fig. 3). Disordered neurotransmitter function in experimental ischemia was suggested by earlier studies, in which serotonin release from brain in cerebral venous blood was observed after the induction of acute cerebral ischemia in the pri-

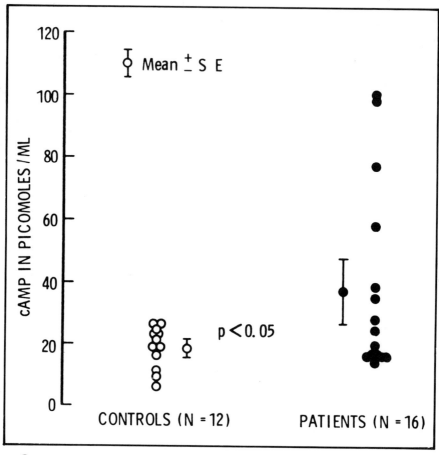

2 Cyclic-AMP in CSF of patients with recent cerebral infarction compared to controls. (Reproduced from Welch et al.: European Neurology, volume 13, pp. 144-154, 1975).

mate (20). Later, catecholamine changes were demonstrated in smaller animal models (21–23). Complementary studies in man have shown elevated levels of norepinephrine and serotonin in CSF after cerebral infarction of recent onset (24). It seems possible from these studies that increase in brain tissue cyclic AMP may be stimulated secondary to increased neurotransmitter release and that this may be reflected by increased extracellular and CSF cyclic

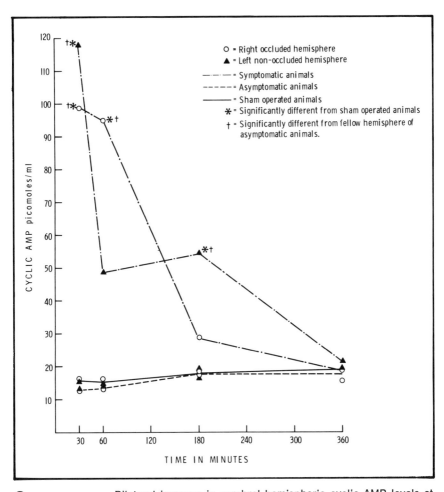

3 Bilateral increase in cerebral hemispheric cyclic AMP levels at various time intervals of unilateral common carotid artery occlusion in gerbils.

AMP levels. Alternatively, or in addition, impaired cell membrane integrity in ischemic brain areas might be responsible for leakage of cyclic AMP into CSF.

The consequences of such changes in central cyclic AMP levels demonstrated in ischemic brain have not been elucidated. Certainly, such changes might be expected to reflect disordered neurotransmission and in this regard it seems interesting that bilateral hemispheric increase in cyclic AMP has been observed in models of unilateral cerebral ischemia (see Fig. 3), suggesting perhaps a generalized neuronal deplorization occurring as a result of focal ischemic insult. This could be related to the phenomenon of diaschisis seen in patients with focal cerebral ischemia, whereby cerebral blood flow and metabolism were found to be altered in brain areas remote from the cerebral ischemia (25). Disordered neurotransmission might be expected since it now appears well established that intraneuronal electroconductivity involves presynaptic neurotransmitter release onto postsynaptic receptors; the resultant change in conductivity and alteration in ion permeability of the postsynaptic cell membrane appears to follow a chain of events starting with adenyl cyclase activation and increase in intracellular cyclic AMP, which in turn activates specific protein kinases leading to phosphorylation via ATP of a protein substrate in the cell membrane (26).

Recent in vivo primate studies have suggested that both cyclic AMP and dibutyryl cyclic AMP result in stimulation of both aerobic and anaerobic glucose metabolism in cerebral tissue (Tagashira et al., J. Neurosurg., in press). Cyclic AMP is thought to be a positive modulator of glycolysis, possibly by an effect on phosphofructokinase activity (27, 28). It therefore seems of interest that cyclic AMP levels are elevated in ischemic brain, where increased anaerobic glycolysis and accumulation of lactate feature prominently (29). Cyclic AMP could perhaps be a mediator of the Pasteur effect seen in normal and ischemic brain and confirmed in patients with cerebral ischemic infarction (30), wherein reduced availability of oxygen results in a shift to anaerobic glycolysis in brain tissue.

Increased brain tissue cyclic AMP may certainly be expected to have behavioral effects. Intracerebral injection of cyclic nucleotide in vivo into various brain regions of experimental animals has produced among other behavioral effects hyperthermia, hyperactivity, mydriasis, piloerection and convulsions (31–33). However, the role of disordered cyclic AMP function in some of the behavioral changes following cerebral ischemia has not been determined.

Schwartz et al. have recently reported that further marked increases in tissue cyclic AMP levels in previously ischemic brain occur upon restoration of blood flow after transient unilateral carotid artery occlusion in the gerbil (34). In the same transient ischemic model, rebound increase in catecholamines, but not serotonin, following removal of arterial occlusion, has also been observed (35). Alteration of cyclic AMP levels in ischemic brain could, therefore, perhaps represent activation of recovery processes. Recent studies utilizing various models of in vivo cerebral ischemia have indicated that energy synthesis remains relatively unimpaired in the early stages of even severe cerebral ischemia (36). Possibly the increased ATP, synthesized during the recovery phase after transient ischemia, is diverted largely to synthesize cyclic AMP, which is known to be involved in the mediation of brain functions apparently more vulnerable to ischemia, such as synaptic transmission and membrane conductance as well as lipid, protein and neurotransmitter synthesis.

In the light of the above it may come as no surprise that cyclic AMP itself has been used as a therapeutic agent in patients with stroke. Gertler et al. administered cyclic AMP intravenously in doses of 2 gm over 2.5 to 3 hr, 3 to 4 times during 7 to 10 days and claimed that eight hemiplegic patients showed increased attentiveness, vocabulary, receptive, and psychologic attainment, improvement in range of motion, gait and endurance, and electroencephalographic improvement (37). Since cyclic AMP infused intravenously is liable to affect many different systems, these results should be interpreted with some caution. However, it may be of some relevance, that after intravenous administration of 10 percent glycerol to patients suffering from recent onset of cerebral infarction, narrowing of cyclic AMP cerebral arteriovenous differences and fall in CSF cyclic AMP (18, 38) indicated that resultant improvement in regional cerebral blood flow, decrease in cerebral edema, and improvement in cerebral metabolism coincided with reduction of cyclic AMP released from infarcted brain.

The etiology and treatment of cerebral edema that is frequently a devastating complication of cerebral ischemia and infarction is a major problem to the neurological scientist and clinician. However, some recent beneficial modification of the development of cerebral edema in animal models of cerebral ischemia has been attributed to the use of cyclic AMP phosphodiesterase inhibitors (39, 40), although their mechanisms are unproved.

Subarachnoid Hemorrhage

In vitro studies of the functional role of cyclic AMP in vasomotor control have suggested that alteration in vascular tone of systemic vascular beds, which occurs secondarily to receptor stimulation by neurogenic or humoral influences, may be mediated in part by changes in intracellular cyclic AMP levels (41–43). Studies of the role of cyclic AMP in the cerebral circulation have been surprisingly limited. Intracarotid and intracisternal administration of cyclic AMP and dibutyryl cyclic AMP has been found to increase cerebral blood flow, probably secondary to an effect on vascular receptors, as well as secondary to stimulation of cerebral metabolism (Tagashira et al., J. Neurosurg., in press). These results seem in accord with the observations of Peterson et al. on experimental subarachnoid hemorrhage in the cat and monkey, wherein alleviation of basilar artery spasm was produced when dibutyryl cyclic AMP was directly applied to the vessel wall (44). Cerebral vascular spasm is a serious complication in patients suffering from subarachnoid hemorrhage and the actual mechanisms for its production remain to a large measure obscure. Peterson and his colleagues have raised the possibility that in subarachnoid hemorrhage there arises a series of topical events, associated with direct application of blood to the vascular wall, which are associated with a more widespread vasoconstriction of cerebral blood vessels in the subarachnoid space and that this in some way involves the intracellular activity of cyclic AMP and its destruction by one or more substances present in the clot or unformed elements of blood (45). Support for this hypothesis is culled from the observation that papaverine, a cyclic AMP phosphodiesterase inhibitor (46), is partially effective in alleviating cerebral vascular spasm after intravenous administration or topical application to the vessel wall. Furthermore, adrenergic receptor blocking agents have shown a partial effectiveness in reducing cerebrovascular spasm (47). Peterson's work, which has included the direct application of dibutyryl cyclic AMP to the vessel, has demonstrated that under this influence the relief of spasm has a prolonged effect, and when vessels are further subjected to stimulation, either mechanical or by blood, vascular spasm is no longer demonstrated. Since dibutyryl cyclic AMP is able to pass across cell membranes more effectively than cyclic AMP (48), and since it is also catabolized to a lesser extent than cyclic AMP itself by cyclic AMP phosphodiesterase (49), there may be some potential for this agent in the alleviation of cerebral vasospasm in clinical patients with subarachnoid hemorrhage.

Migraine

Cyclic AMP is the second messenger of a number of circulating vasoactive agents, e.g., serotonin, that have been implicated in the

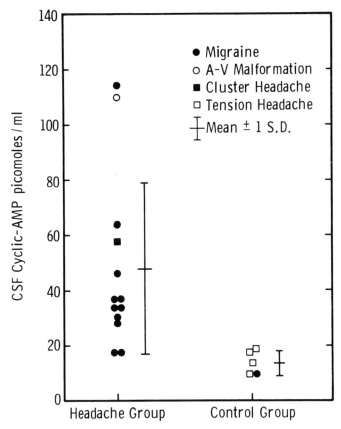

4 Levels of cyclic AMP in CSF obtained from patients during or within 48 hr of a vascular headache of migraine type and from controls asymptomatic for vascular headache or with tension headache. The case of cerebral A-V malformation presented with typical classical migraine attacks. The mean values are significantly different at a $p < 0.05$ level.

etiology of migraine headache (50). Despite evidence that vasoreceptor stimulation may be reflected by alteration in circulating plasma cyclic AMP content (51), in our laboratory no significant differences were observed between plasma levels in migraineurs studied at the time of a migraine attack and levels measured during a headache free interval. Evidence for disordered cyclic AMP metabolism in migraine was, however, obtained from CSF measurement (Fig. 4), cyclic AMP levels becoming elevated, possibly secondary to the transient cerebral ischemia of the migraine prodrome and explained by mechanisms referred to in earlier sections of this chapter. This finding raises the possibility that the effectiveness of propranolol in migraine prophylaxis (52) may be due in part to inhibition of central receptor-stimulated generation in intracellular cyclic AMP, thereby diminishing ischemic metabolic shifts that might otherwise lead to tissue lacticacidosis, dilatation of intraparenchymatous resistance vessels and consequent increase in cerebral blood flow associated with the headache phase of the migraine attack (53).

Brain Trauma

Severe concussive and contusive blows to the cerebral cortex produce slowing of the EEG, reduction in cerebral oxygen consumption ($CMRO_2$) and reduction in regional cerebral blood flow (rCBF), the severity of change in each parameter being dependent on the severity of injury (54). The most severe degree of brain injury, laceration, produces profound decrease in $CMRO_2$, rCBF, cerebral venous oxygen tension, and cerebral venous pH with lactacidosis of brain tissue. Recently, Watanabe and Passoneau reported that unilateral laceration of the cerebral cortex in rats caused marked (aproximately 731 percent) and rapid (within 1 min) bilateral increase in cortical cyclic AMP levels (55). Inhibition of the cyclic AMP rise by pretreatment with theophylline has led these workers to suggest that the cyclic AMP elevation resulted from adenylate cyclase activation by adenosine, which is present in increased amounts in cerebral tissue secondary to decreased phosphorylation of adenine nucleotides under conditions of ischemic anoxia that accompany trauma. Bilateral alteration in cortical cyclic AMP levels seems in keeping with generalized EEG, rCBF, and metabolic changes that follow brain injury. Watanabe and Passoneau also reported that hypothermia, which may have a protective effect on brain function after injury, prevented the increase in cortical cyclic AMP induced by laceration.

Search of the literature has failed to provide evidence of altered cyclic AMP metabolism in clinical patients with brain trauma although other evidence exists of disordered catecholamine and serotonin neurotransmitter metabolism (56), which would, however, represent a further mechanism via which brain tissue cyclic AMP levels could become altered. Correlation of the profound changes in cyclic AMP with the pathogenetic mechanisms of brain trauma therefore remains to be established.

Seizures

Clinical evidence of altered cyclic AMP metabolism in epilepsy appears limited to the report of Myllylä et al. (57) who measured increased CSF levels in patients with seizures of recent onset compared to a group of control patients. Significant elevation in brain levels of cyclic AMP has been measured in association with the development of focal epileptogenic activity following cortical freezing (58). Such cyclic AMP increase could, of course, be a nonspecific effect of agents and events which produce repetitive neuronal depolarization. However, the dibutyryl derivative of cyclic AMP is a potent epileptogenic agent when injected intracerebrally in the rat and cat (33), and topical application of the drug also induces paroxysmal convulsive activity (59). The physiologic property of cyclic AMP, that may account for its propensity for producing seizure activity, may lie in its ability to activate specific cyclic-AMP–dependent membrane-associated protein kinases, thereby inducing in the membrane changes which allow altered ion permeability characteristics (26).

It would be a mistake to imply, from measurement of elevated CSF cyclic AMP levels after seizures, that the nucleotide is a primary factor in the etiology of seizure mechanisms. Massive neuronal depolarization as well as disordered cerebral energy metabolism are all known to occur during and after cortical seizure activity (60) and both might secondarily cause CSF cyclic AMP increase. In addition, catecholamines are known to modulate seizure threshold in experimental animals (61–63). Alteration in cyclic AMP metabolism may therefore occur secondarily to changes in catecholamine function. In this regard, studies showing reduced levels of HVA in CSF of patients with epilepsy may have some relevance (64, 65).

Movement Disorders

The basal ganglia are intimately involved in extrapyramidal control of movement and muscle tone. Intensive studies during the last decade have identified these structures as comprised of interlinking neuronal systems, each apparently specific for certain neurotransmitters. According to current information, the dopaminergic neuronal system seems to be of major importance to extrapyramidal function and an increasing number of movement disorders is now associated with disorder of this system.

Abiotrophy specifically affecting the melanin containing cells of the substantia nigra and neurons of the nigrostriatal system is now well known to produce features of bradykinesia, ridigity and tremor that have been identified as Parkinson's syndrome. These clinical features are the result of dopamine deficiency in the nigrostriatal neurons with consequent reduced delivery of the neurotransmitter at receptor sites (66). Although it is a controversial subject and an hypothesis as yet unproven, the work of Kebabian et al. has suggested that the dopamine receptor for nigrostriatal neurons may be adenylate cyclase (67). The neurotransmitter function of dopamine and its influence on extrapyramidal function therefore may be mediated via alteration in postsynaptic cellular cyclic AMP. This deduction is supported by a number of studies, for example, the observation that dibutyryl cyclic AMP injected into the striatum of normal rats causes contralateral turning of the body in a way similar to that seen after injection of dopamine (68). Furthermore, when administered in excess, dopamine receptor antagonists, such as the neuroleptic drugs, which also prevent postsynaptic generation of cyclic AMP, can produce a near-identical pattern of clinical features to that seen in the degenerative form of Parkinson's syndrome (69).

Unfortunately, clinical studies of cyclic AMP metabolism in patients with Parkinson's disease have not substantiated the theory that disordered cyclic AMP metabolism plays a role in this condition. Cramer (70) demonstrated that basal CSF cyclic AMP levels in patients with Parkinson's disease did not differ from other neurologic patients (Fig. 5). Furthermore, no difference in the rate of cyclic AMP rise was observed after probenecid treatment in these and other neurologic patients (Fig. 6). Neither did L-dopa treatment alter the probenecid-induced accumulation of cyclic AMP in lumbar CSF (see Fig. 6). Nevertheless, such results equally do not rule out the possibility of associated cyclic nucleotide disorder but probably indicate that measurement of cyclic AMP in CSF is not a sufficiently sensitive index of selective neuronal system dysfunction, since the CSF cyclic AMP pool is also contributed to by other neuronal systems which are likely to be intact.

Dopamine receptor activation also seems to be indirectly de-

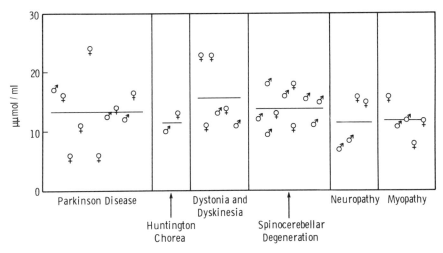

5 Cyclic AMP levels in lumbar CSF of untreated patients with various neurologic disorders. Horizontal bars denote mean values. (Reproduced from Cramer et al.: Arch. Neurol., 29:197-199, 1973.)

pendent upon the activities of the GABAminergic and cholinergic neuronal systems, both of which, on the evidence of recent biochemical estimations in post mortem brain, appear to be impaired in patients with Huntington's disease (71, 72). A resultant decreased inhibitory influence on dopaminergic receptors has been postulated to explain the hyperkinetic movements associated with this disease. Cramer (70) was unable however, to discover in basal CSF cyclic AMP levels any differences that could distinguish patients with the hyperkinetic movement disorder of Huntington's disease from the hypokinetic disorder of Parkinson's syndrome.

Dopamine receptor hyperactivity has also been postluated to explain the hyperkinetic involuntary movements that sometimes develop in patients with Parkinson's disease treated with L-dopa (69). Bernheimer et al. have suggested that the clinical features of Parkinson's disease appear at a relatively late stage of nigrostriatal degeneration since the striatum can compensate to some degree for dopamine deficiency (73). This is perhaps accomplished by increased sensitivity of the dopaminergic receptors, so that although the total amount of dopamine released from remaining intact neurons is reduced, transmission is unimpaired because the

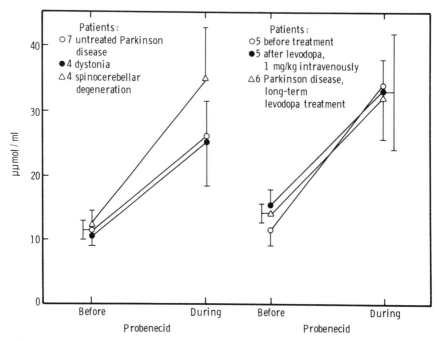

6 Effect of (left) disease or (right) levodopa on the probenecid-induced accumulation of cyclic AMP in lumbar CSF. Probenecid (2 gm) was given orally immediately after the initial lumbar puncture, and again three and six hours later. At right, the five-patient group comprised two with Huntington's chorea, two with dystonia, and one with Parkinson's disease. (Reproduced from Cramer et al.: Arch. Neurol., 29:197–199, 1973.)

neurotransmitter reacts with supersensitive receptor sites. Some workers have speculated that the receptor supersensitivity is brought about by induction of striatal dopamine-sensitive adenylate cyclase (69). The induction of "new DA receptors" might also explain the development of choreiform movements that are signs of striatal dopaminergic hyperactivity in patients receiving chronic neuroleptic therapy. Such hypotheses, however, still require proof, both in the experimental and clinical setting. Nevertheless, possible involvement of cyclic AMP in the production of these hyperkinetic states has been supported by studies of Goldstein et al. (74), who determined that treatment of animals with the phosphodiesterase inhibitor, theophylline, enhanced and prolonged the involuntary movements induced by agents that stimulate the dopamine receptor.

Essential tremor is a 6 to 12 cps tremor of action fixation seen in various parts of the body, especially the hand, which is a familial as well as an isolated disorder coming on particularly late in life, thought to be due to a physiologic tremor that is accentuated because of a genetically determined functional deficit of the nervous system. The tremor is made worse probably as a result of peripheral catecholamine liberation (75). While no measurements of cyclic AMP have been made in patients with this condition, the consistent effectiveness of the beta-adrenergic blocking agent propranolol in alleviation of the symptoms (76) suggests that cyclic AMP could be involved in the chain of neuronal events that bring about the syndrome.

Dementia

Despite the high incidence and apparently increasing prevalence of dementia, little is known of its etiology. There have been recent studies, however, that implicate disordered neurotransmitter function associated with this disorder. The specific neuronal abiotrophic process of Parkinson's disease, sometimes accompanied by dementia (77), has now been subjected to detailed clinical, morphologic and neurochemical analysis leading to clearly effective therapy. The major deficit in this disorder as discussed earlier appears to be dysfunction and depletion of the essential neurotransmitter of the dopaminergic system. Huntington's disease is another form of apparently specific neuronal abiotrophy accompanied by dementia in which there appears to be a major biochemical defect. Biochemical analysis of autopsied brain specimens has provided evidence of disordered GABA metabolism and perhaps associated GABA–dopamine imbalance responsible for hyperkinetic features of the syndrome (71, 72).

Until recently, little attention has been focused on the investigation of any possible underlying disorder of neurotransmitter function in patients with the primary degenerative dementia of Alzheimer's type (previously termed presenile dementia). Studies in autopsied brain of patients with the disease have, however, shown decreased activity of the enzymes dopa decarboxylase (DOPAD) and glutamic acid decarboxylase (GAD) (78). This decrease could not be explained on the basis of general cortical atrophy but appeared more likely to be due to accelerated and specific fallout of the neurotransmitter neuronal system. Measurements of CSF monoamine metabolites by Gottfries et al. have consistently demonstrated low levels of HVA and 5-HIAA in patients with Alzheimer's disease (79). A reduced turnover of dopamine and less significantly serotonin was demonstrated in the same patients by means of the probenecid test. These studies have been supported by measurements which showed low levels of CSF HVA and 5-HIAA in Alzheimer's disease patients with an associated rigid-akinetic type of Parkinsonism (80).

The hypothesis of possible neurotransmission disorder in patients with Alzheimer's disease is backed up by certain experimental and morphological observations. Ultrastructural study of neurofibrillary tangles, found on examination of autopsied brain samples, suggests that they may be formed by a degeneration of neuronal elements essential for intracellular transport processes (81). In specific neurotransmitter neuronal systems, transport failure can result in impaired transfer of synaptic vesicles, enzymes essential for neurotransmitter synthesis and breakdown, as well as failure of transport of the neurotransmitters themselves (82). In accord with this, enlargement of presynaptic terminals, accumulations of vacuolar material, and reduced numbers of synaptic vesicles have been observed in biopsied tissue from Alzheimer's disease patients (83).

Independent of whether the severity of dementia in Alzheimer's disease is related to gross loss of cerebral tissue, or is due to selective neuronal abiotrophy, as suggested by measurement of GABA and monoamine metabolites, or is due even to a primary synaptic functional defect, some associated disorder of cyclic AMP function might be expected. Few studies, however, have evaluated cyclic AMP metabolism in dementia states. As referred to in an earlier section, Cramer reported basal levels of CSF cyclic AMP to be within the range of controls in two patients with dementia due to Huntington's disease (70). When patients with Alzheimer's disease were studied in our laboratory, CSF cyclic AMP levels after probenecid directly correlated with the severity of dementia as assessed by neurophychological measurements (Table 1), the more demented the patient the lower the CSF values observed.

The interpretation of the aforementioned finding is complex. The problems of interpreting CSF cyclic AMP levels after probenecid as representative of central metabolic change in the nervous system have been discussed in an earlier section, and as yet the procedure has not been extensively evaluated. Furthermore,

Table 1
Correlation of Cyclic AMP Accumulation in Lumbar CSF After Probenecid (4.5 gm in 9 hours) with Performance on Neuropsychological Testing in Patients with Alzheimer's Disease

Case No.	MQ	PIQ	Cyclic AMP** (picomoles/ml)
1	40	40	22.4
2	68	58	20.0
3	69	76	62.0
4	96	91	27.4
5	85	103	48.0
6	40	40	23.8
7	97	90	44.5
8	64	40	19.5

Significance on Spearman rank correlation*:
- MQ: $p < 0.001$
- PIQ: $p < 0.05$
- Cyclic AMP: $p < 0.05$

*one-tailed t-test
**CSF values after probenecid
MQ: Memory quotient
PIQ: Performance intelligence quotient

normal control series have not been studied, making it impossible to state whether the cyclic AMP rise after probenecid in dementia patients is in fact reduced compared to normal, a finding implied by this limited study. Another objection is that lower values after probenecid simply may reflect reduction in cerebral tissue mass, that contributes to the CSF cyclic AMP pool. Myllylä et al. reported reduction in basal CSF cyclic AMP values in patients with cerebral atrophy demonstrated on pneumoencephalography as compared to values in patients without demonstrated atrophy (8). Cyclic AMP values in the series under discussion here, however, bore no correlation to brain weights calculated from regional cerebral blood flow measurement. It is concluded that results in the small series of patients reported here seem in keeping with the hypothesis of disordered cyclic AMP metabolism in dementia, although other supportive evidence is unavailable.

Neuromuscular Disease

Drugs which elevate cyclic AMP levels by stimulating its formation (e.g., norepinephrine, ephedrine) are capable of facilitating neuromuscular transmission, whereas drugs which inhibit cyclic AMP breakdown by phosphodiesterase inhibition (e.g., theophylline, caffeine, and papaverine) augment the tension and contraction of striated muscle (84). Possibly, cyclic AMP is involved in acetylcholine release from nerve terminals, since calcium is essential for acetylcholine quanta released or nerve impulse and the importance of cyclic AMP in cellular calcium mobilization and availability seems well-established.

These theoretical and experimental observations are borne out in part by a small number of clinical studies. Hokkanen et al. reported clinical benefit in 9 of 13 patients with myasthenia gravis treated with papaverine (85). Takanori et al., however, using electromyographic techniques in patients with classical myasthenia gravis, were unable to substantiate any effect of methylxanthines on neuromuscular transmission (86). Nevertheless, in patients with myasthenic myopathy, the Eaton-Lambert syndrome, in which calcium-dependent acetylcholine release by nerve impulses is well documented to be defective, distinct improvement of neuromuscular transmission was achieved after epinephrine as well as methylxanthine treatment. These workers further demonstrated that the favorable presynaptic modulation of function after epinephrine was produced via the alpha-adrenergic receptor. Despite the aforementioned studies the use of agents that modify cyclic AMP metabolism at the neuromuscular junction does not appear to have gained a place in the routine clinical management of patients with myasthenia states.

AFFECTIVE DISORDERS

Distrubance of central neurotransmitter function has been postulated to occur in the affective disorders (87). The depressive phase of this condition is thought to be associated with relative deficiency of neurotransmitter (particularly catecholamine) at the postsynaptic neuronal receptor, while the manic phase is associated with excess transmitter at similar sites. Since cyclic AMP in its second messenger role is intimately involved in a reaction chain of

Table 2
Twenty-Four-Hour Cyclic AMP Excretion

Group	Micromoles Cyclic AMP Excreted per 24 hours ± S.E.M.*	Number of Samples**	Number of Patients	p value***
Normal	5.64 ± 0.68	34	10	$>.05 <.10$
Psychotic depression	3.64 ± 0.19	17	7	$>.05$
Manic	9.94 ± 1.88	10	5	$<.01$
Neurotic depression	6.70 ± 0.37	47	25	$>.10$

*This is the mean of the patient means such that each patient is represented by a single number S.E.M. = Standard error of the mean.
**The number of samples is given for information only.
***The p values represent comparison between normal and patient groups; they are computed using an approximation of the Tukey method, an approximation necessary because of differing number of patients in the sample.

(Reproduced from Paul et al: American Journal of Psychiatry, volume 126, pp. 1493-1497, 1970).

synaptic and neuronal events concerned with neurotransmission, this biochemical parameter has been an obvious candidate for study in the substantiation of the neurotransmitter hypothesis of affective illness.

Studies by Paul et al. (88) revealed that 24 hour urinary excretion of cyclic AMP in a cohort of patients studied during the depressive stage of the affective illness was reduced below control values and elevated in patients evaluated during a manic phase of the illness (Table 2). When these studies were extended in relation to the response of individuals to therapy, it was shown that, in patients treated with lithium, urinary cyclic AMP excretion increased with alleviation of depressive symptomatology (89). Similarly, values decreased from previously high levels in manic patients as their symptoms normalized in response to Lithium therapy. Depressed patients treated with L-dopa also showed normalization of cyclic AMP excretion values as successful therapeutic response was obtained.

Pursuing their studies further, Paul et al. (90) demonstrated marked elevations of urinary cyclic AMP excretion on the day of rapid "switch" from depressed to manic phases (Fig. 7) and

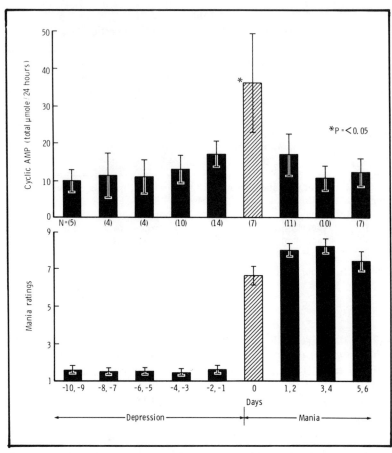

7 Increase in urinary cyclic AMP associated with the switch from depression to mania (N = 7 episodes). (Reproduced from Paul et al.: Amer. J. Phychiatr., 126:1493-1497, 1970. Copyright 1970, the American Psychiatric Association. Reprinted by permission.)

further postulated this to be possible trigger factor to central catecholamine release and the behavioral changes of mania.

The work of Abdulla and Hamada (91), independently and coincidentally performed, closely replicated the results of the above studies, patients with affective illness again showing decreased urinary cyclic AMP excretion during the depressive phase of the illness with a recovery of levels towards control values as symptoms and signs resolved after therapy. Naylor et al. (92) in a smaller series, but one in which patients were used as their own control, confirmed increase of cyclic AMP excretion as patients recovered from their depressed illness due to therapy with tricyclic antidepressants or electroconvulsive therapy. Although their studies provided neither correlation of cyclic AMP levels to the severity of symptomatology or evidence of a relationship between the rate of increase of urinary cyclic AMP excretion and the alleviation of symptomatology by treatment, they did show that increase in urine volume was not related to the observed increase in cyclic AMP excretion found in their own and other studies.

Although a few limited case series have failed to confirm alteration of cyclic AMP metabolism (93), it seems reasonably established that urinary cyclic AMP excretion is altered in affective illness. The interpretation of these findings, however, is the subject of some debate.

Assuming that the cyclic AMP findings in the series quoted above do reflect alteration in central nervous system turnover of the cyclic nucleotide, how can they be interpreted in terms of the etiology of affective illness? Neurotransmitter depletion at the presynaptic and postsynaptic receptors sites certainly might lead to decreased intracellular cyclic AMP generation but in addition, enhanced activity of the cyclic AMP catabolic enzyme, cyclic AMP phosphodiesterase, might also militate towards low central activity. Although far from being unequivocal, studies of biogenic amine and amine metabolites in urine, CSF, and autopsied brain samples have supported the concept of reduced neurotransmitter availability at receptor sites, in association with depressive symptomatology (94–96). Prange and associates have used an alternative approach to prove disorder of neurotransmission in affective states. They evaluated the physiological response of depressed subjects to pharmacological agents and correlated recovery from depressive symptomatology with enhanced pressor response to infused norepinephrine (97). This was interpreted to indicate a diminished catecholamine receptor responsivity. There is also considerable evidence that drugs such as the tricyclic antidepressants may not only increase availability of neurotransmitters at postsynaptic receptor sites but may also elevate intracellular cyclic AMP levels by inhibiting cyclic AMP phosphodiesterase (98).

Involvement of cyclic AMP in affective illness is indirectly supported by the studies of Bunney et al., who in 10 patients showed sharp increase in urinary norepinephrine excretion on the day before onset of mania (94). During the manic episode urinary catecholamine excretion remained increased and was associated with decreased REM and total sleep time. Shildkraut, during several longitudinal studies of individual patients, observed that urinary 3-methoxy-4-hydroxyphenylglycol (MHPG) excretion was higher during mania, intermediate during asymptomatic intervals and lower during depression in patients with manic-depressive illness (99). MHPG is a norepinephrine metabolite which provides an additional index of synthesis and metabolism of norepinephrine in brain.

Although it seems established that fluctuation in urinary cyclic AMP levels may occur in affective illness, there are reasonable objections to the assumption that the values reflect altered central cyclic AMP metabolism with resultant behavioral change. This objection was emphasized by the work of Robison et al. (7), who measured concentrations of cyclic AMP in CSF of both manic and depressed patients and found no statistical difference in values between these groups or between a control group of neurological patients. Further, Cramer et al. have reported no difference in CSF values between manic patients and controls (100). On the contrary, these workers found elevation of cyclic AMP in depressed patients compared to a group of neurologic patients. Both the studies quoted above unfortunately made comparison using patients with neurological disease as controls. CSF cyclic AMP levels may also be affected in a variety of neurologic diseases so that derival of an hypothesis based on such studies can reasonably be questioned. Indeed, Cramer et al. found in two patients with mania that there was a more marked rise in cyclic AMP after probenecid than in other patients (100).

The interpretation of alteration in urinary cyclic AMP values as an indication of alteration in central nervous system cyclic AMP turnover raises the most serious objection to the studies that have used this approach to support the neurotransmitter hypothesis of affective illness. Approximately 40 percent of cyclic AMP in urine is produced by the kidney itself; the remainder is contributed by glomerular filtration. Since cyclic AMP has been implicated as an important metabolic regulator of target organs of endocrinal systems and is intimately involved in many systemic metabolic pathways, alteration in circulating humoral levels, e.g., glucagon, parathyroid hormone, or change in systemic metabolism may affect plasma, renal and hence urinary cyclic AMP levels (for review see Broadus et al. [101]). For example, Jenner and his colleagues (102) have argued that alteration in urinary cyclic AMP excretion occurs secondary to increased ADH secretion and the influence this has on ADH sensitive renal adenylate cyclase. Nevertheless, since ADH secretion is probably under hypothalamic control, this could indicate the fundamental origin of the syndrome.

Since urinary cyclic AMP excretion is so related to systemic humoral or metabolic change it is necessary to exclude fluctuation, in both stress and physical activity, which occurs during manic and depressive phases of affective illness, as being responsible for the fluctuation of urinary cyclic AMP levels. For example, stress is known to promote hypothalamic corticotrophin releasing factor (CRF) leading to anterior pituitary ACTH release. The subsequent stimulation of corticosteroid production is mediated by increased cyclic AMP production in adrenal tissues (103). There is similar controversy concerning the effects of physical activity on cyclic AMP production. While Paul concluded that after prolonged physical activity in normal subjects urinary cyclic AMP excretion was unaltered (87), Eccleston et al. have produced evidence to the contrary (104).

An interesting alternative explanation for the findings of Paul and others has been put forward by Hansen (105) who (in keeping with the early work of Waelsch and Weil-Malherbe [106] who demonstrated reduced systemic glucose utilization in depressed patients) found a relationship between the degree of elevated whole blood uridine diphosphate glucose (UDPG) levels with severity of depressive symptomatology in his patients with affective disorders. Hansen also found decreased levels of ATP in the same patients, in keeping with the initial property of UDPG to increase the rate of ATP hydrolysis. Reduced cyclic AMP levels might be expected in these patients as a result of reduced ATP, as was indeed the case. Hence, at least in the depressive phase of affective illness, decreased urinary cyclic AMP excretion could be explained on this basis.

Some workers have claimed the validity of platelet receptor studies as an indication of central neuronal activity. Although interesting results have been obtained by this approach in some neurological diseases, existing studies of platelet cyclic AMP function in patients with psychiatric illness have proved disappointing. Wang et al. (107) studied platelets obtained from depressed patients and normal controls. No differences were noted between either group in the ability of prostaglandin E_1 to stimulate cyclic AMP formation or in the inhibition of this response by norepinephrine, thus suggesting absence of a generalized defect of adrenergic responses in depression. Murphy et al. (108) extended these studies and arrived at identical conclusions.

In conclusion, alteration of urinary cyclic AMP excretion related to manic-depressive illness seems established. However, it is unknown whether these changes are mediated centrally or are simply a reflection of peripheral metabolic change. It seems impossible to interpret urinary cyclic AMP excretion as an index of alteration in any specific central neuronal system. Nevertheless, with certain reservations, the studies reviewed in this section have contributed some support for the hypothesis of an underlying organic basis for affective illness.

SUMMARY

The human brain has a great capacity for formation of cyclic AMP, the function of the synthetic enzyme, adenylate cyclase, and the catabolic enzyme, cyclic AMP phosphodiesterase, being similar to that in the central nervous system of other mammalian organisms. Cyclic AMP is also measurable in CSF.

Possible involvement of cyclic AMP in various neurological diseases was reviewed. Altered cyclic AMP metabolism during cerebral ischemia associated with cerebrovascular occlusive disease and migraine may be responsible for disordered neurotransmission, shifts in cerebral energy metabolism and adverse behavorial effects, although it may also serve to initiate recovery processes in ischemic brain. Cyclic AMP relieves cerebral vasospasm associated with subarachnoid hemorrhage. Cyclic AMP was increased in traumatized brain tissue. Increased levels of cyclic AMP in brain tissue and CSF after seizure indicate that cyclic AMP is involved in epileptogenic activity, although the primary or secondary nature of this involvement cannot be determined. Although research in experimental models suggests disorder of cyclic AMP metabolism both in the hypokinetic and hyperkinetic movement disorders, clinical investigation reveals no supportive evidence for these findings. Preliminary results, which suggest diminished cyclic AMP synthesis in the brain of patients with dementia, are presented. Drugs that stimulate cyclic AMP synthesis at the neuromuscular junction facilitate neuromuscular transmission, although the use of such agents does not appear to have gained a place in the clinical management of patients with defective neuromuscular transmission.

With certain reservations, the studies which have described disordered cyclic AMP metabolism in affective illness support the theory that this condition has an organic basis.

Clinical study of cyclic AMP metabolism in neurologic and affective disorders is still extremely limited. Although in the physiologic state the "second messenger" role of cyclic AMP seems well established, its role as a mediator of disease processes still requires definition.

REFERENCES

1. Sutherland, E. W., and Robison, G. A.: Metabolic effect of catecholamine. Section A. The role of cyclic-3',5'-AMP in response to catecholamines and other hormones. Pharmacol. Rev. 18:145-161, 1966.
2. Sutherland, E. W., Robison, G. A., and Butcher, R. W.: Some aspects of the biological role of adenosine 3',5'-monophosphate (cyclic AMP). Circulation 37:279-306, 1968.
3. Klainer, L. M., Chi, Y. M., Friedberg, S. L., Rall, T. W., and Sutherland, E. W.: Adenyl cyclase. IV. The effects of neurohormones on the formation of adenosine 3',5'-phosphate by preparations from brain and other tissues. J. Biol. Chem. 237:1239-1243, 1962.
4. Rall, T. W., and Gilman, A. G.: The role of cyclic AMP in the nervous system. N.R.P. Bull. 8:219-323, 1970.
5. Williams, R. H., Little, S. A., Beug, A. G., and Ensinck, J. W.: Cyclic nucleotide phosphodiesterase activity in man, monkey, and rat. Metabolism 20:743-748, 1971.
6. Kodama, T., Matsukado, Y., and Shimizu, H.: The cyclic AMP system of human brain. Brain Res. 50:135-146, 1973.
7. Robison, G. A., Coppen, A. J., Whybrow, P. C., and Prance, A. J.: Cyclic AMP in affective disorders. Lancet 2:1028-1029, 1970.
8. Myllylä, V. V., Vapaatalo, H., Hokkanen, E., and Heikkinen, E. R.: Cerebrospinal fluid concentration of cyclic adenosine-3',5'-monophosphate and pneumoencephalography. Eur. Neurol. 12:28-32, 1974.
9. Heikkinen, E. R., Myllylä, V. V., Vapaatalo, H., and Hokkanen, E.: Urinary excretion and cerebrospinal fluid concentration of cyclic adenosine-3',5'-monophosphate in various neurological diseases. Eur. Neurol. 11:270-280, 1974.
10. Cramer, H., Ng, N. K. Y., and Chase, T. N.: Effect of probenecid on levels of cyclic AMP in human cerebrospinal fluid. J. Neurochem. 19:1601-1603, 1972.
11. Sebens, J. B., and Korf, J.: Cyclic AMP in cerebrospinal fluid: Accumulation following probenecid and biogenic amines. Exp. Neurol. 46:333-344, 1975.
12. Hikada, H., Shibuya, M., Asano, T., and Hara, F.: Cyclic nucleotide phos-

phodiesterase of human cerebrospinal fluid. J. Neurochem. 25:49-53, 1975.
13. Berzin, Y. E., Auna, Z. P., and Brezhinskiy, G. Y.: The significance of blood serotonin and CSF for the clinical picture and pathogenesis of acute cerebral circulatory disturbance. Zh. Nevropatol. Psikhiatrii. 69:1011-1015, 1969.
14. Stoica, E., Pausescu, E., and Trandafirescu, E.: Variations in plasma serotonin and catecholamines induced by cold stress in patients with cerebrovascular accidents. Rev. Roum. Neurol. 5:267-277, 1968.
15. Tomomatsu, T., Ueba, Y., Matsumoto, T., Ikoma, S., Kondo, N., Ijiri, S.. and Oda, M.: Electrocardiographic observations and urinary excretion of catecholamines in cardiovascular disorders. Jap. Circ. J. 28:905-992, 1964.
16. Toyoda, M., and Meyer, J. S.: Effects of cerebral ischemia, seizures, and anoxia on arterial concentrations of catecholamines. Cardiovasc. Res. Cent. Bull. 8:59-74, 1969.
17. Meyer, J. S., Stoica, E., Pascu, I., Shimazu, K., and Hartmann, A.: Catecholamine concenrations in CSF and plasma of patients with cerebral infarction and hemorrhage. Brain 96:277-288, 1973.
18. Welch, K. M. A., Meyer, J. S., and Chee, A. N. C.: Evidence for disordered cyclic AMP metabolism in patients with cerebral infarction. Eur. Neurol. 13:144-154, 1975.
19. Steiner, A. L., Ferrendelli, J. A., and Kipnis, D. M.: Radioimmunoassay for cyclic nucleotides. III. Effect of ischemia, changes during development and regional distribution of adenosine 3',5'-monophosphate and guanosine 3',5'-monophosphate in mouse brain. J. Biol. Chem. 247:1121-1124, 1972.
20. Welch, K. M. A., Meyer, J. S., Teraura, T., Hashi, K., and Shimazu, S.: Ischemic anoxia and cerebral serotonin levels. J. Neurol. Sci. 16:85-92, 1972.
21. Zervas, N. T., Hori, H., Negoro, M., Wurtman, R. J., Larin, F., Lavyne, M. H.: Reduction in brain dopamine following experimental cerebral ischemia. Nature 247:283-284, 1974.
22. Welch, K. M. A., Meyer, J. S., Chabi, E., and Buckingham, J.: The effect of ischemia on catecholamine and 5-hydroxytryptamine levels in the cerebral cortex of gerbils: Influence of parachlorophenylalanine (PCPA). In Blood Flow and Metabolism in the Brain, A. M. Harper, W. B. Jennett, J. D. Miller, and J. O. Rowan (Eds.) pp. 3.18-3.22, Churchill Livingstone, Edinburgh, London and New York, 1975.
23. Kogure, K., Scheinberg, P., Matsumoto, A., Busto, R., and Reinmuth, O. M.: Catecholamines in experimental brain ischemia. Arch. Neurol. 32: 21-24, 1975.
24. Meyer, J. S., Welch, K. M. A., Okamoto, S., and Shimazu, K.: Disordered neurotransmitter function. Demonstration by measurement of norepinephrine and 5-hydroxytryptamine in CSF of patients with recent cerebral infarction. Brain 97:655-664, 1974.
25. Meyer, J. S., Shinohara, Y., Kanda, T., Fukuuchi, Y., Ericsson, A. D., and Kok, N. K.: Diaschisis resulting from acute unilateral cerebral infarction: quantitative evidence for man. Arch. Neurol. 23:241-247, 1970.
26. Greengard, P., McAfee, D. A., and Kebabin, W.: On the mechanism of action of cyclic AMP and its role in synaptic transmission. In Advances in Cyclic Nucleotide Research, P. Greengard, and G. A. Robison, (Eds.) p. 337, Vol. 1, Raven Press, New York, 1972.

27. Eshler, C. J., and Ammon, H. P. T.: The influence of the sympatheticolytic agent, propranolol, on glycogenolysis and glycolysis in muscle, brain and liver of white mice. Biochem. Pharmacol. 15:2031-2035, 1966.
28. Lowry, O. H., and Passoneau, J. V.: Kinetic evidence for multiple binding sites on phosphofructokinase. J. Biochem. 241: 2268-2279, 1966.
29. Meyer, J. S., Sawada, T., Kitamura, A., and Toyoda, M.: Cerebral oxygen, glucose, lactate, and pyruvate metabolism in stroke. Therapeutic considerations. Circulation 37:1036-1048, 1968.
30. Meyer, J. S., Ryu, T., Toyoda, M., Shinohara, Y., Wiederholt, I., and Guiraud, B.: Evidence for a Pasteur effect regulating cerebral oxygen and carbohydrate metabolism in man. Neurology 19:954-962, 1969.
31. Asakawa, T., and Yoshida, H.: Studies on the functional role of adenosine 3',5'-monophosphate, histamine and prostaglandin E_1 in the central nervous system. Jap. J. Pharmacol. 21:569-583, 1971.
32. Breckenridge, B. M., and Lisk, R. D.: Cyclic adenylate and hypothalamic regulatory functions. Proc. Soc. Exp. Biol. Med. 131:934-935, 1969.
33. Gessa, G. L., Krishna, G., Forn, J., Tagliamonti, A., and Brodie, B. B.: Behavioral and vegetative effects produced by dibutyryl cyclic AMP injected into different areas of the brain. In Advances in Biochemical Psychopharmacology, P. Greengard, and E. Costa (Eds.) pp. 371-381, Vol. 3, Raven Press, New York, 1970.
34. Schwartz, J. P., Mrsulja, B. B., Mrsulja, B. J., Passoneau, J. V., and Klatzo, I.: Alteration in the enzymes involved in cyclic nucleotide metabolism following unilateral ischemia and recovery in gerbils. In Neuroscience Abstracts, 5th Annual Meeting of the Society for Neurosciences, Vol. 1, Society for Neuroscience, Bethesda, Maryland, p. 349, 1975 (abstract).
35. Gaudet, R., Chabi, E., Welch, K. M. A., and Wang, B.: Effect of transient ischemia on monoamine levels in the cerebral cortex of gerbils. In Neuroscience Abstracts, 5th Annual Meeting of the Society for Neurosciences, Vol. 1, Society for Neuroscience, Bethesda, Maryland, p. 397, 1975 (abstract).
36. Yatsu, F. M., Lee, L-W., and Liao, C-L.: Energy metabolism during brain ischemia. Stroke 6:678-683, 1975.
37. Gertler, M. M., Saluste, E., Leetma, H. E., and Guthrie, R. G.: The use of cyclic AMP (CAMP) in congestive heart failure and stroke. In Advances in Cyclic Nucleotide Research, G. I. Drummond, P. Greengard, and G. A. Robison (Eds.) p. 574, Vol. 5, 2nd International Conference on Cyclic AMP, Raven Press, New York, 1975.
38. Welch, K. M. A., Meyer, J. S., Okamoto, S., Itoh, Y., Sakaki, S., Miyakawa, Y., Ericsson, A. D., Chabi, E., and Chee, A. N.: Beneficial effects of glycerol in cerebral infarction—osmotic or metabolic? Trans. Amer. Neurol. Assoc. 99:17-20, 1974.
39. Kogure, K., Scheinberg, P., Busto, R., and Reinmuth, O. M.: Influence of functional activity of neurons on the fate of ischemic brain tissue. In Blood Flow and Metabolism in the Brain, A. M. Harper, W. B. Jennett, J. D. Miller, and J. O. Rowan (Eds.) pp. 12.19-12.21, Churchill Livingston, Edinburgh, London and New York, 1975.
40. Ganser, V., and Boksay, I.: Effect of pentoxifylline on cerebral edema in cats. Neurology 5: 487-493, 1974.
41. Triner, L., Nahas, G. G., Vulliemoz, Y., Overweg, N. A., Verosky, M.,

Habif, D. V., and Ngai, S. H.: Cyclic AMP and smooth muscle function. Ann. N.Y. Acad. 185:458-476, 1971.
42. Volicier, L., and Hynie, S.: Effect of catecholamines and angiotensin on cyclic AMP in rat aorta and tail artery. Eur. J. Pharmacol. 15:214-220, 1971.
43. Andersson, R.: Role of cyclic AMP and Ca++ in mechanical and metabolic events in isometrically contracted vascular muscle. Acta Physiol. Scand. 87:84-95, 1973.
44. Peterson, E. W., Searle, R., Francis, F., Mandy, B., and Leblanc, R.: The reversal of experimental vasospasam by dibutyryl 3',5' adenosine monophosphate. J. Neurosurg. 39:730-734, 1973.
45. Peterson, E. W., Leblanc, R., Searle, R., and Mandy, F.: Some considerations on cerebral vasospasm. Am. Heart J. 89:124-126, 1975.
46. Kukovetz, W. R., and Poch, G.: Inhibition of cyclic 3',5' nucleotide phosphodiesterase as a possible mode of action of papaverine and similarly acting drugs. Naunyn. Schniedebergs Arch. Pharmacol. 267:189-194, 1970.
47. Fiamm, E. S., Yasargil, M. G., and Ransohoff, J.: Control of cerebral vasospasam by parenteral phenoxybenzamine. Stroke 3:42, 1972.
48. Robison, G. A., Butcher, R. W., and Sutherland, E. W.: Some actions of cyclic AMP. In Cyclic AMP, G. A. Robison, R. W. Butcher and E. W. Sutherland (Eds.) pp. 91-144. Academic Press, New York, 1971.
49. Posternak, Th., Sutherland, E. W., and Henion, W. F.: Derivatives of cyclic 3',5'-adenosine monophosphate. Biochim. Biophys. Acta 65:558-560, 1962.
50. Anthony, M., Hinterberger, H., and Lance, J. W.: Plasma serotonin in migraine and stress. Arch. Neurol. 16:544-552, 1967.
51. Kaminsky, N. I., Ball, J. H., Broadus, A. E., Hardman, J. G., Sutherland, E. W., and Liddle, G. W.: Hormonal effects on extracellular cyclic nucleotides in man. Trans. Assoc. Amer. Phys. 83:235-244, 1970.
52. Weber, R. B., and Reinmuth, O. M.: The treatment of migraine with propranolol. Neurology 22:366-369, 1972.
53. Skinhøj, E.: Hemodynamic studies within the brain during migraine. Arch. Neurol. 29:95-98, 1973.
54. Meyer, J. S., Kondo, A., Nomura, F., Sakamoto, K., and Teraura, T.: Cerebral hemodynamics and metabolism following experimental head injury. J. Neurosurg. 22:304-319, 1970.
55. Watanabe, H., and Passoneau, J. V.: Cyclic adenosine monophosphate in cerebral cortex. Arch. Neurol. 32:181-184, 1975.
56. Vecht, C. J., van Woerkom, Th. C .A. M., Teelken, A. W., and Minderhoud, J. M.: Homovanillic acid and 5-hydroxyindoleacetic acid cerebrospinal fluid levels. Arch. Neurol. 32:792-797, 1975.
57. Myllylä, V. V., Heikkinen, E. R., Vapaatalo, H., and Hokkanen, E.: Cyclic AMP concentration and enzyme activities of cerebrospinal fluid in patients with epilepsy or central nervous system damage. Eur. Neurol. 13: 123-130, 1975.
58. Walker, J. E., Lewin, E., Sheppard, J. R., and Cromwell, R.: Enzymatic regulation of adenosine 3',5'-monophosphate (cyclic AMP) in the freezing

epileptogenic lesion of rat brain and in homologous contralateral cortex. J. Neurochem. 21:79-85, 1973.
59. Purpura, D. P.,˙ and Shofer, R. J.: Excitatory action of dibutyryl cyclic adenosine monophosphate on immature cerebral cortex. Brain Res. 38: 179-181, 1972.
60. Meyer, J. S., Gotoh, F., and Favale, E.: Cerebral metabolism during epileptic seizures in man. Electroencephalog. Clin. Neurophysiol. 21:10-22, 1966.
61. Azzaro, A. J., Wenger, G. R., Craig, C. R., and Stitzel, R. E.: Reserpine-induced alterations in brain amines and their relationship to changes in the incidence of minimal electroshock seizures in mice. J. Pharmacol. Exp. Ther. 180:558-568, 1972.
62. Wenger, G. R., Stitzel, R. E., and Craig, C. R.: The role of biogenic amines in the reserpine-induced alteration of minimal electroshock seizure thresholds in the mouse. Neuropharmacology 12:693-703, 1973.
63. Kilian, M., and Frey, H-H.: Central monoamines and convulsive thresholds in mice and rats. Neuropharmacology 12:681-692, 1973.
64. Papeschi, R., Molino-Negro, P., Sourkes, T. L., and Erba, G.: The concentration of homovanillic and 5-hydroxyindoleacetic acids in ventricular and lumbar CSF: Studies in patients with extrapyramidal disorders, epilepsy and other diseases. Neurology 22:1151-1159, 1972.
65. Shaywitz, B. A., Cohen, D. J., and Bowers, M. B., Jr.: Reduced cerebrospinal fluid 5-hydroxyindoleacetic acid and homovanillic acid in children with epilepsy. Neurology 25:72-79, 1975.
66. Ehringer, H., and Hornykiewicz, O.: Verteilung von noradrenalin und dopamin (3-hydroxytyramin) im gehirn des menschen und ihr verhalten bei erkrankungen des extrapyramidalen systems. Klin. Wschr. 38:1236, 1960.
67. Kebabian, J. W., Petzold, G. L., and Greengard, P.: Dopamine-sensitive adenylate cyclase in caudate nucleus of rat brain and its similarity to the "dopamine receptor." Proc. Natl. Acad. Sci. U.S.A. 69:2145-2149, 1972.
68. Iversen, L. L., Horn, A. S., and Miller, R. J.: Actions of dopaminergic agonists on cyclic AMP production in rat brain homogenates. In Advances in Neurology, Vol. 9, D. B. Caine, T. N. Chase, and A. Barbeau (Eds.) pp. 197-212, Raven Press, New York, 1975.
69. Hornykiewicz, O.: Parkinsonism induced by dopaminergic antagonists. In Advances in Neurology, Vol. 9, D. B. Caine, T. N. Chase, and A. Barbeau (Eds.) pp. 155-164, Raven Press, New York, 1975.
70. Cramer, H., Ng, L. K. Y., and Chase, T. N.: Adenosine 3',5'-monophosphate in cerebrospinal fluid. Arch. Neurol. 29:197-199, 1973.
71. Bird, E. D., and Iversen, L. L.: Huntington's chorea: Post-mortem measurement of glutamic acid decarboxylase, choline acetyltransferase and dopamine in basal ganglia. Brain 97:457-472, 1974.
72. Perry, T. L., Hansen, S., and Kloster, J.: Huntington's chorea: Deficiency of gamma aminobutyric acid in brain. New Eng. J. Med. 288:337-342, 1973.
73. Bernheimer, H., Birkmayer, W., Hornykiewicz, O., Jellinger, K., and Seitelberger, F.: Brain dopamine and the syndromes of Parkinson and

Huntington: Clinical, morphological and neurochemical correlations. J. Neurol. Sci. 20:415-455, 1973.
74. Goldstein, M., Battista, A. F., and Miyamoto, T.: Modification of involuntary movements by central acting drugs. In Advances in Neurology, Vol. 9, D. B. Caine, T. N. Chase, and A. Barbeau (Eds.) pp. 299-305, Raven Press, New York, 1975.
75. Marsden, C. D., and Owen, D. A. L.: Mechanisms underlying emotional variation in Parkinsonian tremor. Neurology 17:711-715, 1967.
76. Dupont, E., Hansen, H. J., and Dalby, M. A.: Treatment of benign essential tremor with propranolol. A controlled clinical trial. Acta Neurol. Scandinav. 49:75-84, 1973.
77. Pollock, M., and Hornabrook, R. W.: The prevalence, natural history and dementia of Parkinson's disease. Brain 89:429, 1966.
78. Bowen, D. M., White, P., Flack, R. H. A., Smith, C., and Davison, A. N.: Brain decarboxylase activities as indices of pathological change in senile dementia. Lancet 1:1247-1249, 1974.
79. Gottfries, C. G., Gottfries, I., and Roos, B. E.: Homovanillic acid and 5-hydroxyindoleacetic acid in the cerebrospinal fluid of patients with senile dementia, presenile dementia and Parkinsonism. J. Neurochem. 16: 1341-1345, 1969.
80. Parkes, J. D., Marsden, C. D., Rees, J. S., Curzon, G., Kantamaneni, B. D., Knill-Jones, R., Akbar, A., Das, S., and Kataria, M.: Parkinson's disease, cerebral arteriosclerosis and senile dementia. Quart. J. Med. 63: 49-61, 1974.
81. Terry, R. D., and Wísniewski, H. M.: Ultrastructure of senile dementia and of experimental analogs. In Aging and the Brain, C. M. Gaitz (Ed.) pp. 89-116, Plenum Press, New York, 1972.
82. Hökfelt, T., and Dahlström, A.: Effects of two mitosis inhibitors (colchicine and Vinblastine) on the distribution and axonal transport of noradrenaline storage particles, studied by fluorescence and electron microscopy. Zellforsch. 119:460-482, 1971.
83. Gonatas, N. K., Anderson, W., and Evangelista, I.: The contribution of altered synapses in the senile plaque: An electron microscopic study in Alzheimer's dementia. J. Neuropath. Exp. Neurol. 26:25, 1967.
84. Singer, J. J., and Goldberg, A. L.: Cyclic AMP and transmission at the neuromuscular junction. In Advances in Biochemical Psychopharmacology, Vol. 3, P. Greengard and E. Costa (Eds.) pp. 335-348, Raven Press, New York, 1970.
85. Hokkanen, E., Vapaatalo, H., and Anttila, P.: On the possible role of cyclic AMP in the neuromuscular transmission and in the treatment of myasthenia gravis. Acta Neurol. Scand. Suppl. 51:315-316, 1972.
86. Takamori, M., Ishii, N., and Mori, M.: The role of cyclic 3',5'-adenosine monophosphate in nueromuscular transmission. Arch. Neurol. 29:420-424, 1973.
87. Schildkraut, J. J.: Catecholamine hypothesis of affective disorders: Review of supporting evidence. Am. J. Psychiatr. 122:509-522, 1965.
88. Paul, M. I., Ditzion, B. R., Paul, G. L., and Janowsky, D. S.: Urinary adenosine 3',5'-monophosphate excretion in affective disorders. Amer. J. Psychiat. 126:1493-1497, 1970.

89. Paul, M. I., Cramer, H., and Goodwin, F. K.: Urinary cyclic AMP excretion in depression and mania. Arch. Gen. Psychiat. 24: 327-333, 1971.
90. Paul, M. I., Cramer, H., and Bunney, W. E.: Urinary adenosine 3',5'-monophosphate in the switch process from depression to mania. Science 171:300-303, 1971.
91. Abdulla, Y. H., and Hamadah, K.: 3',5' cyclic adenosine monophosphate in depression and mania. Lancet 1:378-381, 1970.
92. Naylor, G. J., Stansfield, D. A., Whyte, S. F., and Hutchinson, F.: Urinary excretion of adenosine 3':5'-cyclic monophosphate in depressive illness. Brit. J. Psychiat. 125:275-279, 1974.
93. Brown, B. L., Salway, J. G., Albano, J. D. M., Hullin, R. P., and Ekins, R. P.: Urinary excretion of cyclic AMP and manic-depressive psychosis. Brit. J. Psychiat. 120:405-408, 1972.
94. Bunney, W. E., Murphy, D. L., and Goodwin, F. K.: The switch process from depression to mania: Relationship to drugs which alter brain amines. Lancet 1:1022-1027, 1970.
95. Post, R. M., Gordon, E. K., Goodwin, F. K., and Bunney, W. E.: Central norepinephrine metabolism in affective illness: MHPG in the cerebrospinal fluid. Science 179:1002-1003, 1973.
96. Pare, C. M. B., Yeung, D. P. H., Price, K., and Stacey, R. S.: 5-hydroxytryptamine, noradrenaline, and dopamine in brainstem, hypothalamus, and caudate nucleus of controls and of patients committing suicide by coal-gas poisoning. Lancet 2:133-135, 1969.
97. Prange, A. J., Jr., Wilson, I. C., Nox, A. E., McClane, T. K., Breese, G. R., Marting, B. R., Alltop, L. B., and Lipton, M. A.: Thyroid-imipramine clinical and chemical interaction: evidence for a receptor deficit in depression. J. Psychiatr. Res. 9:187-205, 1972.
98. Beer, B., Chasin, M., Clody, D. E., Vogel, J. R., and Horovitz, Z. P.: Cyclic adenosine monophosphate phosphodiesterase in brain: Effect on anxiety. Science 176:428-430, 1972.
99. Schildkraut, J. J.: Catecholamine metabolism and affective disorders: Studies of MHPG excretion. In Frontiers in Catecholamine Research, pp. 1165-1171, Pergamon Press, Great Britain, 1973.
100. Cramer, H., Goodwin, F. K., Post, R. M., and Bunney, W. E.: Effects of probenecid and exercise on cerebrospinal-fluid cyclic A.M.P. in affective illness. Lancet 1:1346-1347, 1972.
101. Broadus, A. E., Hardman, J. G., Kaminsky, N. I., Ball, J. H., Sutherland, E. W., and Liddle, G. W.: Extracellular cyclic nucleotides. Ann. N.Y. Acad. Sci. 185:50-66, 1971.
102. Jenner, F. A., Sampson, G. A., Somerville, A. R., Beard, N. A., and Smith, A. A.: Manic-depressive psychosis and urinary excretion of cyclic AMP. Brit. J. Psychiat. 121:236-237, 1972.
103. Paul, M. I., Cramer, H., and Goodwin, F. K.: Urinary cyclic A.M.P. in affective illness. Lancet 1:996, 1970.
104. Eccleston, D., Loose, R., Pullar, I. A., and Sugden, R. F.: Exercise and urinary excretion of cyclic AMP. Lancet 1:612-613, 1970.
105. Hansen, O.: Blood nucleoside and nucleotide studies in mental disease. Brit. J. Psychiat. 121:341-350, 1972.
106. Waelsch, H., and Weil-Malherbe, H.: Neurochemistry and psychiatry.

In Psychiatrie der Gegenwart, H. W. Gruhle, R. Jung, W. Mayer-Gross, and M. Müller (Eds.) p. 27, Springer-Verlag, Berlin, Göttingen, Heidelberg, 1964.
107. Wang, Y-C., Pandey, G. N., Mendels, J., and Frazer, A.: Platelet adenylate cyclase responses in depression: Implications for a receptor defect. Psychopharmacologia 36:291-300, 1974.
108. Murphy, D. L., Donnelly, C., and Moskowitz, J.: Catecholamine receptor function in depressed patients. Am. J. Psychiat. 131:1389-1391, 1974.

13

Role of Cyclic Nucleotides in Drug Addiction and Withdrawal

L. VOLICER
S. K. PURI
B. P. HURTER

Drug addiction and occurrence of withdrawal symptoms after discontinuation of drug administration were until recently thought to be associated only with centrally acting drugs of abuse. However, there is recent evidence that abrupt discontinuation of some cardiovascular drugs also produces withdrawal symptoms (1, 2), and that effects produced by alterations of a neurotransmitter function may be considered as phenomena associated with dependence on the endogenous substances (3). It is very likely that future development of new drugs will increase the probability of occurrence of withdrawal symptoms after discontinuation of therapy.

The mechanism of withdrawal symptoms for some of these drugs is relatively simple; for others it is only partly elucidated. Thus rebound increase of blood pressure occurring in some hypertensive patients after discontinuation of clonidine therapy is due to increased sympathetic activity (1) and we know several factors which might participate in the exacerbation of coronary artery disease after abrupt cessation of antianginal therapy with propranolol (2). However, despite the effort of many investigators the mechanism of classical withdrawal syndromes, namely, morphine and ethanol withdrawals, is not well understood.

In contrast to the withdrawal syndromes occurring after cessation of clonidine or propranolol therapy, morphine and ethanol withdrawals are much more complex and have a wider variety of symptoms. This is probably due to the fact that both morphine and ethanol are relatively nonselective and affect many more physiologic functions than clonidine and propranolol. Therefore it is more difficult to investigate the mechanism of there withdrawal syndromes since it is not clear which system is responsible for a specific symptom.

Much attention has been devoted to changes of neurotransmitters during withdrawal syndromes. Many changes have been detected but at present they do not provide an acceptable explanation of the mechanisms involved. Our knowledge of neurotransmitter functions is incomplete and some of the neurotransmitter changes found during withdrawal syndromes may be secondary to a modified function of another, and possibly unknown, mediator.

Discovery of cyclic nucleotides and development of the concept of cyclic nucleotides as second messengers of hormone and mediator effects (4) provided a new tool for investigation of withdrawal syndromes. This approach assumes that a cell function is modified by the intracellular level of cyclic nucleotides and therefore changes of cyclic nucleotides during withdrawal syndromes indicate that the cell function might be affected by withdrawal. Investigation of withdrawal-induced changes of cyclic nucleotides is complicated by the multiple cell types present in the brain and by the fluctuation of cyclic nucleotides in response to various stimuli but it nevertheless brought new insights into the mechanism of withdrawal syndromes.

This review will be limited to changes of cyclic nucleotides induced by chronic administration of morphine or ethanol and changes detected during morphine or ethanol withdrawal. The effects of acute ethanol administration on the adenosine 3',5'-monophosphate (cyclic AMP) system were recently reviewed elsewhere (5).

CYCLIC NUCLEOTIDES IN NARCOTIC DEPENDENCE AND WITHDRAWAL

Narcotic dependence has been attributed to a latent hyperexcitability of the central nervous system which is induced by repeated administration of narcotic drugs (6, 7). Many modifications have been proposed to a latent hyperexcitability theory of dependence (8, 9, 19, 11). However, it is now widely accepted that drugs producing physical dependence might interfere in some way with synaptic neurotransmission and that the persistent blockade of synaptic neurotransmission during chronic administration produces supersensitivity of the affected neuroreceptors when the blockade is removed by withdrawing the drug. There have been many difficulties in obtaining direct evidence for the nature and mechanism of the changes in receptor activity because of the lack of suitable central nervous system receptor models. More recently cyclic nucleotide systems have been utilized as useful biochemical models to test systematically the changes in CNS-receptor sensitivity in narcotic dependence and withdrawal.

It has been suggested that adenylate cyclase activity is closely associated with the receptors involved in synaptic neurotransmission in the central nervous system. If the chronic administration of an opiate alters central nervous system synaptic activity, then measurement of nucleotide cyclase in vitro and determination of its sensitivity to various neurotransmitters may provide information on the modification of central receptor activity during dependence and withdrawal.

Chronic administration of narcotic drugs was reported to increase adenylate cyclase activity in whole rat brain homogenate (12). A similar increase (13) and no change (14) in the mouse cerebral cortex have also been repotred. No changes were observed in rat cerebral cortex, cerebellum, or thalamus-hypothalamus (15, 16). We have found a twofold increase in the basal adenylate cyclase activity in the striatum of rats made dependent on morphine as compared to saline-treated controls. This increase was observed for at least 96 hr after the last morphine injection (Fig. 1). A similar elevation in the basal activity in the striatum has also been reported elsewhere (17). Increased basal activity of adenylate cyclase might be responsible for the increase of cyclic AMP levels in the striatum of morphine-dependent rats observed by us (Table 1) and

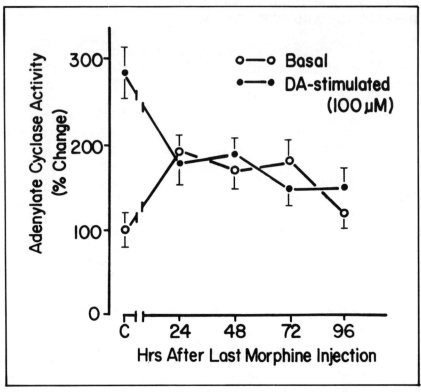

1 Effect of chronic administration of morphine and morphine withdrawal on basal- and dopamine-sensitive adenylate cyclase activity in the striatum.

others (17, 19). Recently, dopamine-sensitive adenylate cyclase has been found in the striatum, and it was suggested that the physiologic effect of dopamine may be mediated through cyclic AMP (20). Therefore, the increase of cyclic AMP formation in the striatum might be mediated by an increase in dopamine receptor sensitivity. Although Iwatsubo and Clouet (21) showed that chronic administration of narcotic drugs to rats significantly increased the response of adenylate cyclase to dopamine in comparison with nondependent rats, our data, which are in agreement with Merali et al. (15) and Van Inwegen et al. (16), indicate that the dopamine sensitivity of adenylate cyclase was significantly lowered in morphine-dependent rats (Fig. 1).

Table 1
Effect of Chronic Administration of Morphine and Morphine Withdrawal on Cyclic AMP and Cyclic GMP Levels in Striatum, Hypothalamus, Cerebral Cortex and Midbrain-Thalamus of Rat Brain

	Striatum	Hypothalamus	Midbrain-Thalamus	Cerebral Cortex
	Cyclic AMP (nmoles/g of tissue) Mean ± S.E. (N)			
Non-addicted Control	0.83 ± 0.13 (9)	1.27 ± 0.11 (10)	1.56 ± 0.13 (5)	1.21 ± 0.16 (5)
Morphine-addicted[1]				
1	1.23 ± 0.16* (5)	2.15 ± 0.23* (5)	2.22 ± 0.09* (5)	0.97 ± 0.10 (5)
72	1.29 ± 0.08* (10)	2.13 ± 0.19* (9)	2.42 ± 0.21* (5)	1.07 ± 0.03 (5)
	Cyclic GMP (pmoles/g of tissue)			
Non-addicted Control	24.6 ± 3.4 (6)	37.6 ± 6.5 (10)	41.4 ± 6.1 (5)	176 ± 9.8 (5)
Morphine-addicted				
1	38.5 ± 6.6 (4)	22.8 ± 4.6 (4)	38.0 ± 14 (5)	159 ± 19 (4)
72	48.0 ± 8.0* (10)	118.0 ± 12.4* (4)	46.0 ± 13 (5)	246 ± 46 (5)

[1]Time in hours after last maintenance dose of morphine (100 mg/kg)
*Significantly different from non-dependent control ($p < 0.05$)

The decreased sensitivity of adenylate cyclase to dopamine is in contrast to behavioral studies indicating that dopamine receptors are supersensitized during chronic administration of morphine (22, 23). Supersensitivity to apomorphine—a dopamine agonist—also has been demonstrated with respect to behavior and dopamine turnover (22–24). To determine if the decreased sensitivity of dopa-

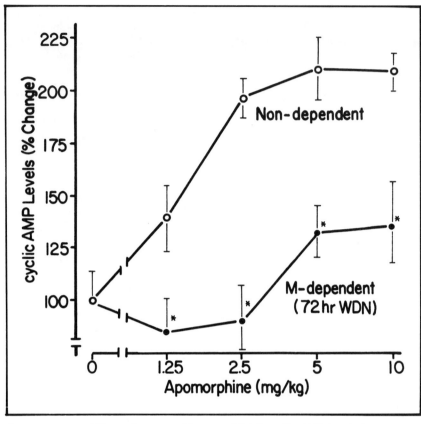

2 Effect of apomorphine on striatal cyclic AMP levels in nondependent control and 72-hr-withdrawn morphine-dependent rats. The animals were sacrificed 15 min after apomorphine injection. Asterisks indicate levels significantly different from levels in similarly treated nondependent rats ($p < 0.05$).

mine-stimulated adenylate cyclase also occurs in vivo, we measured changes of cyclic AMP levels induced by apomorphine in the striatum. We found that apomorphine was less potent in increasing cyclic AMP levels in morphine-dependent rats than in control animals (Fig. 2). At present it is not clear why morphine dependence increases apomorphine potency on dopamine turnover and decreases apomorphine potency in affecting cyclic AMP levels. It is possible that apomorphine may have two sites of action and that these sites are affected differentially by morphine dependence and withdrawal. The nature and localization of these sites need further investigation.

Many observations in the literature suggest that the degree of dependence and the abstinence syndrome may be directly related to changes in cyclic nucleotide systems. An increase in cyclic AMP levels has been demonstrated in striatum (17, 18, 25), thalamus, and substantia nigra (18) after chronic administration of morphine. A moderate increase in striatal and cerebellar cyclic AMP levels was also observed in naloxone-precipitated withdrawal (19). Data from our laboratory indicate that chronic administration of morphine increases cyclic AMP levels not only in striatum but also in hypothalamus and midbrain-thalamus. When morphine is withdrawn, cyclic AMP levels remain elevated in striatum, hypothalamus, and midbrain-thalamus. On the other hand, the cyclic AMP level in the cerebral cortex is unchanged (Table 1). Increased cyclic AMP levels might be responsible for some of the symptoms of morphine withdrawal since it was observed that repeated administration of exogenous cyclic AMP or its dibutyryl derivates to mice implanted with a morphine pellet increases the development of tolerance to and dependence on morphine (26). The administration not only of cyclic AMP but also of theophylline—a phosphodiesterase inhibitor—enhances markedly tolerance and dependence development in mice implanted with a morphine pellet (27). It is interesting to note that phosphodiesterase inhibitors, like theophylline, not only intensify most of the signs of precipitated abstinence in morphine-dependent rats, but also produce an almost perfect replica of the abstinence syndrome when given to naive rats shortly before naloxone (28, 29).

Very few data are available on the changes in brain cyclic GMP after chronic administration of morphine. Recently Bonnet (18) reported a decrease in cyclic GMP in caudate, substantia nigra, hypothalamus and thalamus. We have not been able to reproduce this finding. Our data indicate that cyclic GMP is not altered either after chronic administration of morphine or after withdrawal in cerebral cortex, or midbrain-thalamus. Only in the caudate and hypothalamus was there an increase in cyclic GMP levels after morphine withdrawal. These observations suggest that one of the biochemical processes of abstinence might be related to the activation of the cyclic AMP system and that the cyclic GMP system is less affected. This interpretation is further supported by the finding that cyclic AMP, but not cyclic GMP, intensifies the abstinence syndrome (30).

Administration of prostaglandin E_1 (PGE_1) elicits pain or hyperalgesia (31–33). It was therefore proposed that morphine may act by antagonizing the stimulation of adenylate cyclase by PGE_1 (34). Although there is some evidence for this antagonism (34, 35) other investigators did not observe a consistent stimulation of adenylate cyclase by PGE_1 (16, 36), and Dismukes and Daly (37) found potentiation of PGE_1 effect in rat brain slices by morphine. It was reported that in heroin-dependent rats, PGE_1 usually failed to increase the adenylate cyclase activity in brain homogenates (12).

Recently, a neuroblastomaxglioma hybrid cell culture model has been utilized to study the effect of opiates on cyclic nucleotides, because these cells possess morphine receptors (38). Opiate-like drugs decreased adenylate cyclase activity (38, 39), increased the accumulation of cyclic GMP (40, 41) and reversed the stimulation of adenylate cyclase by PGE_1 (39) in these cells. When the cells were exposed chronically to morphine the basal activity of adenylate cyclase increased and they became tolerant to the inhibitory effect of morphine on PGE_1 stimulation of adenylate cyclase (40). The displacement of morphine by naloxone in dependent cells increased basal, PGE_1-stimulated and adenosine-stimulated adenylate cyclase activity with concomitant increase of cyclic AMP levels (42). These results further support the notion that elevated levels of cyclic AMP during withdrawal may be a biochemical substrate of the abstinence syndrome.

CYCLIC NUCLEOTIDES IN ETHANOL DEPENDENCE AND WITHDRAWAL

The role of cyclic nucleotides in ethanol addiction has been investigated less extensively than their role in narcotic addiction. However, available evidence indicates that cyclic nucleotides are affected by chronic ethanol administration and might participate in the mechanism of some withdrawal symptoms.

After chronic ethanol administration, tolerance develops to the acute effect of ethanol on the cyclic AMP system. Although acute administration of ethanol decreases cyclic AMP levels in cerebral cortex, pons, medulla oblongata and cerebellum (43, 44), it was reported that cyclic AMP levels in cerebral cortex are higher in mice chronically given ethanol in a liquid diet than in control animals (45). We have observed that there is no difference in brain cyclic AMP levels between control and dependent rats (46) 1 hr after the last ethanol administration (44). The mechanism of tolerance is not clear. It is possible that it is mediated by an increased basal activity of adenylate cyclase which was found in mice kept on liquid ethanol-containing diet (45) but the tolerance development to the depressant effect of ethanol on nerve activity might also participate.

Chronic administration of ethanol in a liquid diet changes sensitivity of adenylate cyclase to norepinephrine. It was reported that adenylate cyclase in cerebral cortex is less sensitive to norepinephrine stimulation in mice and rats kept on the liquid diet (47, 48). This decrease of sensitivity might be specific for norepinephrine because histamine-induced stimulation was unaffected by chronic ethanol administration (47). The decreased sensitivity of adenylate cyclase to norepinephrine might play a role in the development of seizures during ethanol withdrawal since norepinephrine limits the spread of audiogenic seizure discharge in rats (49), and the intensity of the syndrome is increased by drugs which deplete catecholamines or block adrenergic receptors (50). However, we have observed that cyclic AMP levels are actually increased in the cerebral cortex during an early phase of ethanol withdrawal (Fig. 3). This indicates that increased norepinephrine turnover, which occurs during ethanol withdrawal (51, 52, 53), compensates for the decreased adenylate cyclase sensitivity. It is likely that the increase in norepinephrine turnover and in cyclic AMP levels is a consequence rather than a cause of the withdrawal syndrome.

French et al. (48) have found supersensitivity of adenylate cyclase from the cerebral cortex to norepinephrine and other neurotransmitters (54) at 72 hr after withdrawal of rats from an ethanol-containing diet. This withdrawal stage corresponds behaviorally with the development of delirium tremens in man (55) and increased aggressiveness in rats (56). It is possible that dopamine is involved in this withdrawal phase since aggressiveness may be induced by apomorphine, a dopamine agonist (56). Dopamine probably does not participate in the early phase of withdrawal since Goldstein (50) found no effect on withdrawal hyperexcitability with drugs affecting the dopamine system and Ahtee (53) has reported no change in dopamine turnover during early ethanol with-

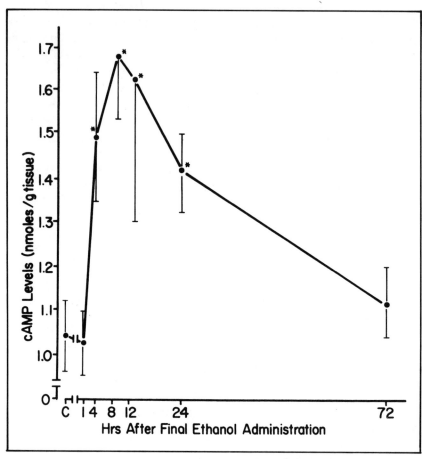

3 Changes of cyclic AMP levels in rat cerebral cortex during ethanol withdrawal. Rats were made ethanol-dependent by three daily administrations of ethanol per os (46). The bars represent ± SEM and the asterisks indicate $p < 0.05$ compared to control rats.

drawal. We found no change in cyclic AMP levels in any brain area at three days after the final dose of ethanol. However, a selective change in cyclic AMP levels in dopaminergic structures cannot be excluded since we did not assay these brain areas separately.

There is also some evidence that guanosine 3',5'-monophosphate (cyclic GMP) might be involved in ethanol withdrawal. Intraventricular administration of the dibutyryl derivative of cyclic GMP increases the incidence of head twitches in mice undergoing ethanol withdrawal (57). We have found that tolerance does not develop to the cyclic GMP-decreasing effect of ethanol. Cyclic GMP levels in rats chronically treated with ethanol (46) and sacrificed 1 hr after the last ethanol dose were significantly lower than the levels found in control animals in all brain areas except cerebral cortex (Fig. 4). Cyclic GMP was also decreased in the cerebral cortex but this decrease was not statistically significant ($p < 0.1$). These results are similar to the decrease of cyclic GMP levels observed after acute ethanol administration (44).

Cyclic GMP levels were unchanged during ethanol withdrawal in the cerebral cortex and subcortex. In the cerebellum there was a significant increase of cyclic GMP levels which started 8 hr after the last ethanol dose and was still present 3 days later. Cyclic GMP in pons and medulla oblongata also increased during ethanol withdrawal but this increase was significant only at 24 hr after the last ethanol dose and the cyclic GMP level returned to normal two days later (Fig. 4).

Since cyclic GMP might be connected with both cholinergic and GABA-nergic transmission, both systems might be involved in the mechanism of ethanol effect on cyclic GMP levels. However, Goldstein (50) has reported that agents modifying acetylcholine action do not alter the withdrawal syndrome. On the other hand, blockade of GABA receptors by picrotoxin was shown to increase the severity of convulsions induced during alcohol withdrawal. An increase in GABA levels induced by inhibition of GABA transaminase by amino-oxyacetic acid decreases the severity of withdrawal. Patel and Lal (58) have demonstrated a decrease in GABA levels in the brain at 8 hr after removal from an ethanol vapor chamber. We have found a decrease in GABA levels in the cerebellum and pons medulla during ethanol withdrawal at the time of cyclic GMP increase (Volicer, Williams and Puri, in preparation).

Cerebellar changes in cyclic GMP and GABA during the hyperexcitability phase of ethanol withdrawal correlated well with the model of cyclic GMP involvement in seizure activity proposed by Costa et al. (59). According to this model, cyclic GMP levels may be altered by an interaction of GABA and an unknown neurotransmitter in the Purkinje fibers. It is possible that a decrease in GABA levels may result in an increase in cyclic GMP levels concomitantly with the increase of neuronal activity during ethanol withdrawal.

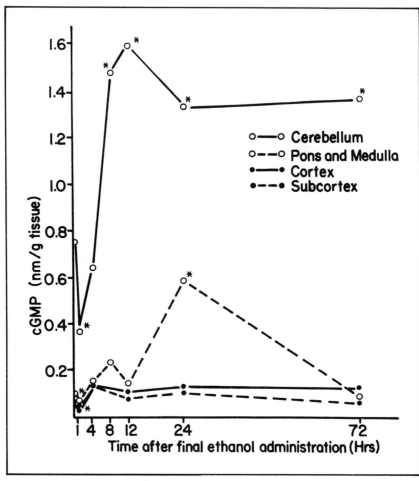

4 Changes of cyclic GMP levels in various areas of rat brain during ethanol withdrawal. Rats were made ethanol-dependent by three daily administrations of ethanol per os (46). The asterisks indicate levels significantly different ($p < 0.05$) from levels in control rats.

The cyclic GMP-decreasing effect of diazepam, a drug used for treatment of ethanol withdrawal supports the role of cyclic GMP in ethanol withdrawal. It should be noted that in the cerebellum, levels of cyclic GMP remain elevated up to 72 hr after the last ethanol dose. Perhaps cyclic GMP plays a role in the increased sensitivity to apomorphine-induced aggression found to be maximal three days after ethanol withdrawal (56), since we found that apomorphine increases the cyclic GMP level in rat cerebellum (Puri and Volicer, unpublished observation).

SUMMARY

Investigation of drug-induced changes in cyclic nucleotides was found to be useful in the search for mechanisms underlying drug dependence and withdrawal syndromes. Chronic administration of morphine increases cyclic AMP levels in the brain. This increase might be responsible for some of the symptoms of morphine withdrawal since elevation of cyclic AMP levels in the brain increases development of tolerance to morphine and intensifies withdrawal signs in morphine-dependent animals. Chronic administration of morphine also increases basal activity of adenylate cyclase in the straitum and possibly in some other brain areas. However, sensitivity of striatal adenylate cyclase to dopamine is decreased by chronic morphine administration which is in contrast to the increased behavioral effects caused by dopamine receptor stimulation in morphine-withdrawn animals. Exposure of neuroblastoma glioma hybrid cells to morphine increases basal activity of adenylate cyclase and naloxone-induced withdrawal increases the cyclic AMP level in these cells.

While morphine withdrawal seems to affect mainly the cyclic AMP system, ethanol withdrawal leads to pronounced changes in the cyclic GMP system. Cyclic GMP levels are markedly increased in the cerebellum and pons-medulla oblongata at the time of the behavioral withdrawal syndrome, and in the cerebellum, this increase lasts for at least 72 hr. The increase of cyclic GMP together with reciprocal changes of GABA levels might be related to increased susceptibility to seizures during ethanol withdrawal. There is also an increase of cyclic AMP levels in the cerebral cortex during ethanol withdrawal which might be caused by increased norepinephrine turnover.

During chronic administration of ethanol a tolerance develops to the acute cyclic AMP-lowering effect of ethanol and the basal activity of adenylate cyclase in the cerebral cortex is increased. Simultaneously, sensitivity of adenylate cyclase to norepinephrine is decreased while sensitivity to histamine remains unchanged. In contrast, sensitivity of adenylate cyclase to several neurotransmitters is increased in the cerebral cortex of rats three days after ethanol withdrawal.

REFERENCES

1. Hansson, L., Hunyor, S. N., Julius, G. and Hoobler, G. W.: Blood pressure crisis following withdrawal of clonidine with special reference to arterial and urinary catecholamine levels and suggestion for acute management. Amer. Heart J. *85*:605-610, 1973.
2. Miller, R. R., Olson, H. G., Amsterdam, E. A. and Mason, D. T.: Propranolol withdrawal rebound phenomenon. New Engl. J. Med. *293*:416-418, 1975.
3. Wyatt, R. J., and Gillin, J. C.: Development of tolerance to and dependence on endogenous neurotransmitters. In: Neurobiological Mechanisms of Adaptation and Behavior. ed. Mandell, A. J. Raven Press, New York, 1975.
4. Robison, G. A., Butcher, R. W. and Sutherland, E. W.: Cyclic AMP. Academic Press, New York and London, 1971.
5. Volicer, L. and Gold, B. I.: Interactions of ethanol with cyclic AMP. In: Biochemical Pharmacology of Ethanol. ed. Majchrowicz, E. Plenum Press, New York and London. pp. 211-237, 1975.
6. Seevers, M. H. and Deneau, G. A.: A critique of the "Dual Action" hypothesis of morphine dependence. Arch. Int. Pharmacodyn. *140*:514-517, 1962.
7. Jaffe, J. H. and Sharpless, S. K.: Pharmacological denervation supersensitivity in the central nervous system. A theory of physical dependence. *In:* Addictive States, ed. Wikler, A. William and Wilkins, Baltimore. Asso. Nerv. Ment. Dis. *46*:226-246, 1968.
8. Collier, H. O. J.: Humoral transmitters, supersensitivity, receptors and dependence. In: Scientific Basis of Drug Dependence, ed. Steinberg, H. Grune and Stratton, Inc. New York. pp. 49-66, 1969.
9. Goldstein, D. B. and Goldstein, A.: A possible role of enzyme inhibition and repression in drug tolerance and addiction. Biochem. Pharmacol. *8*: 48-52, 1961.
10. Schulz, R., Cartwright, C. and Goldstein, A.: Reversibility of morphine tolerance and dependence in guinea pig brain and myenteric plexus. Nature *251*:329-331, 1974.
11. Shuster, L.: Tolerance and physical dependence. In: Narcotic Drugs, Biochemical Pharmacology, ed. Clouet, D. G. Plenum Press, New York. pp. 408-423, 1971.
12. Collier, H. O. J., Francis, D. L., McDonald-Gibson, W. J., Roy, A. C. and Saeed, S. A.: Prostaglandins, cyclic AMP and the mechanism of opiate dependence. Life Sci. *17*:85-90, 1975.
13. Naito, K. and Kuriyama, K.: Effect of morphine administration on adenyl cyclase and 3',5'-cyclic nucleotide phosphodiesterase activities in the brain. Jap. J. Pharmacol. *23*:274-276, 1973.
14. Chou, W. S., Ho, A. K. S., and Loh, H. H.: Effect of acute and chronic morphine and norepinephrine on brain adenyl cyclase activity. Proc. West. Pharmacol. Soc. *14*:42-46, 1971.

15. Singhal, R. L., Kacew, S. and Lafreniere, R.: Brain adenyl cyclase in methadone treatment of morphine dependency. J. Pharm. Pharmacol. *25*: 1022-1024, 1973.
16. Van Inwegen, R. G., Strada, S. J., and Robison, G. A.: Effects of prostaglandins on brain adenylyl cyclase. Life Sci. *16*:1875-1876, 1975.
17. Merali, Z., Singhal, R. L., Hrdina, P. D. and Ling, G. M.: Changes in brain cyclic AMP metabolism and acetylcholine and dopamine during narcotic dependence and withdrawal. Life Sci. *16*:1889-1894, 1975.
18. Bonnet, K. A.: Regional alterations in cyclic nucleotide levels with acute and chronic morphine treatment. Life Sci. *16*:1877-1882, 1975.
19. Mehta, C. S. and Johnson, W. E.: Possible role of cyclic AMP and dopamine in morphine tolerance and physical dependence. Life Sci. *16*:1883-1888, 1975.
20. Iverson, L. L.: Dopamine receptors in the brain. Science *188*:1084-1089, 1975.
21. Iwatsubo, K. and Clouet, D. H.: Dopamine-sensitive adenylate cyclase of the caudate nucleus of rats treated with morphine or haloperidol. Biochem. Pharmacol. *24*:1499-1503, 1975.
22. Puri, S. K., O'Brien, J. and Lal, H.: Potentiation of morphine-withdrawal aggression by d-amphetamine, DOPA or apomorphine. Pharmacologist *13*: 280, 1971.
23. Puri, S. K. and Lal, H.: Effect of dopaminergic stimulation or blockade on morphine-withdrawal aggression. Psychopharmacologia *32*:113-120, 1973.
24. Kuschinsky, K.: Dopamine receptor sensitivity after repeated morphine administration to rats. Life Sci. *17*:43-48, 1975.
25. Clouet, D. H., Gold, G. J. and Iwatsubo, K.: Effects of narcotic analgesic drugs on the cyclic adenosine 3',5'-monophosphate adenylate cyclase system in rat brain. Br. J. Pharmacol. *54*:541-548, 1975.
26. Ho, I. K., Loh, H. H. and Way, E. L.: Effects of cyclic 3',5'-adenosine monophosphate on morphine tolerance and physical dependence. J. Pharmacol. Exp. Therap. *185*:347-357, 1973.
27. Ho, I. K., Loh, H. H., Bhargava, H. N. and Way, E. L.: Effects of cyclic nucleotides and phosphodiesterase inhibition on morphine tolerance and physical dependence. Life Sci. *16*:1895-1900, 1975.
28. Collier, H. O. J.: The concept of Quasi-Abstinence effect and its use in the investigation of dependence mechanism. Pharmacology *11*:58-61, 1974.
29. Collier, H. O. J., Francis, D. L., Henderson, G. and Schneider, C.: Quasi morphine-abstinence syndrome. Nature (London) *249*:471-473, 1974.
30. Collier, H. O. J. and Francis, D. L.: Morphine abstinence is associated with increased brain cyclic AMP. Nature *255*:159-162, 1975.
31. Solomon, L. M., Juhlin, L. and Kirschenbaum, M. B.: Prostaglandin on cutaneous vasculature. J. Invest. Dermatol. *51*:280-285, 1968.
32. Collier, H. O. J. and Schneider, C.: Nociceptive response to prostaglandins and analgesic actions of aspirin and morphine. Nature *236*:141-143, 1972.
33. Ferreira, S. H.: Prostaglandins, aspirin-like drugs and analgesia. Nature *240*:200-203, 1972.
34. Collier, H. O. J. and Roy, A. C.: Morphine-like drugs inhibit the stimulation of E prostaglandins of cyclic AMP formation by rat brain homogenates. Nature *248*:24-27, 1974.

35. Roy, A. C. and Collier, H. O. J.: Prostaglandins, cyclic AMP and the biochemical mechanism of opiate agonist action. Life Sci. *16*:1857-1862, 1975.
36. Tell, G. P., Pasternak, G. W. and Cuatrecasas, P.: Brain and caudate nucleus adenylate cyclase: Effects of dopamine, GTP, E prostaglandins and morphine. FEBS Letters *51*:242-245, 1975.
37. Dismukes, K. and Daly, J. W.: Accumulation of adenosine 3',5'-monophosphate in rat brain slices: Effect of prostaglandins. Life Sci. *17*:199-210, 1975.
38. Klee, W. A. and Nirenberg, M.: A neuroblastoma X glioma hybrid cell line with morphine receptors. Proc. Natl. Acad. Sci. U.S.A. *71*:3474-3477, 1974.
39. Traber, J., Fischer, K., Latzin, S. and Hamprecht, B.: Morphine antagonizes action of prostaglandin in neuroblastoma and neuroblastoma X glioma hybrid cells. Nature *253*:120-122, 1975.
40. Gullis, R., Traber, J. and Hamprecht, B.: Morphine elevates levels of cyclic GMP in a neuroblastoma X glioma hybrid cell line. Nature *256*: 57-59, 1975.
41. Klee, W. A., Sharma, S. K. and Nirenberg, M.: Opiate receptors as regulators of adenylate cyclase. Life Sci. *16*:1863-1868, 1975.
42. Sharma, S. S., Klee, W. A. and Nirenberg, M.: Dual regulation of adenylate cyclase accounts for narcotic dependence and tolerance. Proc. Nat. Acad. Sci. U.S.A. *72*:3092-3096, 1975.
43. Volicer, L. and Gold, B. I.: Effect of ethanol on cyclic AMP levels in the rat brain. Life Sci. *13*:269-280, 1973.
44. Volicer, L. and Hurter, B.: Effect of acute and chronic ethanol administration and ethanol withdrawal on cyclic nucleotide levels in rat brain. J. Pharmacol. Exp. Therap. (in press).
45. Kuriyama, K. and Israel, M. A.: Effect of ethanol administration on cyclic 3',5'-adenosine monophosphate metabolism in brain. Biochem. Pharm. *22*: 2919-2922, 1973.
46. Majchrowicz, E.: Induction of physical dependence upon ethanol and the associated behavioral changes in rats. Psychopharmacologia *43*:245-254, 1975.
47. Israel, M. A., Kimura, H. and Kuriyama, K.: Changes in activity and hormonal sensitivity of brain adenyl cyclase following chronic ethanol administration. Experientia *28*:1322-1323, 1972.
48. French, G. W., Palmer, D. L., Narod, M. E., Reid, P. E. and Ramey, C. W.: Noradrenergic sensitivity of the cerebral cortex after chronic ethanol ingestion and withdrawal. J. Pharmacol. Exper. Ther. *194*:319-326, 1975.
49. Jobe, P. C., Picchioni, A. L. and Chin, L.: Role of brain NE in audiogenic seizure in the rat. J. Pharmacol. Exper. Ther. *184*:140, 1973.
50. Goldstein, D. B.: Alcohol withdrawal reactions in mice: Effects of drugs that modify neurotransmission. J. Pharmacol. Exper. Ther. *186*:1-9, 1973.
51. Hunt, W. A., and Majchrowicz, E.: Alterations in turnover of brain norepinephrine and dopamine in alcohol-dependent rats. J. Neurochem. *23*: 549-552, 1974.
52. Pohorecky, L. A., Jaffe, L. S. and Berkeley, H. A.: Ethanol withdrawal in the rat: Involvement of noradrenergic neurons. Life Sci. *15*:427-437, 1974.
53. Ahtee, L. and Svartstrom-Fraser, M.: Effect of ethanol dependence and withdrawal on the catecholamines in rat brain and heart. Acta Pharmacol. Toxicol. *36*:289-298, 1975.

54. French, G. W., Palmer, D. L. and Narod, M. E.: Effect of withdrawal from chronic ethanol ingestion on the cAMP response of cerebral cortical slices using the agonists histamine, serotonin and other neurotransmitters. Can. J. Physiol. Pharmacol. *53*:248-255, 1974.
55. Victor, M. and Adams, R. D.: The effect of alcohol on the nervous system. Res. Publs. Ass. Res. Nerv. Ment. Dis. *32*:526-673, 1953.
56. Mann, S. and Lal, H.: Enhancement of drug (apomorphine or amphetamine)-induced aggression in rats withdrawn from chronic exposure to ethanol inhalation. Fed. Proc. *34*:720, 1975.
57. Hammond, M. D. and Schneider, C.: Cyclic nucleotides and ethanol withdrawal head-twitches in mice. Br. J. Pharm. *52*:138, 1974.
58. Patel, G. J., and Lal, H.: Reduction in brain gamma-aminobutyric acid and in barbital narcosis during ethanol withdrawal. J. Pharmacol. Exper. Ther. *186*:625-629, 1973.
59. Costa, E., Guidotti, A. and Mao, C. C.: A GABA hypothesis for the action of benzodiazepines. In: GABA in Nervous System Function. eds. Roberts, E. Chase, T. N. and Tower, D. N. Raven Press, New York, pp. 413-426, 1976.

14

Roles for Cyclic Nucleotides in Diseases of the Eye

ARTHUR H. NEUFELD

The major efforts to investigate cyclic nucleotides in the eye began within this decade, but already there are areas of research which are of potential significance, clinically. These areas include the investigations of the cyclic nucleotide phosphodiesterase in the retina, which is associated with retinal degeneration in an animal model system, and the mediation by adenosine 3',5'-monophosphate (cyclic AMP) of the response to adrenergic compounds to decrease intraocular pressure, the goal of the medical therapy for glaucoma. The nature of the experiments in these two areas are quite different: the observations on the retina have been largely biochemical and little is known about the role of cyclic nucleotides in the functioning of this tissue; whereas, the observations on aqueous humor dynamics have been primarily physiological and can be related to the control of intraocular pressure. Thus, the review of these areas will be qualitatively different; that of cyclic nucleotides and the retina is descriptive while that of cyclic AMP and aqueous humor dynamics is more integrative. In addition, recent observations concerning the cyclic nucleotide system in the cornea will be reviewed because they too may soon be clinically relevant.

RETINA

The retina is a complicated nervous tissue that can be fractionated, to some extent, for biochemical studies. The retinal pigment epithelium, which is posterior and in close association with the photopigment-containing cells, can be dissected away from the neural retina. Furthermore, the highly specialized rods consist of two quite different and separable parts of the same cell: the outer segment, which houses the stacks of membranes containing the visual pigment, and the inner segment, which contains the nucleus, mitochondria, protein synthetic apparatus, and makes synaptic connections with the bipolar and horizontal cells. More anterior layers of the retina contain bipolar cells, horizontal cells, ganglion cells, and amacrine cells, all forming a multitude of synapses. Upon gentle agitation of the neural retina, the rod outer segments pinch off and can be further purified. These types of preparations, as well as the degeneration of photoreceptor cells to be discussed, have allowed investigators to study cyclic nucleotides in specific layers of the retina. Some of the current investigations into cyclic nucleotides in the retina are summarized in Table 1.

In the retinal pigment epithelium, cyclic nucleotides regulate cell growth (1). When normal, chick retinal pigment epithelium is cloned and grown in tissue culture media, dibutyryl cyclic AMP retards the increase in cell number and decreases the final number of cells per dish, while increasing the rate of cell adhesion to the substratum. Interestingly, addition of dibutyryl cyclic AMP decreases pigmentation, normally considered an expression of increased differentiation, but this is perhaps related to the artificial situation of incubation in high concentrations of cyclic AMP for prolonged periods (G. Chader, personal communication).

In the neural retina, cyclic AMP and guanosine 3',5'-monophosphate (cyclic GMP) are active at many levels. Cyclic AMP added to the media of the chick embryo retina in tissue culture reduces de novo synthesis of glutamine synthetase (2). A dopamine-stimulated, adenylate cyclase is found in retinal homogenates (3, 4) and dopamine and apomorphine increase the cyclic AMP concentrations of intact retinas (5). The retina has dopamine containing neurons

Table 1. Retina

Tissue	Physiology	Cyclic Nucleotides
Pigment Epithelium	Retinal Nutrition Phagocytosis of Rods	Control Cell Division and Differentiation
Photoreceptor Layer Rod Outer Segment	Rhodopsin in Disk Membranes Photon Capture Sodium Conductance Changes	Guanylate Cyclase Light Inactivated Phosphodiesterase Light Activated $K_m = 2\text{-}6 \times 10^{-4}$M cGMP; $K_m = 2\text{-}8 \times 10^{-3}$M cAMP High cGMP Levels
Rod Inner Segment	Synapses Synthesis of Rhodopsin, Disk Membranes	

Inner Retinal Layers
  Bipolar Cells
  Horizontal Cells
  Ganglion Cells
  Amacrine Cells
	Information Processing	Adenylate Cyclase
		Dopamine Activated
		Phosphodiesterase
		$K_m = 7 \times 10^{-5}$M cGMP;
		$K_m = 2 \times 10^{-4}$M cAMP

(6) and the dopamine-sensitive, adenylate cyclase probably participates in postsynaptic events, following release of dopamine as a neurotransmitter.

The most active and controversial area of research on cyclic nucleotides in the eye is the investigation into cyclic nucleotide activity in photoreceptors and its relation to the mechanism of visual excitation. A review of this work is beyond the scope of this chapter, but the current concensus indicates that in rod outer segments there is a light-inactivated guanylate cyclase (7, 8), a light-activated, cyclic GMP phosphodiesterase (to be discussed) and very high levels of cyclic GMP (9). Upon illumination, in vivo, one suspects that a large decrease in the level of cyclic GMP occurs in rod outer segments. The relevance of cyclic GMP metabolism to any mechanism involved in visual excitation has not been demonstrated.

Because of the considerable interest in the phosphodiesterase in the photoreceptor cells, a chronological description of developments in this area is pertinent. In 1971, Bitensky, Gorman and Miller (10) suggested that cyclic AMP might be a link in the molecular events that regulate sodium permeability in light and in darkness in the vertebrate photoreceptor. Quite understandably, this hypothesis stimulated much work on the system. In 1972, Pannbacker et al. (11) described the presence of a cyclic nucleotide phosphodiesterase in mammalian photoreceptors. The enzyme described for bovine rod outer segments shows simple kinetics, has a high Km for cyclic AMP, estimated to be about 7mM, and is present in extremely high levels compared to chick brain, rat brain or whole retina. An enzyme was also found that hydrolyzes cyclic GMP and the similarity between the Km for cyclic GMP of 1.8×10^{-4}M and the Ki of 1.5×10^{-4}M for inhibition by cyclic GMP of the hydrolysis of cyclic AMP suggested that the same enzyme hydrolyzes cyclic AMP or cyclic GMP and that the preferred substrate is cyclic GMP. Dibutyryl cyclic AMP, caffeine, theophylline and papaverine all inhibit the hydrolysis of cyclic AMP or cyclic GMP; papaverine is the most effective. In addition, the assay of whole homogenates of retina for phosphodiesterase demonstrated complex kinetics, suggesting another enzyme with a lower Km for cyclic AMP.

As research continued on the role of cyclic AMP in photoreception, it gradually became clear that light regulates the system not through adenylate cyclase activity but rather, at least in part, through phosphodiesterase activity. In late 1973, Chader et al. (12) pointed out that there were difficulties interpreting measurements of cyclic AMP levels in preparations of rod outer segments which have high phosphodiesterase activity. They were able to confirm that in the light, there appeared to be less cyclic AMP made by the preparation, but there was some question as to the linearity of formation of cyclic AMP in the dark and in the light. In addition, they demonstrated that the formation of 5′ AMP, the metabolite of cyclic AMP, increased in light bleached preparations. This suggested that increased cyclic AMP degradation occurs after light bleaching and when tested, this hypothesis was found to be true. Thus, ^{3}H-cyclic AMP was degraded twice as fast in light bleached samples than in dark samples. These workers also mentioned the difficulty of recovering cyclic GMP in the assay for guanylate cyclase as the concentration of protein was increased in the incubation tube; this too was attributed to the extremely active phosphodiesterase.

At about the same time, Bitensky and co-workers (13) were reworking their hypothesis in terms of light-activated cyclic nucleotide phosphodiesterase being the key to regulating cyclic nucleotide levels. They demonstrated that the activation by light of the phosphodiesterase in frog rod outer segments requires the presence of ATP, with a half-maximal concentration of 0.2mM. Half-maximal stimulation of phosphodiesterase is achieved with a mixture of only 0.6 percent bleached material and 99.4 percent unbleached material. The Km for cyclic AMP is the same in the light and the dark, approximately 8mM, but the Vmax increases fivefold upon illumination only in the presence of ATP. The enzyme has a lower Km for cyclic GMP, approximately 1.6×10^{-4}M in the light and the dark, and illumination increases the Vmax for this cyclic nucleotide as well. At 10^{-7}M cyclic nucleotides, the ratio of activities for cyclic GMP/cyclic AMP is 23:1.

Soon thereafter, Chader et al. (14) published an extensive study of cyclic nucleotide phosphodiesterase in bovine retina, further purifying phosphodiesterase from rod outer segments, and also demonstrating that the hydrolytic activity for cyclic GMP is greater than that for cyclic AMP. Once purified the enzyme is no longer influenced by light. The soluble fraction of retinal homogenates demonstrates complex kinetics for cyclic AMP with Km's of 2.5×10^{-5}M and 3.9×10^{-6}M. For cyclic GMP, the kinetics are simple with a Km of 1×10^{-5}M. According to these workers, in rod outer segments, phosphodiesterase has a Km for cyclic AMP of 5×10^{-5}M and is partially soluble and partially membrane bound. The Vmax is three times greater for cyclic GMP than cyclic AMP in this preparation and the hydrolysis of cyclic AMP is inhibited by cyclic GMP. The assay for phosphodiesterase in these studies used cyclic nucleotides at lower concentrations than that used by others and this probably explains the discrepancies between the Km values obtained by this group compared to higher reported values.

Goridis and Virmaux (15) also confirmed in bovine rod outer segments the presence of a cyclic nucleotide phosphodiesterase that prefers cyclic GMP and that loses its ability to be activated by light upon further purification. As others were able to show, light activation returns upon addition of dark adapted rod outer segment material. Noteworthy in this study is the apparent absence of ATP in the assay system when light activation of phosphodiesterase is obtained. These authors were also able to show decreased levels of cyclic GMP, formed from added hypoxanthine, in whole, incubated, bovine retinas, when incubated in the light without inhibitors of phosphodiesterase. Phosphodiesterase activity, isolated from other tissue, was not altered when added to light or dark adapted material from retinas. Thus, the specificity for light activation lies with the peculiar properties of phosphodiesterase from photoreceptors.

Chader et al. (16, 17) also showed that there is light activation of phosphodiesterase only in the presence of ATP (0.1mM) and that higher concentrations of ATP (1mM) inhibit the hydrolysis of both cyclic AMP and cyclic GMP in light bleached samples. Calcium does not activate the rod outer segment phosphodiesterase; thus, this phosphodiesterase is not similar to the brain enzyme, which exhibits regulation by calcium. Because the hydrolytic activity for cyclic AMP is not separable from that of cyclic GMP, the authors tentatively concluded that a single enzyme hydrolyzes either nucleotide and postulated that fluctuating levels of cyclic nucleotides in rod outer segments are regulated, at least in part, through activation of phosphodiesterase.

Pannbacker (18) demonstrated the presence of the phosphodiesterase in a preparation of human photoreceptors at comparable levels of activity to bovine photoreceptors. Thus, in human tissue, the relative activity of phosphodiesterase for cyclic GMP is also higher than that for cyclic AMP and the Km for cyclic AMP is 4 mM with no evidence of a form of the enzyme with a lower Km.

Using a histochemical, lead precipitate method for localizing phosphodiesterase, Robb (19) demonstrated reaction product on the lamellae of rod outer segments when incubated in media containing cyclic AMP and also a light precipitate, occasionally present, on the outer segment plasma membrane. Little or no reaction product is found in the photoreceptor inner segments or nucleus, or when the preparation is incubated with cyclic AMP and papaverine. With media containing cyclic GMP, the precipitate is much less dense and shows great variability. These experiments were done with mice retina but preliminary experiments with cattle and monkey retinas reportedly gave similar results. The author claims that these results are specific for cyclic AMP and do not suffer from the usual artifacts of histochemical techniques using lead precipitation. The lack of formation of reaction product with cyclic GMP is attributed to greater susceptibility of cyclic GMP phosphodiesterase during glutaraldehyde fixation, thus implying separate enzymes for cyclic AMP and cyclic GMP hydrolysis, a suggestion not favored by most of the published biochemical data.

Goridis et al. (20) demonstrated very high levels of cyclic GMP in incubated, intact, whole retinas of both calf and frog and reported that brief exposures to light decreases the level of cyclic GMP markedly. Thus, the light sensitive mechanisms for decreasing the level of cyclic GMP are active in an intact in vitro preparation. Krishna et al. (9) incubated preparations of rod outer segments and measured the levels of cyclic AMP and cyclic GMP. In the presence of GTP or ATP, hydrolysis of exogenous cyclic GMP is markedly stimulated by exposure of rod outer segments to light. In addition, dark-adapted, bovine rod outer segments have very high levels of endogenous cyclic GMP, approximately 100 times that of cyclic AMP. Using dark adapted, rod outer segments from the frog, Fletcher and Chader (71) demonstrated a tenfold decrease in the level of endogenous cyclic GMP (with no change in cyclic AMP) upon exposure to light.

Miki et al. (21) have further characterized the activation of the phosphodiesterase. They have demonstrated that polyanions, such as heparin or sonicated silica gel, activate the enzyme in a manner resembling that produced by light and ATP. Activation by polyanions, or by light and ATP, is prevented by prior treatment of disk membranes with organic mercurials, and this is reversed by the addition of dithiothreitol. These workers suggested that light and ATP may cause an anionic transformation in the native disk membrane, thus activating phosphodiesterase. The phosphodiesterase can be eluted from disk membranes with buffers of low ionic strength containing EDTA, but eluted phosphodiesterase is no longer responsive to light and ATP. Upon addition of magnesium, the eluted phosphodiesterase can be reabsorbed onto the disk membrane and the sensitivity to light is partially restored.

These authors have also purified the phosphodiesterase to homogeneity, representing a 185-fold increase in specific activity over that seen in a rod outer segment preparation. The enzyme has a molecular weight of 240,000 daltons and during sodium dodecyl sulfate, polyacrylamide gel electrophoresis separates into two subunits of 120,000 and 110,000 daltons. The Km for the purified phosphodiesterase with cyclic GMP as substrate is $7 \times 10^{-5}M$, similar to that reported by Chader et al. (14).

The physiological role of cyclic GMP in photoreceptor cells is as yet unknown; however, the system is impressive in that dark-adapted, bovine rod outer segments contain 550 pmoles cyclic GMP/mgm protein (9). The potential of the system appears to be to decrease rapidly the level of cyclic GMP upon illumination by decreasing synthesis (through light-inactivated guanylate cyclase) and by increasing degradation (through light-activated phosphodiesterase). Most workers now believe that fluctuations in the concentration of cyclic GMP would be too slow to be involved in the initial mechanism of visual excitation. However, regeneration of rhodopsin, following extensive bleaching, could be stimulated by changing levels of cyclic GMP. Thus, when going from an illuminated room into a darkened room, cyclic GMP levels may increase in rod outer segments, stimulating regeneration of rhodopsin and promoting, over several minutes, the dark-adapted state.

Alternatively, if one considers the hypothesis that relative levels of cyclic nucleotides influence differentiation and mitotic rate, the photoreceptor cells, with their high ratio of cyclic GMP to cyclic AMP should be undergoing rapid cell division, which they are not. These cells are specialized, showing a high degree of differentiation, adapted exquisitely for their function. The rod outer segment, the part of the cell that contains the high levels of cyclic GMP, is, however, turning over rapidly. The disk membranes are constantly renewed by de novo synthesis of rhodopsin and membrane material at an area proximal to the cell body of the inner segment and phagocytosed at the distal end by the retinal pigment epithelium. One wonders if the cyclic GMP system is somehow involved in regulating the turnover of rod disk membranes.

As work on the biochemistry of the phosphodiesterase was continuing, Lolley and his colleagues were studying the fate of this enzyme in the retina of rodents with inherited blindness. C3H/HeJ mice undergo an autosomal recessive mutation which results in the degeneration of all the photoreceptor cells during the second postnatal week, precisely the time in the development of the retina when the photoreceptor and bipolar cells normally differentiate. Death of the cells begins within the central retina and spreads to the periphery. Presumably, some aspect of photoreceptor cell differentiation is adversely affected by the mutant gene, but Lolley (22) previously reported that for the first 10 days of postnatal life, glucose, lactate, ATP and P-creatinine are all normal in the photoreceptor cell layer of the retinas of these mice. Thus, the lesion is not at the level of glucose metabolism or energy production.

Comparing retinas from normal mice and mutant mice, Schmidt and Lolley (23) demonstrated that during development of the retina of C3H mice, a specific phosphodiesterase, with a high Km for cyclic nucleotides and greater affinity for cyclic GMP (Km = 6.6×10^{-4}M) than for cyclic AMP (2×10^{-3}M), is never observed.

In the normal mouse retina, the appearance of this high Km phosphodiesterase, a rapid, eight-fold increase in activity, coincides with the differentiation and growth of the photoreceptor outer segments. A similar increase in cyclic GMP phosphodiesterase occurs in the developing chick retina (24). When various regions of the central nervous system of the mice are assayed, for example the hypothalamus or cortex, the phosphodiesterase is normal in animals with retinal degeneration. A low Km phosphodiesterase for cyclic AMP, 2×10^{-4}M, is also present in the retina of the normal mouse and the mouse that develops retinal degeneration. This phosphodiesterase is localized in the inner layers of the retina, remains in both strains of animals and is not appreciably affected by the degeneration of the photoreceptor layer.

Farber and Lolley (25) have examined the proteins in developing mice retina with SDS polyacrylamide gel electrophoresis. Of the approximately 30 bands of protein observed from adult retina in the normal animals, the retinas from adult C3H mice are deficient in three bands of protein. The authors identified two of the bands: one is opsin and the other is cyclic nucleotide phosphodiesterase.

Lolley, Schmidt, and Farber (26) demonstrated that cyclic AMP is not the important factor in photoreceptor cell degeneration. In the C3H mouse, as development of the retina continues, adenylate cyclase activity increases similarly to the normal retina for the first week; afterwards, adenylate cyclase activity becomes greater than normal and remains so until adulthood. Cyclic AMP levels are abnormally high after 10 days of age and this persists during and after photoreceptor cell degeneration. The characteristics of the enhanced activity of adenylate cyclase, such as stimulation by dopamine or fluoride and the Km values, are similar in the normal and mutant mice. Thus, the elevated level of cyclic AMP occurs in the surviving cells of the inner layer of the retina and the photoreceptor layer actually has a low level of enzyme activity and cyclic AMP.

Having ruled out cyclic AMP, Farber and Lolley (27) determined the level of cyclic GMP in retinas of normal and C3H mice. Cyclic GMP in the normal retina increases postnatally and approximately doubles between days 6 and 12. In the developing retina of the C3H mouse, the level of cyclic GMP is normal for 6 days postnatally and then increases to a peak value, at least threefold higher than normal, at day 15. By microdissecting freeze-dried retinas and measuring cyclic GMP in separated layers, the authors demon-

Table 2.
Mouse Retina, 10 Days After Birth
(from references 36, 39, 40)

	Normal (DBA)	Dystrophic (C3H)
Phosphodiesterase		
Vmax (nmoles cyclic AMP/mg prot/min)	18	7
Km (mM cyclic AMP)	2.5	
	0.13	0.18
*(mM cyclic GMP)	0.65	0.55
Adenylate Cyclase		
Activity (pmoles cyclic AMP/mg prot/min)		
Basal	50	44
Dopamine-Stimulated	116	110
Cyclic AMP		
(pmoles/mg prot)	19	30
Cyclic GMP		
(pmoles/mg prot)	150	350

*determined in adult retinae

strated that most of the cyclic GMP elevation occurs in the photoreceptor layer of C3H mice. After degeneration occurs, the cyclic GMP level decreases to a value that approaches the normal retina. Thus, the initial increase in the level of cyclic GMP in the degenerating retina at day 6 is due to the deficiency of the high Km phosphodiesterase in the photoreceptor layer and, therefore, the inability to hydrolyze cyclic GMP. The subsequent decrease in cyclic GMP in these retinas is due to degeneration and death of the photoreceptor cells. Presumably, the elevated cyclic GMP causes disturbances in the metabolism or function of these cells, prior to the onset of their degeneration, and this alteration of the homeostatic balance produces changes that result in degeneration. The results from the developing mouse retina, on day 10, are summarized in Table 2.

Recently, Robb (28) has provided histochemical evidence that C57B1/6J mice that experience retinal degeneration have the same amount of cyclic AMP phosphodiesterase as do normal littermates. This argues against generalizing that all lesions leading to retinal degeneration involve a phosphodiesterase. However, a further understanding of the lesion in this strain will come with biochemical techniques which are quantifiable and with the consideration of the fate of cyclic GMP.

Lolley and Farber (29) have extended their work to the inherited retinal dystrophy of another species, the Royal College of Surgeons (RCS) rats. In these rats photoreceptor cell degeneration is somewhat different than in mice, occurring later, between 20 and 40 days postnatally. The differentiated photoreceptor cells produce rod outer segment material which is not phagocytosed by the retinal pigment epithelium, leading to the accumulation of debris, rod outer segment and pigment epithelium material, between the two layers. These investigators determined the activities of cyclic AMP and cyclic GMP phosphodiesterase in the inner layers of the retina (the bipolar and ganglion cell layers) and the photoreceptor layer. The cyclic AMP phosphodiesterase of the photoreceptor layer has a Km of 2×10^{-3}M cyclic AMP, first appears on day 8, and increases in activity as the rod outer segments increase in length and debris accumulates in the retina of the RCS rats. By day 23, when photoreceptor debris is the thickest, the phosphodiesterase activity is greater than in normal retinas and afterwards, as the cells further degenerate, the cyclic AMP phosphodiesterase activity decreases. This pattern of cyclic AMP phosphodiesterase activity is very similar to the changes seen in rhodospin content of these degenerating retinas, as expected if the two are in the same cell and changing similarly. The cyclic GMP phosphodiesterase of the inner and photoreceptor cell layers show complex changes in the RCS rat retina, unlike the C3H mouse retina. At about 18 days postnatally, when the photoreceptor debris is accumulating, the inner retinal layer cyclic GMP phosphodiesterase is no longer demonstrable, and the cyclic GMP phosphodiesterase of the photoreceptor layer has a Km value lower than that previously measured, both apparently due to the presence of a macromolecular inhibitor. Following photoreceptor degeneration, the cyclic GMP phosphodiesterase of the inner retinal layer reappears. The authors suggest that prior to degeneration of the photoreceptor of RCS rats, cyclic GMP is elevated due to decreased activity of cyclic GMP phosphodiesterase.

Recently, at least partial confirmation of this work has been obtained by Dewar, Barron and Richmond (30). Although they did not measure activity with cyclic GMP as substrate, these authors did show that in the retinas of Campbell rats with retinal dystrophy there is an apparent defect in a phosphodiesterase with a high Km for cyclic AMP. This enzyme will undoubtedly prove to be the cyclic GMP phosphodiesterase of the photoreceptor cells. In Hunter rats, a pigmented strain with retinal dystrophy, the changes in retinal cyclic AMP phosphodiesterase appear to be secondary to degeneration of the photoreceptor cells. It remains to be seen what other photoreceptor cell degenerations, particularly in primates, are related to a missing phosphodiesterase, an inhibited phosphodiesterase, or some other alteration of the metabolism of cyclic nucleotides.

AQUEOUS HUMOR DYNAMICS

The intraocular pressure in most, and perhaps all, conscious mammals is 15 to 18 mm Hg and is the steady state value resulting from the relative rates of inflow and outflow of aqueous humor. The turnover rate of aqueous humor is approximately 1 percent per minute; thus, in 1 hr, most of the aqueous humor, which supplies the lens, cornea, and trabecular meshwork with nutrients, is renewed. The mechanisms governing the movement of fluid through the anterior portion of the eye are only partially understood, primarily due to the difficulty of studying these processes. Nevertheless, detailed information concerning the mechanisms of inflow and outflow of aqueous humor is crucial, because small changes in the fluid dynamics may be disastrous to the eye, leading to atrophy of the optic nerve and blindness.

The formation of aqueous humor, e.g., at a rate of 2 μl/min in primates, has two components, secretion and ultrafiltration, and occurs primarily in the ciliary processes, the fingerlike projections extending radially into the posterior chamber, behind the iris (Figure 1). The secretory cells, the nonpigmented epithelial cells which border the posterior chamber, have numerous invaginations on this surface, similar to secretory epithelia in other tissues. The secretion of aqueous humor is active and probably requires the movement of sodium pumped into the nascent fluid in the posterior chamber. Chloride and bicarbonate follow to maintain electrical neutrality and there may also be an active component for transport of these ions. The net flux of ions causes water movement, producing aqueous humor approximately isotonic with plasma as it flows from the posterior chamber, through the pupil, into the anterior chamber.

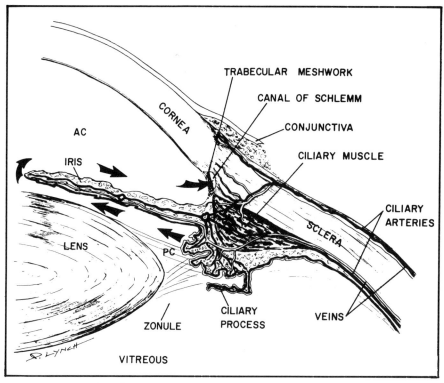

1 Anterior segment of the eye. The broad arrows indicate the flow of aqueous humor through the posterior chamber (PC) and the anterior chamber (AC).

The other mechanism for the formation of aqueous humor, either in series or in parallel to secretion, is ultrafiltration. This provides a pressure-dependent flow of an ultrafiltrate of plasma and the final result is an aqueous humor relatively free of plasma proteins. The maintenance of this blood-aqueous barrier to plasma proteins ensures a clear media in the eye for the passage of light to the retina. Not only do we know little about the mechanisms for the formation of aqueous humor, but much uncertainty exists concerning even the relative contributions of secretion and ultrafiltration to this process. (For further discussions of formation, as well as all aspects of aqueous humor dynamics, see Refs. 31–33.)

Since the endothelium of the capillaries of the iris are joined by tight junctions and the vessels are relatively impermeable, there

Table 3.
Influences on Intraocular Pressure

		Inflow				Outflow	
		Secretion		Ultrafiltration			
		aff	eff	aff	eff	aff	eff
Vascular	Constriction	↓	↑	↓	↑	↓	↑
	Dilation	↑	↑	↓	↓	↑	↓
Cellular	cAMP action	↑ or ↓*		?		↓	

arrows indicate effect on intraocular pressure
aff = afferent blood vessels
eff = efferent blood vessels
* indirect evidence in different species

is little net flow across the iridial vessels. However, the capillaries of the heavily vascularized ciliary processes are fenestrated and, therefore, any consideration of the formation of aqueous humor must take into account the state of these blood vessels. Table 3 summarizes vascular and other factors that may influence intraocular pressure. Vasoconstriction of efferent vessels would favor an increase in ultrafiltration (34), due to the increased transluminal pressure difference between the vessels and the eye, while vasoconstriction of afferent blood vessels in the ciliary body would decrease ultrafiltration (35). Vasodilation, which leads to a decrease in blood pressure, would decrease ultrafiltration slightly. Vascular effects on secretion would be appreciable only if blood flow was limiting. Then, afferent vasoconstriction would decrease, and efferent vasoconstriction would increase, secretion (36); whereas, vasodilation might favor a small increase in secretion, if blood flow increased. Also, effects of agents on secretion may be exerted at the cellular level. Although there is no direct evidence that this is mediated by cyclic AMP, ciliary processes from the rabbit contain an isoproterenol-stimulated, adenylate cyclase (37) and cyclic AMP stimulates the short circuit current in an isolated iris-ciliary process preparation from the rabbit (38). In addition, intracameral isoproterenol increases secretion in the monkey (36) and topical isoproterenol decreases non-pressure dependent flow in normal, human volunteers (39).

The primary mechanism for outflow of aqueous humor is through intrascleral routes in the chamber angle of the eye, just anterior to the junction between the iris and ciliary muscle (Fig. 1). Leaving the anterior chamber, the aqueous humor flows through the trabecular meshwork, first uveal mesh then corneoscleral mesh. The trabecular beams which make up the trabecular meshwork, increase in size towards the canal of Schlemm, gradually decreasing the space between beams. However, these spaces, which are several microns wide, do not offer much resistance until deep into the corneoscleral areas of the trabecular meshwork. The beams are lined by an endothelium that may be coated by a mucopolysaccharide material which perhaps offers resistance, in the inner recesses of the trabecular meshwork, to the flow of aqueous humor. The juxtacanalicular material, the cellular and extracellular material before the canal of Schlemm, does not exhibit any regular patterns of structure, but may, in addition, offer some resistance via extracellular matrices which the aqueous humor must seep through. Finally, the aqueous humor must pass through the anterior, endothelial lining of the canal of Schlemm, a unique endothelium in which the cells are joined by occluding junctions and fluid passes through the cells via transcellular pores or giant vacuoles, the formation of which may offer resistance to flow. For a discussion of the outflow pathway, see Tripathi (40).

Aqueous humor leaves the eye by a bulk flow mechanism which is dependent upon the difference between the intraocular pressure and episcleral venous pressure. The pressure in the episcleral veins probably does not change markedly but vascular changes before (afferent) or after (efferent) the canal of Schlemm, which affect blood flow or pressure in this area, will cause small, transient changes in outflow (41), as indicated in Table 3. The primary resistance to the movement of fluid is in the tissues comprising the outflow pathway. This resistance is often expressed as its reciprocal, or a facility. Outflow facility is, to a first approximation at physiological pressures, a constant, not dependent on pressure, and given by the following equation: $F = C(IOP-P_v)$, where F is the flow of aqueous through the eye in $\mu l/min$, IOP is the intraocular pressure and P_v is the episcleral venous pressure, both in mm Hg, and C is outflow facility in $\mu l/min/mm$ Hg. The value of C is usually about 0.2 to 0.3 $\mu l/min/mm$ Hg and can be determined in laboratory animals, by intraocular perfusion techniques, and in man, by tonography (42). Although the nature of the resistance to the flow of fluid through the outflow channels is not understood, relatively small changes in outflow facility can cause large changes in intraocular pressure. From the above equation, consider what would happen in a normal individual with F=2 μl/min, IOP=17 mm Hg, P_v=9 mm Hg and C=0.25 μl/min/mm Hg, if outflow facility is decreased by 40 percent while flow and episcleral venous pressure remained the same. The intraocular pressure would become 23 mm Hg and the person would be suspected of having glaucoma.

In glaucoma, the intraocular pressure is elevated and the goal of the physician is to reduce the intraocular pressure to a value at which tissue damage will not occur. In adults, two types of primary glaucoma are distinguishable. In narrow angle glaucoma, obstruction of the chamber angle causes the intraocular pressure to increase markedly and the eye becomes painful. Immediate medical or surgical intervention is needed to prevent irreparable damage to the optic nerve. In open angle glaucoma, the chamber angle appears grossly normal but the pressure is elevated because there is increased resistance to the flow of fluid from the eye, measured as a decrease in outflow facility (43). Open angle glaucoma is an insidious disease causing often only mild elevations in intraocular pressure, which chronically over years, possibly through decreased choroidal blood flow or obstruction of axoplasmic flow, leads to degeneration of the ganglion cells and atrophy of the optic nerve. Glaucoma may account for as much as 15 percent of blindness in this country and occurs in 2 percent of individuals over 40 years of age (44).

The medical therapy of open angle glaucoma utilizes drugs which have either one or a combination of three basic types of action (45). One of the most effective ways of reducing the intraocular pressure is to decrease the formation of aqueous humor through the use of systemic acetazolamide. Although the actual mechanism of action, be it carbonic anhydrase inhibition or some other effect, is still the subject of debate, in therapeutic dosages, the drug reduces the inflow of aqueous humor by approximately 50 percent. A second class of drugs are miotics, such as pilocarpine, applied topically, which stimulate contraction of the pupillary sphincter and the ciliary muscle. Because the insertion of the ciliary muscle is the scleral spur, which borders the trabecular meshwork, the contraction of these smooth muscles somehow results in a reduced resistance to flow of fluid through the outflow channels, perhaps due to an enlargement of the cross sectional area.

A third class of drugs are the adrenergic compounds, such as epinephrine, also used topically. Undoubtedly, the catecholamines have vascular effects, causing both vasoconstriction and vasodilation according to dose, route of delivery and time after delivery, and these actions will affect intraocular pressure through influences on both inflow and outflow, as depicted in Table 3. However, after topical epinephrine, the lowered intraocular pressure persists for several hours, presumably, past the time when high concentrations of catecholamines are still free in ocular tissues and long after vascular tone is restored. In addition, the work of Sears (46) indicates that catecholamines, like norepinephrine, act directly on tissues in the outflow channels to decrease the resistance to the outflow of aqueous humor. In the rabbit, this is predominatly an alpha effect of norepinephrine and can be demonstrated 20 hr after superior cervical ganglionectomy when large amounts of norepinephrine, released endogenously into the aqueous humor from the degenerating adrenergic nerve terminals, cause an increase in outflow facility and a decrease in intraocular pressure. In addition, several weeks after denervation, a supersensitivity to norepinephrine, administered intracamerally (directly into the anterior chamber), develops, indicating receptors for norepinephrine in the outflow channels. There are also beta receptors in the rabbit eye which increase outflow (47). Thus, in the rabbit, both alpha and beta stimulation appear to accomplish the same end, increased outflow of aqueous humor. Bill (36) has demonstrated that in the monkey eye, strongly beta drugs, such as isoproterenol, are more effective than strongly alpha drugs, such as norepinephrine, to increase outflow facility when given intracamerally; thus, the nonhuman primate has beta receptors in the outflow channels which can cause an increase in outflow. In the human, intracameral administration of drugs has not been tried, and topical administration of epinephrine to normal eyes, topical epinephrine often causes an improvement in outflow facility. However, in elderly normal individuals, topical isoproterenol increases true outflow facility (48) and, in glaucomatous eyes, topical epinephrine often causes an improvement in outflow facility that persists for some time with prolonged use (49, 50).

The mechanism by which epinephrine acts on the outflow pathway to increase the flow of aqueous humor out of the eye has remained a mystery. We began our investigations into the action of epinephrine for several reasons. First, catecholamines are endogenous compounds and may normally exert a physiological control mechanism on the outflow of aqueous humor. Although the trabecular meshwork and the canal of Schlemm are not heavily adrenergically innervated, there are adrenergic nerve endings sparsely present near these structures (51, 52). However, the absence of structures in this area which can normally be associated with innervation, led us to consider a more cellular and biochemical response to catecholamines, in particular, a response mediated through cyclic AMP. Furthermore, a demonstration of the relevance of cyclic AMP would suggest new medical therapies for the treatment of glaucoma and might provide some insight into the etiology of the disease.

We first determined the concentration of cyclic AMP in aqueous humor, which, after all, is a tissue perfusate that bathes intraocular tissues and may, therefore, reflect intracellular changes. Cyclic AMP is present in the aqueous humor of rabbits, at a concentration of approximately 24 nmoles/liter (slightly lower in primates). This concentration increases three- to fourfold within 2 hr after administration of topical 1 percent epinephrine to the rabbit eye, approximately the same time that the maximum fall in intraocular pressure of 4 to 5 mm Hg occurs (53). There are similar responses, although somewhat reduced, to topical norepinephrine and isoproterenol and an approximate relationship exists between the ability of a compound to reduce intraocular pressure and to increase cyclic AMP in the aqueous humor. Lower concentrations of each drug, for example 0.1 percent epinephrine, cause a smaller decrease in intraocular pressure and a smaller elevation of cyclic AMP in the aqueous humor of the treated eye. Topically or intravenously administered caffeine, theophylline, or aminophylline do not increase cyclic AMP in the aqueous humor or decrease intraocular pressure. Our observations with norepinephrine were confirmed by Radius and Langham (54), who pointed out that cyclic AMP could become elevated in the aqueous humor, in certain instances, without a clear effect on intraocular pressure.

The response to topical epinephrine in the rabbit lasts several hours. The intraocular pressure remains decreased for at least 6 to 7 hr during which time the cyclic AMP in the aqueous humor remains elevated. The effect of epinephrine on pupil size is quite different; the mydriasis (pupil dilation) peaks at about 15 min; then the pupil begins to return slowly to normal size over the next few hours.

In a further attempt to correlate the change in cyclic AMP in the aqueous humor with the change in intraocular pressure, rabbits were made more, or less, sensitive to administered epinephrine and then tested (55). In the rabbit, repeated, daily, topical administration of epinephrine to the eye leads to a progressive decrease in the magnitude of the intraocular pressure response to this drug. Thus, animals were treated with topical 1 percent epinephrine once daily for 5 days while the other eye received the control vehicle. On day 5, in eyes treated for 5 days, a decrease in intraocular pressure of 2 mm Hg occurred, and the increase in cyclic AMP in the aqueous humor was less than twofold. Thus, decreased sensitivity to epinephrine, as far as changes in intraocular pressure, correlate with less of an increase in cyclic AMP in the aqueous humor.

As mentioned above, several weeks after superior cervical ganglionectomy, a denervation supersensitivity develops in the rabbit eye, whereupon a given concentration of topical catecholamine produces a greater fall in intraocular pressure and the eye is more sensitive to a lower concentration of the drug. There is no supersensitivity to isoproterenol. In rabbits, whose eyes had been adrenergically denervated, topical epinephrine elevated cyclic AMP in the aqueous humor six- to sevenfold, shifting the dose-response curve to the left by approximately the same order of magnitude as the supersensitivity for intraocular pressure or outflow facility. Isoproterenol does not cause an enhanced increase in cyclic AMP in the aqueous humor of the supersensitive eye. The mechanism for enhanced levels of cyclic AMP in the denervated eye is still questionable: there may be presynaptic supersensitivity because catecholamines are no longer being inactivated by neuronal tissue or there may be postsynaptic supersensitivity elicited through membrane adenylate cyclase and its receptors for catecholamines. In either event, there is increased sensitivity to topical epinephrine in terms of both intraocular pressure and cyclic AMP in the aqueous humor.

The cyclic AMP in the aqueous humor is presumably an overflow, or spillover, from cells making this cyclic nucleotide, a reflection of intracellular events in response to a stimulus. However, the level of cyclic AMP in the aqueous humor of a non-drug-treated eye is not due to adrenergic neural activity (54) since it has the same value in the adrenergically denervated and normal eye (Neufeld, Chavis, and Sears, unpublished observations). Attempts to determine the tissue of origin of the cyclic AMP in the aqueous humor have not been entirely successful; although, the absence of cyclic AMP in the aqueous humor from the posterior chamber does focus attention on cells bordering the anterior chamber. Ocular tissues from rabbit, monkey and man such as cornea, iris, ciliary body, sclera-trabecular ring (a ring of sclera cut at the limbus, circumferentially around the eye, to include material containing the outflow pathway), and a section of sclera a few millimeters posterior to this ring make and release cyclic AMP, in vitro, in response to catecholamines (56). The possibility that cyclic AMP is made by specific cells in tissues that govern outflow is difficult to test but is still part of our hypothesis.

To prove the causal relationship between intraocular pressure and cyclic AMP, we injected 100 μmoles of cyclic AMP directly into the anterior chamber of one eye of a rabbit, and 5′ AMP into the anterior chamber of the other eye, and measured the intraocular pressure (53). A decrease of 3 to 4 mm Hg occurred in the eye receiving cyclic AMP, with no effect on the eye receiving 5′ AMP. Thus, cyclic AMP reduces intraocular pressure and the magnitude of the effect and the route of delivery, i.e., into the anterior chamber, suggests a direct effect on the outflow pathway. Thus, we measured outflow facility in rabbit eyes using constant pressure perfusion at two steps of pressure above the prevailing intraocular pressure (57). Both cyclic AMP and dibutyryl cyclic AMP, administered directly into the anterior chamber, cause an approximately twofold increase in outflow facility in the rabbit eye, similar to the response to intracameral epinephrine (58). In addition, another active cyclic AMP analog, 8-methylthio cyclic AMP (SQ 80,002) is approximately 10 times more effective than dibutyryl cyclic AMP to increase outflow facility in the rabbit eye (author's unpublished observations). Data summarizing the effect of cyclic AMP, administered intracamerally in the rabbit, is summarized in Table 4.

The outflow pathway in the primate is analogous to, but slightly different than, that in the rabbit; for example, the canal of Schlemm of the primate has its counterpart in the rabbit as a multichambered plexus into which the aqueous humor drains. Therefore, we tested the relevance of mediation by cyclic AMP to the control of outflow in the primate. Outflow facility was measured in the eye of the vervet monkey, *Cercopithecus ethiops*, by intraocular cannulation and constant pressure perfusion. Preliminary experiments indicated that the monkey was less sensitive than the rabbit to administration of cyclic AMP, so 8-methylthio cyclic AMP

Table 4.
Rabbit Eye, Intracameral Delivery
(from references 64, 68)

	IOP (mm Hg)	C (μl/min/mm Hg)
Cyclic AMP		
Before	16.9 ± 0.2 (10)*	0.22 ± 0.02 (11)
After	12.8 ± 0.5 (10)	0.41 ± 0.03 (10)
5' AMP		
Before	16.4 ± 0.3 (10)	0.25 ± 0.02 (12)
After	16.5 ± 0.4 (10)	0.22 ± 0.03 (12)

*mean ± S.E.M. (number of eyes)

was used. This compound, when perfused through the anterior chamber of the monkey eye, in a similar manner to that used by Bill (36) for administering isoproterenol, increases outflow facility approximately twofold (59). Originally, we reported that delivery of 60 μmoles of 8-methylthio cyclic AMP over 30 min increases outflow facility in the vervet monkey without an effect on pupil size. We have since shown that much less cyclic AMP analog is needed, 6 μmoles delivered over two to three minutes, to similarly increase outflow facility in the vervet and macaque. The response is brisk, within ten minutes after the end of the perfusion of cyclic AMP, outflow facility has already increased, in contrast to the relatively slow increase in outflow facility during intracameral perfusion of isoproterenol over a period of 1 to 1.5 hr. However, once an increase in outflow facility is obtained in response to isoproterenol, 8-methylthio cyclic AMP no longer increases outflow facility in that eye. We interpret this to indicate that the two compounds are working at different levels of the same system. Thus, the administered isoproterenol has stimulated production of cyclic AMP, maximally, and administration of an analog of cyclic AMP no longer has an appreciable effect on outflow facility.

Bárány (60) has suggested that, in addition to contraction of specific smooth muscles, pilocarpine increases outflow facility by another, unknown mechanism. We hypothesized that cyclic GMP may mediate this mechanism in the primate eye. In the vervet monkey, the pupil constricts, and outflow facility increases severalfold when pilocarpine is administered intracamerally. However, a potent analog of cyclic GMP, 8-bromo-cyclic GMP (SQ 80,022) administered intracamerally, has no effect on outflow facility. When pilocarpine is then given into the anterior chamber of the same eye, a large increase in outflow occurs. We conclude that the systems are different; the cholinergic effect on outflow facility, if there is another action in addition to smooth muscle contraction, is not exerted through a mechanism mediated by cyclic GMP in the outflow pathway.

Thus, we have demonstrated that cyclic AMP mediates the increase in outflow facility in response to adrenergic stimulation and this explains, at least in part, the decrease in intraocular pressure following adrenergic agonists. This mediation by cyclic AMP is further proof of the presence of beta adrenergic receptors in the outflow pathway which can respond physiologically to endogenous catecholamines, or pharmacologically to administered adrenergic compounds. Also, our observations imply that outflow can be influenced via intracellular events.

At this point, if we apply our knowledge of cyclic AMP to open angle glaucoma, it suggests several possibilities. An adrenergic compound, given topically to a glaucomatous eye, may increase outflow facility through a cyclic AMP mediated mechanism, at a point of resistance, either the same as, or in series, or in parallel to, the area of increased resistance which has caused the glaucoma. Therefore, a cyclic AMP analog, a compound that stimulates cyclic AMP production, or a phosphodiesterase inhibitor might be developed, which, when administered to patients, either topically or systemically, would increase outflow facility or potentiate another drug such as epinephrine. The testing of new antiglaucoma drugs should include their abilities to decrease intraocular pressure as well as to increase cyclic AMP in the aqueous humor of rabbits and monkeys. An adrenergic-type drug should do both to be effective in humans and indirect proof of this point has been obtained recently. d-isoproterenol reduces the intraocular pressure of rabbits but does not increase the cyclic AMP in the aqueous humor. In the primate, this compound does not decrease the intraocular pressure of monkeys or normal, human volunteers (61). However, dipivalyl epinephrine reduces the intraocular pressure and increases the concentration of cyclic AMP in the aqueous humor of rabbits and is effective in hypertensive patients (62).

The demonstration of the regulation by cyclic AMP of the outflow of aqueous humor may lead to an understanding of the etiology of the problem in open angle glaucoma. Cyclic AMP is a good candidate for influencing may of the mechanisms of resistance that have been suggested. This cyclic nucleotide could mediate changes in the intertrabecular beam distance, steroid sensitivity, mucopolysaccharides lining the endothelium of the trabecular beams, the phagocytic activity of the endothelium, the juxtacanalicular matrix, or the permeability of the anterior endothelial lining of the canal of Schlemm, either through an intercellular mechanism or an intracellular mechanism like the giant vacuoles. In addition, there is the possibility that catecholamines or steroids exert a tonic influence on the outflow of aqueous humor through cyclic AMP and that a lesion in this system causes the increase in resistance associated with open angle glaucoma.

CORNEA

There are two areas of research concerning cyclic nucleotides and the cornea which deserve careful attention. The cornea maintains its transparency by controlling the hydration of its stroma, in part through the pumping of chloride and sodium. Cyclic AMP stimulates the corneal epithelium to actively transport chloride from the aqueous humor to the tears in the frog (63) and rabbit cornea (64), and a recent preliminary report indicates that there is a similar mechanism in the human cornea (65). In a preswollen cornea, cyclic AMP causes an increase in transparency (63, 66). Increases in cyclic AMP in the cornea occur via stimulation by catecholamines of a β receptor (56, 67) and perhaps the inhibition of phosphodiesterase by ascorbic acid (68), which is found in high concentrations in the aqueous humor. Thus, in corneal diseases where loss of transparency and edema is often seen, medical therapy involving cyclic nucleotides may help reestablish control of hydration.

Another area of recent interest in the cornea involves mediation by cyclic nucleotides of various aspects of corneal wound healing. When ulcerated rabbit corneas are maintained in tissue culture, dibutyryl cyclic AMP and theophylline partially inhibit the appearance of collagenase activity as well as inhibit degradation of collagen (69). This suggests an effect on the synthesis, secretion, or activity of collagenase and that successful treatment of corneal ulceration might be accomplished through intervention with cyclic nucleotides, pharmacologically. Adrenergic agents might serve as a first messenger to an ulcerating cornea, in vivo, to stimulate the cyclic AMP formation and regulate collagenase production.

In addition, Cavanagh (70) has suggested that one of the problems in corneal wound healing may be an epithelial defect involving an upset in the relative balance of cyclic nucleotides. Ulcerating human corneas might therefore be improved by compounds which increase cyclic GMP levels in the cells, bringing the cyclic GMP to cyclic AMP ratio back to a state associated with normal growth of this tissue. Initial successes have been reported in treating patients with corneal defects, such as herpetic corneal disease, with compounds that may increase the level of cyclic GMP such as topical carbachol, phospholine iodide, or phorbolymyristate acetate.

SUMMARY

Involvement of cyclic nucleotides in various aspects of ocular biochemistry and physiology suggests mechanisms for the etiology, and applications for the treatment, of several diseases of the eye. In the retina, a light-activated cyclic GMP phosphodiesterase is present in the photoreceptor cells. In strains of mice and rats that develop retinal dystrophy soon after birth, this phosphodiesterase is missing, or inactive, prior to degeneration of the photoreceptor cells, and elevated levels of cyclic GMP occur. Cyclic AMP mediates the action of adrenergic compounds to decrease intraocular pressure by decreasing the resistance to the flow of aqueous humor out of the normal rabbit and monkey eye. This mechanism may be important to the treatment and understanding of open angle glaucoma, a disease of increased intraocular pressure due to increased outflow resistance. In the cornea, cyclic AMP regulates transparency by stimulating a chloride pump, and an imbalance of the cyclic AMP to cyclic GMP ratio may contribute to persistent epithelial defects.

REFERENCES

1. Newsome, D. A., Fletcher, R. T., Robison, W. G. Jr., Kenyon, K. R. and Chader, G. J.: Effects of cyclic AMP and sephadex fractions of chick embryo extract on cloned retinal pigmented epithelium in tissue culture. J. Cell Biol. 61:369-382, 1974.

2. Chader, G. J.: Hormonal effects on the neural retina: Induction of glutamine synthetase by cyclic-3',5'-AMP. Biochem. Biophys. Res. Comm. 43: 1102-1105, 1971.
3. Brown, J. H. and Makman, M. H.: Stimulation by dopamine of adenylate cyclase in retinal homogenates and of adenosine 3',5'-cyclic monophosphate formation in intact retina. Proc. Nat. Acad. Sci. USA 69:539-543, 1972.
4. Brown, J. H. and Makman, M. H.: Influence of neuroleptic drugs and apomorphine on dopamine-sensitive adenylate cyclase of retina. J. Neurochem. 21:477-479, 1973.
5. Bucher, M. B. and Schorderet, M.: Apomorphine-induced accumulation of cyclic AMP in isolated retinas of the rabbit. Biochem. Pharmacol. 23:3079-3082, 1974.
6. Häggendal, J. and Malmfors, T.: Identification and cellular localization of the catecholamines in the retina and the choroid of the rabbit. Acta Physiol. Scand. 64:58-66, 1965.
7. Pannbacker, R. G.: Control of guanylate cyclase activity in the rod outer segment. Science 182:1138-1140, 1973.
8. Bensinger, R. E., Fletcher, R. T. and Chader, G. J.: Guanylate cyclase: Inhibition by light in retinal photoreceptors. Science 183:86-87, 1974.
9. Krishna, G., Krishnan, N., Fletcher, R. T. and Chader, G.: Effects of light on cyclic GMP metabolism in retinal photoreceptors. J. Neurochem. In Press.
10. Bitensky, M. W., Gorman, R. E. and Miller, W. H.: Adenyl cyclase as a link between photon capture and changes in membrane permeability of frog photoreceptors. Proc. Nat. Acad. Sci. USA 68:561-562, 1971.
11. Pannbacker, R. G., Fleischman, D. E. and Reed, D. W.: Cyclic nucleotide phosphodiesterase: High activity in a mammlian photoreceptor. Science 175:757-758, 1972.
12. Chader, G. J., Bensinger, R., Johnson, M. and Fletcher, R. T.: Letter: Phosphodiesterase: An important role in cyclic nucleotide regulation in the retina. Exp. Eye Res. 17:483-486, 1973.
13. Miki, N., Keirns, J. J., Marcus, F. R., Freeman, J. and Bitensky, M. W.: Regulation of cyclic nucleotide concentrations in photoreceptors: An ATP-dependent stimulation of cyclic nucleotide phosphodiesterase by light. Proc. Nat. Acad. Sci. USA 70:3820-3824, 1973.
14. Chader, G., Johnson, M., Fletcher, R. and Bensinger, R.: Cyclic nucleotide phosphodiesterase of the bovine retina: Activity, subcellular distribution and kinetic parameters. J. Neurochem. 22:93-99, 1974.
15. Goridis, C. and Virmaux, N.: Light-regulated guanosine 3',5'-monophosphate phosphodiesterase of bovine retina. Nature 248:57-58, 1974.
16. Chader, G. J., Herz, L. R. and Fletcher, R. T.: Light activation of phosphodiesterase activity in retinal rod outer segments. Biochim. Biophys. Acta 347:491-493, 1974.
17. Chader, G., Fletcher, R., Johnson, M. and Bensinger, R.: Rod outer segment phosphodiesterase: Factors affecting the hydrolysis of cyclic-AMP and cyclic-GMP. Exp. Eye Res. 18:509-515, 1974.
18. Pannbacker, R. G.: Cyclic nucleotide metabolism in human photoreceptors. Invest. Ophthalmol. 13:535-538, 1974.
19. Robb, R. M.: Histochemical evidence of cyclic nucleotide phosphodiesterase in photoreceptor outer segments. Invest. Ophthalmol. 13:740-747, 1974.

20. Goridis, C., Virmaux, N., Cailla, H. L. and Delaage, M. A.: Rapid, light-induced changes of retinal cyclic GMP levels. FEBS Lett. 49:167-169, 1974.
21. Miki, N., Baraban, J. M., Keirns, J. J., Boyce, J. J. and Bitensky, M. W.: Purification and properties of the light activated cyclic nucleotide phosphodiesterase of rod outer segments. J. Biol. Chem. 250:6320-6327, 1975.
22. Lolley, R. N.: Changes in glucose and energy metabolism *in vivo* in developing retinae from visually-competent (DBA/1J) and mutant (C3H/HeJ) mice. J. Neurochem. 19:175-185, 1972.
23. Schmidt, S. Y. and Lolley, R. N.: Cyclic-nucleotide phosphodiesterase: An early defect in inherited retinal degeneration of C3H mice. J. Cell Biol. 57:117-123, 1973.
24. Chader, G. J., Fletcher, R. T. and Newsome, D. A.: Development of phosphodiesterase activity in the chick retina. Dev. Biol. 40:378-380, 1974.
25. Farber, D. B. and Lolley, R. N.: Proteins in the degenerative retina of C3H mice: Deficiency of a cyclic-nucleotide phosphodiesterase and opsin. J. Neurochem. 21:817-828, 1973.
26. Lolley, R. N., Schmidt, S. Y. and Farber, D. B.: Alterations in cyclic AMP metabolism associated with photoreceptor cell degeneration in the C3H mouse. J. Neurochem. 22:701-707, 1974.
27. Farber, D. B. and Lolley, R. N.: Cyclic guanosine monophosphate: elevation in degenerating photoreceptor cells of the C3H mouse retina. Science 186:449-451, 1974.
28. Robb, R. M.: Electron microscopic histochemical studies of cyclic 3'.5'-nucleotide phosphodiesterase in the developing retina of normal mice and and mice with hereditary retinal degeneration. Trans. Am. Ophthalmol. Soc. 72:650-669, 1974.
29. Lolley, R. N. and Farber, D. B.: Cyclic nucleotide phosphodiesterases in dystrophic rat retinas: Guanosine 3',5' cyclic monophosphate anomalies during photoreceptor cell degeneration. Exp. Eye Res. 20:585-597, 1975.
30. Dewar, A. J., Barron, G. and Richmond, J.: Retinal cyclic AMP phosphodiesterase activity in two strains of dystrophic rat. Exp. Eye Res. 21:299-306, 1975.
31. Davson, H.: The intraocular fluids. The intraocular pressure. In: The Eye, Vol. 1, H. Davson (ed.), Academic Press, New York, 1969.
32. Kinsey, V. E. and Reddy, D. V. N.: Chemistry and dynamics of aqueous humor. In: The Rabbit in Eye Research, J. H. Prince (ed.), Thomas, Springfield, Illinois, 1964.
33. Bill, A.: Blood circulation and fluid dynamics in the eye. Physiol. Rev. 55:383-417, 1975.
34. Macri, F. J., Cevario, S. J. and Ballintine, E. J.: The arterial pressure dependency of the increased aqueous humor formation induced by Ach and eserine. Invest. Ophthalmol. 13:153-155, 1974.
35. Macri, F. J. and Cevario, S. J.: A possible vascular mechanism for the inhibition of aqueous humor formation by ouabain and acetazolamide. Exp. Eye Res. 20:563-569, 1975.
36. Bill, A.: Effects of norepinephrine, isoproterenol and sympathetic stimulation on aqueous humor dynamics in vervet monkeys. Exp. Eye Res. 10:31-46, 1970.
37. Waitzman, M. B. and Woods, W. D.: Some characteristics of an adenyl

cyclase preparation from rabbit ciliary body tissue. Exp. Eye Res. 12: 99-111, 1971.
38. Cole, D. F. and Nagasubramanian, S.: The effect of natural and synthetic vasopressins and other substances on active transport in ciliary epithelium of the rabbit. Exp. Eye Res. 13:45-57, 1972.
39. Gaasterland, D., Kupfer, C., Ross, K., and Gabelnick, H. L.: Studies of aqueous humor dynamics in man. III. Measurements in young normal subjects using norepinephrine and isoproterenol. Invest. Ophthalmol. 12:267-279, 1973.
40. Tripathi, R. C.: Comparative physiology and anatomy of the outflow pathway. In: The Eye, Vol. 5, H. Davson and L. T. Graham Jr. (eds.), Academic Press, New York, 1974.
41. Langham, M. E., Kitazawa, Y. and Hart, R. W.: Adrenergic responses in the human eye. J. Pharmacol. Exp. Therap. 179:47-55, 1971.
42. Grant, W. M.: Clinical measurements of aqueous outflow. Arch. Ophthalmol. 46:113-131, 1951.
43. Kolker, A. E. and Hetherington, J., Jr.: Becker-Shaffer's Diagnosis and Therapy of the Glaucomas, C. V. Mosby Co., St. Louis, 1970.
44. Kahn, H. A.: The prevalence of chronic simple glaucoma in the United States. Amer. J. Ophthalmol. 74:355-359, 1972.
45. Grant, W. M.: Action of drugs on movement of ocular fluids. Ann. Rev. Pharmacol. 9:85-94, 1969.
46. Sears, M. L.: Catecholamines in relation to the eye. In: Handbook of Physiology-Endocrinology, VI, Chapter 35, E. Astwood and R. Creep (eds.), Amer. Physiol. Soc., Baltimore, Maryland, pp. 553-590, 1975.
47. Gnädinger, M. C. and Bárány, E. H.: Die wirkung der β-adrenergischen substanz isoprenalin auf die ausflub-fazilitat des kaninchenauges. v. Graefes Archiv für Ophthalmologie 167:483-492, 1964.
48. Kupfer, C., Gaasterland, D., and Ross, K.: Studies of aqueous humor dynamics in man. V. Effects of acetazolamide and isoproterenol in young and old normal volunteers. Invest. Ophthalmol. 15:349-356, 1976.
49. Ballintine, E. J. and Garner, L. L.: Improvement of the coefficient of outflow in glaucomatous eyes. Arch. Ophthalmol. 66:314-317, 1961.
50. Becker, B., Pettit, T. H. and Gay, A. J.: Topical epinephrine therapy of open-angle glaucoma. Arch. Ophthalmol. 66:219-225, 1961.
51. Laties, A. M. and Jacobwitz, D.: A histochemical study of the adrenergic and cholinergic innervation of the anterior segment of the rabbit eye. Invest. Ophthalmol. 3:592-600, 1964.
52. Nomura, T. and Smelser, G. K.: The identification of adrenergic and cholinergic nerve endings in the trabecular meshwork. Invest. Ophthalmol. 13:525-532, 1974.
53. Neufeld, A. H., Jampol, L. M. and Sears, M. L.: Cyclic-AMP in the aqueous humor: The effects of adrenergic agents. Exp. Eye Res. 14:242-250, 1972.
54. Radius, R. and Langham, M. E.: Cyclic-AMP and the ocular responses to norepinephrine. Exp. Eye Res. 17:219-229, 1973.
55. Neufeld, A. H., Chavis, R. M. and Sears, M. L.: Cyclic-AMP in the aqueous humor: The effects of repeated topical epinephrine administration and sympathetic denervation. Exp. Eye Res. 16:265-272, 1973.

56. Neufeld, A. H. and Sears, M. L.: Cyclic-AMP in ocular tissues of the rabbit, monkey and human. Invest. Ophthalmol. 13:475-477, 1974.
57. Bárány, E. H.: Simultaneous measurement of changing intraocular pressure and outflow facility in the vervet monkey by constant pressure infusion. Invest. Ophthalmol. 3:135-143, 1964.
58. Neufeld, A. H., Dueker, D. K., Vegge, T. and Sears, M. L.: Adenosine 3',5'-monophosphate increases the outflow of aqueous humor from the rabbit eye. Invest. Ophthalmol. 14:40-42, 1975.
59. Neufeld, A. H. and Sears, M. L.: Adenosine 3',5'-monophosphate analogue increases the outflow facility of the primate eye. Invest. Ophthalmol. 14: 688-689, 1975.
60. Bárány, E. H.: The mode of action of pilocarpine on outflow resistance in the eye of a primate (Cercopithecus ethiops). Invest. Ophthalmol. 1: 712-727, 1962.
61. Kass, M. A., Reid, T. W., Neufeld, A. H., Bausher, L. P. and Sears, M. L.: The effect of d-isoproterenol on intraocular pressure of the rabbit, monkey and man. Invest. Ophthalmol. 15:113-118, 1976.
62. Kaback, M. B., Podos, S. M., Harbin, T. S., Jr., Mandell, A. and Becker, B.: The effect of dipivalyl epinephrine on the eye. Amer. J. Ophthalmol. 81:768-773, 1976.
63. Chalfie, M., Neufeld, A. H. and Zadunaisky, J. A.: Action of epinephrine and other cyclic AMP-mediated agents on the chloride transport of the frog cornea. Invest. Ophthalmol. 11:644-650, 1972.
64. Klyce, S. D., Neufeld, A. H. and Zadunaisky, J. A.: The activation of chloride transport by epinephrine and DB cyclic-AMP in the cornea of the rabbit. Invest. Ophthalmol. 12:127-139, 1973.
65. Fischer, F. and Wiederholt, M.: Sodium and chloride transport in the isolated human cornea. Presented at ARVO Spring Meeting, 1975, Sarasota, Florida.
66. Klyce, S. D.: Transport of Na, Cl, and water by the rabbit corneal epithelium at resting potential. Amer. J. Physiol. 228:1446-1452, 1975.
67. Montoreano, R., Candia, O. A. and Cook, P.: α and β adrenergic receptors in the regulation of ionic transport in the frog cornea. Amer. J. Physiol. 230:1487-1494, 1976.
68. Buck, M. G. and Zadunaisky, J. A.: Stimulation of ion transport by ascorbic acid through inhibition of 3',5'-cyclic-AMP phosphodiesterase in the corneal epithelium and other tissues. Biochim. Biophys. Acta 389:251-260, 1975.
69. Berman, M. B., Cavanagh, H. D., and Gage, J.: Regulation of collagenase activity in the ulcerating cornea by cyclic AMP. Exp. Eye Res. 22:209-219, 1976.
70. Cavanagh, H. D.: Inflammation and anti-inflammatory therapy in herpetic ocular disease. In: Ocular Viral Diseases, D. P. Langston (ed.), Inter. Ophthalmol. Clin. 15:67-88, 1975.
71. Fletcher, R. T., and Chader, G. J.: Cyclic GMP: Control of concentration by light in retinal photoreceptors. Biochem. Biophys. Res. Comm. 70:1297-1302, 1976.

15

The Role of Cyclic Nucleotides in the Pathogenesis of Psoriasis

DONALD A. CHAMBERS
CYNTHIA L. MARCELO
JOHN J. VOORHEES

Normal epidermis is highly ordered (1) and contains basal cells characterized by low rates of proliferation (2), as well as cells which are terminally differentiated (1). This quiescent steady state is markedly altered in psoriasis such that the epidermis shifts to a metabolically active state in which cellular proliferation and glycogen accumulation increases (3–5) and a population of cells with incomplete terminal differentiation appears (6). This review concerns itself with the role of cyclic nucleotides in the pathophysiology of psoriasis. Since psoriasis is a prototype of human proliferation skin disease, this information may be of general significance to other cellular proliferative diseases as well.

ANATOMY OF THE SKIN RELATIVE TO CYCLIC NUCLEOTIDE STUDIES AND PSORIASIS

A brief review of cutaneous anatomy and function will be given here. (For more detailed reviews, see Refs. 7, 8.) Skin is

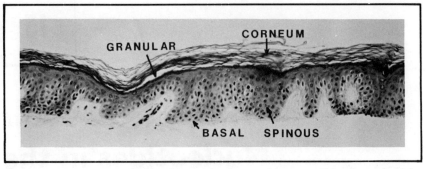

1 Light micrograph of normal human epidermis, stained by hematoxylin and eosin.

composed of three major layers: the epidermis (outer layer), the dermis, and underlying fat. The epidermis originates from the embryonic ectoderm, is avascular but is nourished by diffusion from the underlying dermal vasculature. The epidermis itself is composed of layers; the outer one-fourth contains the horny and granular layers which are made up of heavily keratinized or dying cells, whereas the inner three-quarters of the epidermis (basal and spinous layers) is composed of living cells which are not yet terminally differentiated. In the lower layers, keratinocytes are the most common cells and comprise about 95 percent of the epidermal volume (9) (see Fig. 1).

The epidermis functions to provide a barrier between the external environment and the human organism. Light microscopy shows that the outer segments of the epidermis (stratum corneum) appears as a dark band and it is likely that these dead cells function as the barrier (1). To maintain this barrier, cells in the lower level of the epidermis must proliferate, differentiate, stratify, keratinize and die. Since psoriasis is typical of proliferative skin disease, it has been used as a model for the study of aberrant epidermal homeostasis. A patient with marked psoriasis is shown in Fig. 2. The microscopic morphology of a psoriatic lesion is depicted in Fig. 3. Some of the features of psoriasis are epidermal hyperplasia (3), incomplete epidermal differentiation (6), enhanced glycogen accumulation (4), decreased stainable surface glycoprotein in lesional epidermis (10), and improvement of a majority of psoriatic lesions by exposure to sunlight (11). As far as is known, psoriasis is confined to Homo sapiens and is rarely fatal. Epidemiology shows that

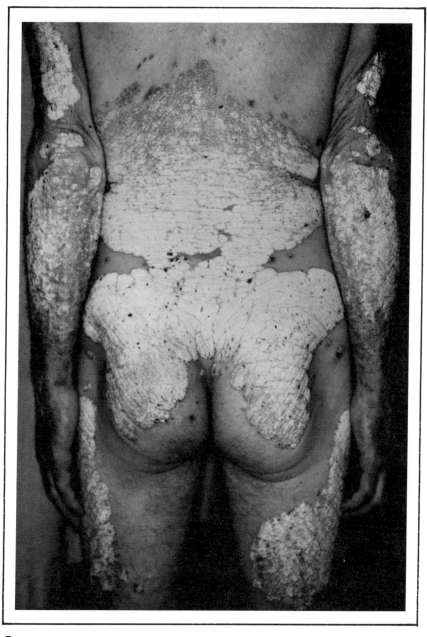

2 A patient with classical psoriasis. Note the sharp margins of the lesional areas and the thick scale.

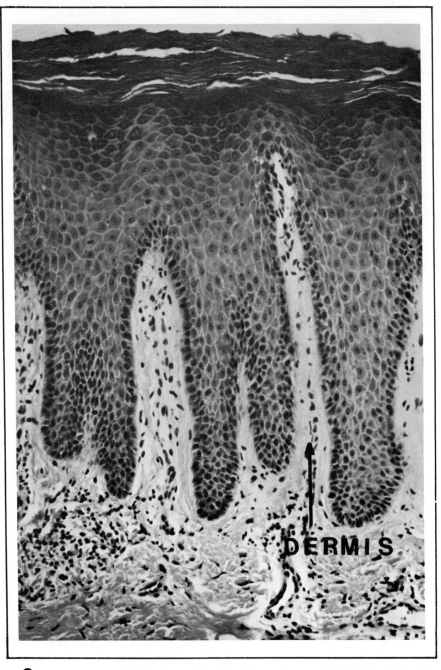

3 Light micrograph of a biopsy of a psoriatic lesion stained by hematoxylin and eosin.

1 to 3 percent of the global population is affected by the disease without sex preference. Psoriasis can occur at any time in a patient's life (11). In most if not all patients, the predisposition to the disease is thought to be genetic and an argument can be made that psoriasis is a multifactorial genetic disease grouping of several different diseases characterized by a similar phenotype. Some evidence exists that psoriatic patients can be classified in terms of HLA antigen specificity (12, 78). However, the ultimate delineation of the genetics of this disease still remains a problem for the future. The primary site for the expression of the genetic abnormality may be in the epidermis but it can also lie in the dermis, serum growth factors, immune system, or elsewhere. Since the activated epidermis represents the major phenotype of clinical expression of the disease, studies of the nature of the regulatory factors involved in epidermal growth affords a rational approach to the investigation of psoriasis. The studies to be described below indicate that cyclic nucleotide metabolism in psoriasis is altered and it is our view that these alterations may be basic to the pathogenesis of the epidermal component of this disease.

MOLECULAR MECHANISMS OF CELL PROLIFERATION

The universality of common physiological mechanisms implies that the *molecular basis of cellular proliferation* has aspects common to all cells. Deciphering these mechanisms is basic to the understanding of cellular proliferation, both normal and abnormal, in all tissues including the epidermis.

Cells during their life time undergo essentially two general processes, (1) proliferation and (2) differentiation. By differentiation we mean the selected activation of specific functions unique to a particular cell type. In modern biologic terms, differentiation means the selective activation of specific genes and the repression of others. That this process occurs has been beautifully described by Gurdon, in his classical experiment showing that a differentiated cell nucleus obtained from a frog epidermal cell still retains all its genetic complement and could be activated in ova cytoplasm to produce an entire frog (13).

Proliferation processes are those that lead to the replication of DNA and the production of progeny cells. Some cells, like neurons,

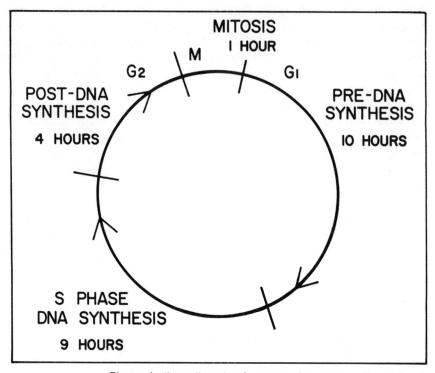

Phases in the cell cycle of a mouse hepatoma cell growing in tissue culture with a generation time of 24 hr. (Redrawn from Watson, J. D. Molecular Biology of the Gene, Third Edition, W. A. Benjamin, Menlo Park, 1976.)

remain "terminally differentiated" and probably never normally proliferate during the life time of the organism whereas other cells such as lymphocytes retain the potential for proliferation.

The Proliferative Program

Cells that proliferate are said to "cycle" (14, 15). The concept of the cell cycle stems from the work of Howard and Pelc and is based on experiments which showed that DNA is synthesized at a specific time in a proliferating eukaryotic cell (16). The time of DNA synthesis is called S phase and all other phases of the cell cycle are related to S phase by definition. Thus G_1 (G is the symbol for gap) represents the time from the end of the previous mitosis until DNA synthesis commences in S phase. G_2 represents the time span after completion of DNA synthesis until the beginning of mitosis; M represents the stage of mitosis (see Fig. 4). Terminally differentiated cells (e.g., epidermal cells in the stratum granulosum) never normally cycle. However, the ability of cells which are differentiated to reenter the cycle may relate to a pathological process (i.e., neurons enter the cell cycle as malignant neuroblastoma cells). In normal nonproliferating cells, the cells are said to be "blocked" and are unable to pass through G_1. Diseases of cellular proliferation may owe their pathology to deranged regulation of the proliferative apparatus. The division of time in each phase of the cell cycle varies according to the cell, but a generalization often made is that most variations occur in the G_1 phase.

Cyclic Nucleotides and Cell Proliferation

Although the molecular nature of the initiating signal for cell proliferation remains unknown, cyclic nucleotides are strongly implicated in either initiation or transmission of that signal. Thus Pastan and his colleagues (17) have noted that malignant fibroblasts contained diminished amounts of *adenosine 3',5'-monophosphate (cyclic AMP)* relative to normal fibroblasts. These workers, and Hsie and Puck, further showed that exogenous addition of cyclic AMP to malignant cells resulted in a phenotypic reversion towards normal. In lymphocytes and lymphosarcoma cells, cell proliferation as a response to a mitogen or oncogen is inhibited by cyclic AMP (19, 20). A current hypothesis which relates cyclic AMP to guanosine 3',5'-monophosphate (cyclic GMP) in terms of cell proliferation and differentiation is that the ability of the cell to enter the cell cycle may be a function of the ratio of cyclic AMP to cyclic GMP such that high amounts of cyclic AMP serve to modulate the "differentiated state" (21).

Recent evidence is accumulating that proliferation may be characterized by increased cyclic GMP levels. Siefert et al. (22) and Moens et al. (23) have noted that cyclic GMP is increased in proliferating fibroblasts. Experiments by Hadden et al. (24) in human peripheral blood lymphocytes and Watson (25) in B lymphocytes obtained from nude mice have shown increases in cyclic GMP after addition of mitogen. However Wedner et al. (26) have been unable to duplicate the observations of Hadden et al. (24), and Miller et al. (27), using a different strain of fibroblasts from that used by the investigators cited above, could not document a rise in cyclic GMP as a function of cell growth.

An alternate approach to studying the role of cyclic GMP in cell growth is by direct addition of the cyclic nucleotide to cells growing in culture. Thus Chambers and his colleagues (19) and Watson (25) have found that cyclic GMP and its more permeable derivatives can act as a mitogenic agent for lymphocytes.

In addition to the effects of cyclic nucleotides per se, researchers are beginning to address themselves to other factors that may interact with cyclic nucleotides in affecting the cell cycle. Several proteins (e.g., fibroblast growth factor, epidermal growth factor, mesenchymal factor and tumor angiogenesis factor) have been isolated from different sources which appear to play some role in cell proliferation in vitro (28). Work on these factors is still in its preliminary stages and must be interpreted with care since serum, most commonly used to maintain cell cultures, contains a heterogeneous mixture of proteins and hormones (e.g., insulin) which can also augment growth. Of particular interest may be the recent work of Chambers et al. (19) in which they described a cyclic GMP-activatible phosphoribosyl pyrophosphate synthetase (PRPP synthetase) induced in the lymphocyte by lectin stimulation. This induction was inhibited by cyclic AMP. PRPP synthetase has been purified free of kinase activity from hepatoma cells by Green and Martin and is still activatible by cyclic GMP (29). It would appear that cyclic GMP acts as an allosteric activator of a new species of enzyme unique to proliferating cells. PRPP is a compound essential to purine and pyrimidine biosynthetic pathways and is necessary to RNA, DNA, nucleotide, and coenzyme synthesis. The induction of such an enzyme particularly specialized to meet the increased demands of proliferation may be exempletive of a specific regulatory phenomena in normal cell physiology. It will be of further interest to see if other enzymes related to proliferation will also be cyclic GMP activatible.

The role of cyclic pyrimidine 3',5'-monophosphates in cell proliferation is still to be defined. Cytidine 3',5'-monophosphate (cyclic CMP) was reported to be present in leukemia cells by Bloch (30) and in bacteria by Ishiyama (31). When added exogenously to a serum-free mouse spleen lymphocyte system, cyclic CMP acted similarly to cyclic GMP in its mitogenic capacity and led to a 10-fold increase in thymidine incorporation into DNA which was inhibited by cyclic AMP (32). In earlier studies, Chambers and Zubay found that cyclic GMP inhibited the ability of cyclic AMP to stimulate a cell-free DNA directed system for the synthesis of β-

galactosidase derived from bacteria (33). Cyclic CMP acted similarly to cyclic GMP in the cell-free DNA directed protein synthesis system. Thus cyclic CMP appears to act analogously to cyclic GMP in both a prokaryotic and a eukaryotic system (32). However, much more work remains to be done before we can ascribe a physiological role to cyclic pyrimidine monophosphates.

Another group of compounds which relate to the cyclic nucleotides are *prostaglandins*. These substances are synthesized from unsaturated essential fatty acids (e.g., *arachidonic acid*). Prostaglandins appear to work in concert with cyclic nucleotides. Prostaglandins of the E series increase cyclic AMP levels and relate to differentiation (34) whereas prostaglandins of the F series can increase cyclic GMP levels (35) and can be associated with proliferation (36).

Finally, one must be concerned with the nature of regulation of cyclic nucleotide concentrations as well as proteins with which the cyclic nucleotides interact. To be a regulatory compound, the synthesis and degradation of such a compound must be carefully controlled. With the cyclic nucleotides, synthesis is the result of activation of adenylate and guanylate cyclases. Many such regulators have been described (e.g., glucogen, epinephrine, prostaglandins, acetylcholine, etc.) (37). Degradation of the cyclic nucleotides is accomplished by cyclic nucleotide phosphodiesterase(s) (38). The action of cyclic nucleotides may be mediated through protein kinases (39) and other mechanisms (e.g., cyclic-GMP–activated enzymes, etc.). A detailed study of the interactions of all these processes will be necessary to demonstrate the molecular specifics of cyclic nucleotides and cell proliferation.

An interesting technique which may prove useful in these studies is the use of ultraviolet radiation to make a covalent bond between a cyclic nucleotide and its receptor protein, pioneered by Antonoff and Ferguson (40). This method can then be utilized to study the nature of the cyclic nucleotide binding proteins present in a cell at a specific time, or when the cell is perturbed by a particular stimulus. In preliminary experiments, Stubbs and Chambers (manuscript in preparation) have utilized this technique to show that cyclic nucleotide binding proteins are present in certain tissues and tumors in apparently normal amounts (e.g., lymphosarcoma cells vs. lymphocytes), but present in other tumors in altered amounts (e.g., hepatoma cells and lymphoma cell mutants). One interpretation of these data is that there are many potential control points for cyclic nucleotide interaction in the proliferative process and that these control points involve not only the regulated synthesis of the cyclic nucleotides but also the availability of proteins with which they interact. Derangements in either area could result in pathology.

CYCLIC NUCLEOTIDES AND PSORIASIS

The characteristics of psoriatic lesional epidermis have been described. Ideally, one would like to define a unifying hypothesis which would explain these abnormalities in molecular terms.

The review of Robison, Butcher, and Sutherland (41) that cyclic AMP could stimulate the breakdown of glycogen in liver and muscle, coupled with the emerging studies of the effect of cyclic AMP on cell proliferation and differentiation (17) suggested to Voorhees and Duell that the cyclic AMP system of homeostatic regulation might be disturbed in psoriatic epidermis (42). This hypothesis was later extended to include cyclic GMP, proposed by Goldberg et al. (21) to have an inverse relationship to cyclic AMP as it relates to proliferation and many other cellular functions.

Several investigators in different laboratories have examined the metabolism of cyclic AMP in psoriatic lesional tissue (involved), "apparently normal tissue" removed from patients possessing psoriatic lesions (uninvolved) and normal epidermis obtained from individuals with no clinical manifestations of the disease. However, despite the intense effort of many workers, no general agreement as to the role of cyclic nucleotides in psoriasis can be arrived at. This situation exists largely because of lack of uniformity or difficulties in sampling technique, method of assay of cyclic AMP, as well as problems with choice of data base. These problems are discussed briefly below and have been detailed "in extenso" elsewhere (43).

Sampling Techniques

Since artifactual errors can often be introduced into experiments as a result of improper sampling techniques, special care must be exercised in terms of the initial epidermal biopsy. Of critical importance is the inclusion of the entire proliferative compartment in the biopsy sample as well as the avoidance of dermal contamination. This point was highlighted by the recent observations of Verma et al. (44) that cyclic AMP was localized in the basal cells of the epidermis. Another choice that the investigator is often forced to make is whether to assay the concentrations of the cyclic nucleotide or concentrations of adenylate cyclase or phosphodiesterase in tissue slices, or subcellular homogenates and fractions. Comparing homogenates of psoriatic tissue with normal or uninvolved tissue is dangerous, because, in the homogenization procedure, one may easily change the situation that prevails in the organized tissue.

Method of Cyclic Nucleotide Assay

The protocol for cyclic nucleotide assay is also critical to obtaining meaningful data. Care must be taken to insure that the assay reflects the true level of the cyclic nucleotide and is not a function of secondary changes in the tissue itself, such as fluctuation in nucleotide pool levels or contamination in tissue preparations which produce artificial assay errors. For example, Duell (personal communication) has found that for assay of epidermal cyclic AMP by the method of Gilman (45), extensive purification of cyclic AMP from other tissue components is necessary.

Data Bases and Population Size

Since cyclic nucleotides are dynamic molecules whose intracellular levels fluctuate as a function of specific transient stimuli, a large number of samples from a randomized population is required to compare the levels of cyclic AMP in normal, involved and uninvolved psoriatic epidermis. In addition, the cyclic nucleotide levels must be normalized to a proper data base to permit such comparisons. The usual data bases that have been considered in normalizing cyclic AMP concentrations to tissue quantity are DNA levels, assayable protein content and wet weight. The differences between normal, uninvolved, and psoriatic lesional epidermis in DNA and assayable protein content, and wet weight (due to variations in scale, edema, and hydration) make it mandatory to use more than one data base for expressing data.

Cyclic AMP and Adenylate Cyclase

The psoriatic lesional epidermis contains approximately fivefold greater glycogen than normal or uninvolved epidermis (4, 46).

Addition of cyclic AMP or its dibutyryl derivative, epinephrine, fluoride or prostaglandin E cause glycogenolysis in psoriatic and normal epidermal slices (47) indicating that the level of glycogen in the epidermis is subject to control by the intracellular levels of cyclic AMP. This fact, coupled with the paradoxical storage of glycogen in tissue that shows an abnormally high proliferative rate and reduced differentiation, suggests that cyclic AMP is decreased in the lesional tissue. A *decrease in cyclic AMP in psoriatic tissue* is consistent with the present theories of the regulation by cyclic AMP of the cell proliferation-differentiation state. Results from this laboratory indicate a probable decrease in the cyclic AMP levels of the lesional psoriatic epidermis when contrasted to uninvolved and normal epidermis. The data obtained from two different cyclic AMP assay systems from a sample of 50 patients showed a highly significant decrease in cyclic AMP concentration in the involved compared to the uninvolved epithelium of psoriatics ($p = 0.001$, 36 percent decrease based on DNA as the data base) (43, 48, 49). From the 50 paired epidermal samples assayed, 42 based on total protein, 39 based on DNA, and 33 based on the wet weight denominator, a decrease in lesional cyclic AMP could be demonstrated. Borderline significance ($p = 0.09$) was obtained when wet weight was used for the data base. This difference probably resulted from the gross weight variations between the normal and pathological tissue and reflects the importance of consideration of the previously mentioned difficulties encountered in dealing with this tissue.

Since the levels of cyclic AMP appear decreased in lesional psoriatic epidermis it is possible that there may be a defect in the adenylate cyclase system and/or cyclic nucleotide phosphodiesterase system of the tissue. An alternate explanation might be the presence of a labile inhibitor of adenylate cyclase or stimulator of phosphodiesterase in the lesion.

Investigations of epidermal adenylate cyclase in homogenates and subcellular fractions and tissue slices have been attempted (50–59). In summary, the accumulated observations imply adenylate cyclase activity may be diminished in psoriatic lesions compared to uninvolved tissue, but because of the technical problems described earlier, no definitive conclusion can be made.

The isoproterenol-mediated stimulation of cyclic AMP in epidermal slices has been examined in a population composed of 16 normals and the involved and uninvolved areas of 23 psoriatic patients by Voorhees, et al. (49). The slices were incubated in the absence of a phosphodiesterase inhibitor. These studies revealed that cyclic AMP concentrations and adenylate cyclase activity after isoproterenol stimulation was similar in both involved and uninvolved tissue.

Cyclic AMP Phosphodiesterase

An alternate means of control of cyclic AMP levels in tissues can be by degradation of the cyclic nucleotide. Fluctuations in cyclic nucleotide concentrations could be accounted for in terms of variations in cyclic phosphodiesterase activity (PDE). Accordingly, the cyclic AMP PDE was studied in this laboratory (60). An assay of the high Km enzyme in involved and uninvolved epidermis from 18 psoriatic patients showed no apparent difference in enzymatic activity. A study of the low Km enzyme, both particulate and soluble fractions, of involved and uninvolved epidermis of six psoriatic patients indicated that the soluble low Km enzyme present in the lesion has an increased Vmax. An increased Vmax could result from the presence of an enzyme activator or an alteration in the enzyme itself. This change in enzyme activity may play some role in maintaining the observed decreased cyclic AMP levels of involved psoriatic tissue.

In summary, the experiments described in this section indicate that cyclic AMP levels are probably decreased in psoriatic lesions. Further confirmatory studies await standardization of the techniques and attention to the problems of measuring cyclic AMP levels discussed earlier.

Cyclic GMP and Psoriasis

Increased intracellular levels of cyclic GMP have been demonstrated in many systems and shown to be associated with physiologic effects antagonistic to those related to elevated cyclic AMP levels (21). Variations of the absoulte levels and the intracellular ratio of both cyclic nucleotides often correlate with the proliferation-differentiation program of a variety of cells in culture, as discussed above. Analysis of the cyclic GMP concentrations of the uninvolved and involved epidermis of 12 psoriatic patients indicated a highly significant increase in the cyclic GMP content of the involved tissue compared to uninvolved areas (49). These studies showed a 94% increase of cyclic GMP in the lesional epidermis versus the unaffected areas (11.8 ± 3.4 vs. 6.1 ± 1.0 femtomol/ μgm DNA). When this 94% cyclic GMP increase is related to cyclic nucleotide ratios an even more dramatic difference is seen. The *cyclic AMP/cyclic GMP ratio* for the *uninvolved* area is 180:1 whereas in the involved epidermis it is 59:1. This threefold increase in ratios supports our present working hypothesis that an abnormal cyclic AMP/cyclic GMP ratio is central to the homeostatic imbalance observed in lesional psoriatic epidermis.

Prostaglandins and Psoriasis

When considering the role of prostaglandins in human skin physiology and pathological states, three interacting spheres of influence should be considered: (1) the endogenous production of prostaglandins by epidermal cells; (2) the reactions of the epidermis to exogenous prostaglandins; and (3) the role of prostaglandins in cutaneous inflammation.

The synthesis of prostaglandins from arachidonic acid by homogenates of human skin (61–63) and the conversion of PGE_2 into a biologically inactive metabolite (62) have been demonstrated. Arachidonic acid cyclo-oxygenase activity is localized in the microsomal fraction of human epidermis and can be inhibited by polyunsaturated fatty acids and nonsteroidal anti-inflammatory drugs (61). To ascertain if an abnormality of the prostaglandin system exists in psoriasis, Hammarström et al. (64) measured the levels of arachidonic acid and the metabolic products of arachidonic acid, 12L-hydroxy-5,8,10,14-eicosatetraenoic acid (HETE), PGE_2, $PGF_{2\alpha}$ in involved and uninvolved psoriatic epidermis. Most interestingly, the involved psoriatic tissue showed a 25-fold increase in free arachidonic acid and of 81-fold in *HETE,* but the levels of *PGE_2* and *$PGF_{2\alpha}$* were only minimally (40 and 82 percent respectively) increased in the diseased tissue. These results indicated that abnormalities of prostaglandin metabolism or its precursors are observable in psoriasis but the significance of the markedly increased levels of free arachidonic acid and HETE remain to be established.

The observations described above are consistent with the recent finding of Penneys et al. (65) of the presence of a prostaglandin synthesis inhibitor in psoriatic lesions but not in normal epidermis. Unfortunately, characterization of this inhibitor awaits further work. Another interesting avenue for investigation in the epidermis is the observation of Newcombe et al. that human synoviocytes can become refractory to prostaglandin stimulation (66). If this situation is also true in the epidermis, it is possible that the psoriatic lesion may be desensitized to prostaglandin effects relative to normal tissue.

Whether the elevated intracellular levels of PGE_2 and $PGF_{2\alpha}$ in the psoriatic lesion are related to the intracellular cyclic AMP and cyclic GMP systems is open to speculation although the reported decrease in cyclic AMP stimulation of psoriatic slices by PGE_2 and $PGF_{2\alpha}$ when compared to uninvolved areas (59, 67) may reflect some interplay between intracellular prostaglandins and adenylate cyclase.

A model that shows great promise in delineating the role of prostaglandins in the epidermis is the essential-fatty-acid deficient (EFA) rat system of Ziboh and Hsia (68). Scaliness of the dorsal skin and of the hind limbs and tail, coupled with decreased cutaneous levels of PGE_2 (69) and other gross systemic defects resulted when the animals were maintained on an EFA free diet. These abnormalities were reversed by either feeding the animals linolenic or arachidonic acid or applying PGE_2 topically to the cutaneous lesions. These results imply relationships between PGE_2 and the homeostatic regulation of epidermal structures. Furthermore, if scaly dermatoses in the rat is analogous to epidermal hyperplasia, these studies suggest that investigation of prostaglandin metabolism and its regulation may be fruitful to increased understanding of hyperplastic diseases such as psoriasis.

Prostaglandins of the E and F series have been related to physically and chemically induced cutaneous inflammation (70-72). These substances may be mediators of the delayed phase of inflammation and it has been strongly suggested that they play an important role in sustained inflammatory response.

The nature of the interactions between the prostaglandin and cyclic nucleotide systems, and epidermal homeostasis is presently unknown. It can be stated, however, that an imbalance in these systems is present in lesional psoriatic epidermis.

MOLECULAR PHARMACOLOGY

A knowledge of mechanisms of disease can often lead to potential therapies as witnessed by the growth of the discipline of molecular pharmacology.

Since the investigations we have discussed in this review lead to the hypothesis that the pathophysiology of psoriasis is significantly related to the imbalance of cyclic nucleotide concentrations, *drug therapies* directed against such imbalances may ameliorate the disease. This conclusion leads to considerations of pharmacological agents which will reverse cyclic nucleotide concentrations or ratios toward normal in the psoriatic lesional tissue.

Drugs that increase cyclic AMP, or cyclic AMP itself, have been demonstrated to decrease the proliferation of cells in culture (17, 73, 74). Clinical bioassays designed to determine whether or not cyclic AMP therapy will be efficacious are in progress. Experimental agents are screened initially by use of a mouse epidermal slice system and those drugs which elevate cyclic AMP levels are considered for further clinical bioassay.

The phosphodiesterase inhibitors papaverine and RO-20-1724 have been tested in double blind assays and appear to cause regression of psoriatic lesions (75, 76). An anecdotal study using 5' AMP injected intramuscularly suggests that this agent may also be therapeutically beneficial (43). An adenosine analog (ethyl adenosine-5'-carboxylate) has been demonstrated to increase cyclic AMP in psoriatic uninvolved and involved epidermal slices (77) and is a potential agent for clinical bioassay.

Thus, results of clinical assays indicate that agents that normalize cyclic AMP/cyclic GMP ratios may, if properly formulated, become future drugs for use in psoriasis therapy and further substantiate the hypothesis that the observed cyclic nucleotide imbalance is central to epidermal homeostasis.

SUMMARY

Psoriasis is a prototype of human proliferation skin disease. In this disease, control of epidermal cell proliferation is deranged, and epidermal cells show altered surface characteristics, enhanced glycogen accumulation, and incomplete differentiation. An investigation of the molecular interactions characteristic of the proliferating state and their control might be expected to illuminate some of the problems in the pathophysiology of psoriasis. Since cyclic nucleotides have been implicated in both cell proliferation and differentiation, these compounds have been studied in normal and abnormal epidermal homeostasis. In summary, it is probable that diminished cyclic AMP levels and increased cyclic GMP levels or imbalanced cyclic nucleotide ratios are characteristic of psoriatic lesions. The metabolism of prostaglandins, regulatory compounds which are metabolically related to cyclic nucleotides, appears altered in psoriasis as well. These studies lead to investigations of potential therapies for psoriasis directed towards regaining the cyclic nucleotide and prostaglandin regulation found in the normal epidermis.

ACKNOWLEDGMENTS

The authors would like to acknowledge the advice of Dr. E. A. Duell and the assistance of M. A. Sevic, B. G. Spencer and C. Rusnak in preparing this manuscript.
Some of the work described herein was supported by grants from the National Institutes of Health, 2P01AM15740 and IR01DE04313-01.

REFERENCES

1. Christophers, E.: Correlation between column formation, thickness, and rate of new cell production in guinea pig epidermis. Virchows Arch. Zell Path. 10:286-292, 1972.
2. Van Scott, E. J. and Ekel, T... M.: Kinetics of hyperplasia in psoriasis. Arch. Dermatol. 88:373-381, 1963.
3. Weinstein, G. D. and McCullough, J. L.: Cytokinetics in diseases of epidermal hyperplasia. Ann. Rev. Med. 24:345-352, 1973.
4. Halprin, K. M. and Ohkawara, A.: Carbohydrate metabolism in psoriasis: an enzymatic study. J. Invest. Dermatol. 46:51-69, 1966.
5. Williams, J. P. G.: Interrelation of epithelial glycogen, cell proliferation, and cellular migration with cyclic adenosine monophosphate in epithelial wound healing. Cell Differentiation. 1:317-323, 1972.
6. Van Scott, E. J.: Tissue compartments of the skin lesion of psoriasis. J. Invest. Dermatol. 59:4-6, 1972.
7. Montagna, W. and Parakkal, P. F.: The Structure and Function of Skin. Third edition, Academic Press, New York, 1974.
8. Breathnach, A. S.: An Atlas of the Ultrastructure of Human Skin Development, Differentiation, and Post-Natal Features. J. and A. Churchill, London, 1971.
9. Zelickson, A. S. and Mottaz, J. H.: Epidermal dendrite cells. A quantitative study. Arch. Dermatol. 98:652-659, 1968.
10. Mercer, E. H. and Maibach, H. I.: Intercellular adhesion and surface coats of epidermal cells in psoriasis. J. Invest. Dermatol. 51:215-221, 1968.
11. Van Scott, E. J. and Farber, E. M.: Chapter 8, Disorders with epidermal proliferation. In: Dermatology in General Medicine. T. B. Fitzpatrick, K. A. Arndt, W. H. Clark, A. Z. Eisen, E. J. Van Scott and J. H. Vaughan (eds.) pp. 228. McGraw-Hill, New York, 1971.
12. Russel, T. J. and Schultes, L. M. and Kuban, D. J.: Histocompatibility

(HL-A) antigens associated with psoriasis. N. Eng. J. Med. 287:738-740, 1972.
13. Gurdon, J. B.: The Control of Gene Expression in Animal Development. Harvard University Press, Cambridge, 1974.
14. Baserga, R.: The Cell Cycle and Cancer. Dekker, Inc., New York, 1971.
15. Mitchison, J. M.: The Biology of the Cell Cycle. Cambridge University Press, London, 1971.
16. Howard, A. and Pelc, S. R.: Synthesis of desoxyribonucleic acid in normal and irradiated cells and its relation to chromosome breakage. Heredity, London (Suppl.). 6:261-273, 1953.
17. Pastan, I. H., Johnson, G. S. and Anderson, W. B.: Role of cyclic nucleotides in growth control. In: Annual Review of Biochemistry, vol. 44, pp. 491-522, 1975.
18. Hsie, A. W. and Puck, T. T.: Morphological transformation of chinese hamster cells by dibutyryl adenosine cyclic 3':5'-monophosphate and testosterone. Proc. Nat. Acad. Sci. USA. 68:358-361, 1971.
19. Chambers, D. A., Martin, D. W. Jr. and Weinstein, Y.: The effect of cyclic nucleotides on purine biosynthesis and the induction of PRPP synthetase during lymphocyte activation. Cell. 3:375-380, 1974.
20. Bourne, H. R., Coffino, P., Melmon, K. L., Tomkins, G. M. and Weinstein, Y.: Genetic analysis of cyclic AMP in a mammalian cell. In: Advances in Cyclic Nucleotide Research, G. I. Drummond, P. Greengard and G. A. Robison (eds.) Vol. 5, pp. 771-786, Raven Press, New York, 1975.
21. Goldberg, N. D., Haddox, M. K., Dunham, E., Lopez, C. and Hadden, J. W.: The yin yang hypothesis of biological control: opposing influences of cyclic GMP and cyclic AMP in the regulation of cell proliferation and other biological processes. In: Control of Proliferation in Animal Cells, B. Clarkson and R. Baserga (eds.) pp. 609-625, Cold Spring Harbor Laboratory, New York, 1974.
22. Seifert, W. E. and Rudland, P. S.: Possible involvement of cyclic GMP in growth control of cultured mouse cells. Nature 248:138-140, 1974.
23. Moens, W., Vokaer, A. and Kram, R.: Cyclic AMP and cyclic GMP concentrations in serum- and density-restricted fibroblast cultures. Proc. Nat. Acad. Sci. USA. 72:1063-1067, 1975.
24. Hadden, J. W., Hadden, E. M., Haddox, M. K. and Goldberg, N. D.: Guanosine 3':5'-cyclic monopsohphate: A possible intracellular mediator of mitogenic influences in lymphocytes. Proc. Nat. Acad. Sci. USA. 69:3024-3027, 1972.
25. Watson, J.: The influence of intracellular levels of cyclic nucleotides on cell prolifieration and the induction of antibody synthesis. J. Exp. Med. 141:97-111, 1975.
26. Wedner, H. J., Danker, R. and Parker, C. W.: Cyclic GMP and lectin-induced lymphocyte activation. J. Immunol. 115:1682-1687, 1975.
27. Miller, Z., Lovelace, E., Gallo, M. and Pastan, I.: Cyclic guanosine monophosphate and cellular growth. Science 190:1213-1215, 1975.
28. Rutter, W. J., Pictet, R. L. and Morris, P. W.: Toward molecular mechanisms of developmental processes. In: Annual Review of Biochemistry, Vol. 42, E. E. Snell (ed.) pp. 601-646, Annual Reviews Inc., 1973.
29. Green, C. D. and Martin, D. W. Jr.: A direct, stimulating effect of cyclic GMP on purified phosphoribosyl pyrophosphate synthetase and its antag-

onism by cyclic AMP. Cell 2:241-245, 1974.
30. Bloch, A.: Cytidine 3',5'-monophosphate (cyclic CMP) I. Isolation from extracts of leukemia L-1210 cells. Biochem. and Biophys. Res. Comm. 58: 652-659, 1974.
31. Ishiyama, J.: Isolation of cyclic 3',5'-pyrimidine mononucleotides from bacterial culture fluids. Biochem. Biophys. Res. Comm. 65:286-292, 1975.
32. Chambers, D. A.: Complementarity in the metabolic code. J. Cell. Biol. 67:60a, 1975.
33. Zubay, G. and Chambers, D. A.: A DNA-directed cell-free system for β-galactosidase synthesis. In: Cold Spring Harbor Symposium on Quantitative Biology, Vol. 34, pp. 753-761, Cold Spring Harbor, New York, 1969.
34. Samuelsson, B., Granström, E., Green, K., Hamberg, M. and Hammarström, S.: Prostaglandins. In: Annual Review of Biochemistry, Vol. 44, pp. 669-695, Annual Reviews Inc., 1975.
35. Dunham, E. W., Haddox, M. K. and Goldberg, N. D.: Alteration of vein cyclic 3':5' nucleotide concentrations during changes in contractility. Proc. Nat. Acad. Sci. USA. 71:815-819, 1974.
36. DeAsua, L. J., Clingan, D. and Rudland, P. S.: Initiation of cell prolifieration in cultured mouse fibroblasts by prostaglandin $F_2\alpha$. Proc. Nat. Acad. Sci. USA. 72:2724-2728, 1975.
37. Hardman, J. G.: Cyclic nucleotides and hormone action. In: Textbook of Endocrinology, R. H. Williams (ed.) pp. 869-880, W. B. Saunders Co., Philadelphia, 1974.
38. Appleman, M. M., Thompson, W. J. and Russell, T. R.: Cyclic nucleotide phosphodiesterases. In: Advances in Cyclic Nucleotide Research, G. I. Drummond, P. Greengard and G. A. Robison (eds.) Vol. 3, pp. 65-98, Raven Press, New York, 1973.
39. Rubin, C. S. and Rosen, O. M.: Protein phosphorylation. In: Annual Review of Biochemistry, E. E. Snell (ed.), Vol. 44, pp. 831-887, 1975.
40. Antanoff, R. S. and Ferguson, J. J.: Photoaffinity labeling with cyclic nucleotides. J. Biol. Chem. 294:3319-3321, 1974.
41. Robison, G. A., Butcher, R. W. and Sutherland, E. W.: Cyclic AMP. In: Annual Review of Biochemistry, P. D. Boyer (ed.) Vol. 37, pp. 149-174, Annual Reviews, Inc., Palo Alto, California, 1968.
42. Voorhees, J. J. and Duell, E. A.: Psoriasis as a possible defect of the adenyl cyclase-cyclic AMP cascade. Arch. Dermatol. 104:352-358, 1971.
43. Voorhees, J. J., Marcelo, C. L. and Duell, E. A.: Cyclic AMP, cyclic GMP, and glucocorticoids as potential metabolic regulators of epidermal proliferation and differentiation. J. Invest. Dermatol. 65:179-190, 1975.
44. Verma, A. K., Dixon, K. E., Froscio, M. and Murray, A. W.: Localization of adenosine 3',5'-monophosphate in mouse epidermis by immunofluorescence. J. Invest. Dermatol. 66:239-241, 1976.
45. Gilman, A. G.: A protein binding assay for adenosine 3':5'-cyclic monophosphate. Proc. Nat. Acad. Sci. USA. 67:305-312, 1970.
46. Halprin, K. M. and Ohkawara, A.: Glucose and glycogen metabolism in the human epidermis. J. Invest. Dermatol. 46:43-50, 1966.
47. Halprin, K. M., Adachi, K., Yoshikawa, K., Levine, V., Mui, M. M. and Hsia, S. L.: Cyclic AMP and psoriasis. J. Invest. Dermatol. 65:170-178, 1975.
48. Voorhees, J. J., Duell, E. A., Bass, L. J., Powell, J. A. and Harrell, E. R.:

Decreased cyclic AMP in the epidermis of lesions of psoriasis. Arch. Dermatol. 105:695-701, 1972.
49. Voorhees, J., Kelsey, W., Stawiski, M., Smith, E., Duell, E., Haddox, M. and Goldberg, N.: Increased cyclic GMP and decreased cyclic AMP levels in the rapidly proliferating epithelium of psoriasis. In: The Role of Cyclic Nucleotides in Carcinogenesis, J. Schultz and H. Gratner (eds.) Vol. 6, pp. 325-373, Academic Press, New York, 1973.
50. Härkönen, M., Hopsu-Havu, V. K. and Raij, K.: Cyclic adenosine monophosphate, adenyl cyclase and cyclic nucleotide phosphodiesterase in psoriatic epidermis. Acta Dermatovener (Stockholm). 54:13-18, 1974.
51. Wright, R. K., Mandy, S. H., Halprin, K. M. and Hsia, S. L.: Defects and deficiency of adenyl cyclase in psoriatic skin. Arch. Dermatol. 107:47-53, 1973.
52. Hsia, S. L., Wright, R. K. and Halprin, K. M.: Adenyl cyclase of human skin and its abnormalities in psoriasis. In: The Role of Cyclic Nucleotides in Carcinogenesis, J. Schultz and H. G. Gratzner (eds.) Vol. 6, pp. 303-323, Academic Press, New York, 1973.
53. Hsia, S. L., Wright, R., Mandy, S. H. and Halprin, K. M.: Adenyl cyclase in normal and psoriatic skin. J. Invest. Dermatol. 59:109-113, 1972.
54. Mui, M. M., Hsia, S. L. and Halprin, K. M.: Further studies on adenyl cyclase in psoriasis. Brit. J. Dermatol. 92:255-262, 1975.
55. Yoshikawa, K., Adachi, K., Levine, V. and Halprin, K. M.: Microdetermination of cyclic AMP levels in human epidermis, dermis and hair follicles. Brit. J. Dermatol. 92:241-248, 1975.
56. Yoshikawa, K., Adachi, K., Halprin, K. M. and Levine, V.: Cyclic AMP in skin: effects of acute ischaemia. Brit. J. Dermatol. 92:249-254, 1975.
57. Yoshikawa, K., Adachi, K., Halprin, K. M. and Levine, V.: On the lack of response to catecholamine stimulation by the adenyl cyclase system in psoriatic lesions. Brit. J. Dermatol. 92:619-624, 1975.
58. Yoshikawa, K., Adachi, K., Halprin, K. M. and Levine, V.: Is the cyclic AMP in psoriatic epidermis low? Brit. J. Dermatol. 93:253-258, 1975.
59. Adachi, K., Yoshikawa, K., Halprin, K. M. and Levine, V.: Prostaglandins and cyclic AMP in epidermis. Brit. J. Dermatol. 92:381-388, 1975.
60. Voorhees, J. J., Colburn, N. H., Stawiski, M., Duell, E. A., Haddox, M. and Goldberg, N. D.: Imbalanced cyclic AMP and cyclic GMP levels in the rapidly dividing, incompletely differentiated epidermis of psoriasis. In: Control of Proliferation in Animal Cells, B. Clarkson and R. Baserga (eds.) 1:635-648, Cold Spring Harbor Lab., New York, 1974.
61. Ziboh, V. A.: Biosynthesis of Prostaglandin E_2 in human skin: subcellular localization and inhibition by unsaturated fatty acids and anti-inflammatory drugs. J. Lipid Res. 14:377-384, 1973.
62. Jonsson, C.-E. and Anggard, E.: Biosynthesis and metabolism of prostaglandin E_2 in human skin. Scand. J. Clin. Lab. Invest. 29:289-296, 1972.
63. Aso, K., Deneau, D. G., Kruling, L., Wilkinson, D. I. and Farber, E. M.: Epidermal synthesis of prostaglandins and their effect on levels of cyclic adenosine 3',5'-monophosphate. J. Invest. Dermatol. 64:326-331, 1975.
64. Hammarström, S., Hamberg, M., Samuelsson, B., Duell, E. A., Stawiski, M. and Voorhees, J. J.: Increased concentrations of nonesterified arachidonic acid, 12L-hydroxy-5,8,10,14-eicosatetranenoic acid, prostaglandin E_2,

and prostaglandin $F_{2\alpha}$ in epidermis of psoriasis. Proc. Nat. Acad. Sci. USA. 72:5130-5134, 1975.
65. Penneys, N. S., Ziboh, V., Lord, J. and Simon, P.: Inhibitor(s) of prostaglandin synthesis in psoriatic plaque. Nature 254:351-352, 1975.
66. Newcombe, D. S., Ciosek, C. P., Ishikawa, Y. and Fahey, J. V.: Human synoviocytes: activation and desensitization by prostaglandins and L-epinephrine. Proc. Nat. Acad. Sci. USA 72:3124-3128, 1975.
67. Aso, K., Orenberg, E. K. and Farber, E. M.: Reduced epidermal cyclic AMP accumulation following prostaglandin stimulation: its possible role in the pathophysiology of psoriasis. J. Invest. Dermatol. 65:375-378, 1975.
68. Ziboh, V. A. and Hsia, S. L.: Effects of prostaglandin E_2 on rat skin: inhibition of sterol ester biosynthesis and clearing of scaly lesions in essential fatty acid deficiency. J. Lipid Res. 13:458-467, 1972.
69. VanDorp, D.: Recent development in the biosynthesis and the analysis of prostaglandins. Ann. N.Y. Acad. Sci. 180:181-199, 1971.
70. Goldyne, M. E.: Prostaglandins and cutaneous inflammation. J. Invest. Dermatol. 64:377-385, 1975.
71. Ziboh, V. A.: Prostaglandins and their biological significance in the skin. Int. J. Dermatol. 14:485-493, 1975.
72. Greaves, M. W. and Kingston, W. P.: Prostaglandins as mediators of sustained inflammation in the skin. Int. J. Dermatol. 14:338-340, 1975.
73. Prasad, K. N. and Sheppard, J. R.: Inhibitors of cyclic-nucleotide phosphodiesterase induce morphological differentiation of mouse neuroblastoma cell culture. Exp. Cell Res. 73:436-440, 1972.
74. Pawelek, J., Sansone, M., Koch, N., Christie, G., Halaban, R., Handee, J., Lerner, A. B. and Varga, J. M.: Melanoma cells resistance to inhibition of growth by melanocyte stimulating hormone. Proc. Nat. Acad. Sci. USA 72:951-955, 1975.
75. Stawiski, M. A., Powell, J. A., Lang, P. G., Schork, A., Duell, E. A. and Voorhees, J. J.: Papaverine: its effects on cyclic AMP in vitro and psoriasis in vivo. J. Invest. Dermatol. 64:124-127, 1975.
76. Stawiski, M., Rusin, L., Schork, M. A., Burns, T., Duell, E. A. and Voorhees, J.: Ro 20-1724 elevates epidermal cyclic AMP levels in vitro and improves psoriasis in vivo. Clin. Res. 24:267A, 1976.
77. Duell, E., Voorhees, J., Bazner, W., Creehan, P., Stawiski, M. and Harrell, R.: Increased cyclic AMP levels in involved and uninvolved psoriasis epidermis after incubation with adenosine analog. Clin. Res. 23:228A, 1975.
78. White, S. H., Newcomer, V. D., Mickey, M. R. and Terasaki, P. I.: Disturbance of HL-A antigen frequency in psoriasis. N. Eng. J. Med. 287: 740-743, 1972.

Cyclic Nucleotide Metabolism in Tumors

WAYNE E. CRISS
FERID MURAD

The cyclic forms of AMP and GMP, perhaps also CMP and UMP, are quite ubiquitous in nature and have been observed in most extracellular fluids and tissues. They seem to have a very general distribution and their concentrations can be altered with a wide variety of hormones, neurohormones, prostaglandins, and drugs. Levels of these cyclic nucleotides in extracellular fluids, particularly urine and plasma, have proved useful in diagnosing some disorders and in studying the pathophysiological states of tissues (1–14). Altered levels of adenosine 3',5'-monophosphate (cyclic AMP) and/or guanosine 3',5'-monophosphate (cyclic GMP) were observed in tissues or in the urine of animals or patients with hyperparathyroidism (5, 7, 9, 10), ectopic PTH-secreting tumors (5, 7), carcinoid tumors, pheochromocytoma*, hepatomas (12–14), renal adenoma and sarcoma (14), tumors associated with the syn-

* Unpublished observations by F. Murad.

drome of inappropriate antidiuretic hormone secretion (7), and Cushing's disease (11). The purpose of this review was to collect and examine the data concerning the cyclic nucleotide system in cancer, specifically in solid tumors. (For a more extensive review of cyclic nucleotides in cancer see *Modified Cellular and Molecular Control in Neoplasia*, ed. by Dr. W. E. Criss, T. Ono, and J. Sabine, in press with Raven Press, N.Y.).

TISSUE CONCENTRATION OF CYCLIC NUCLEOTIDES

A number of recent studies have pointed to important roles for cyclic AMP and cyclic GMP in several experimental cancer systems. Several authors have shown that addition of cyclic AMP or its derivatives to cultured malignant cells will "apparently" restore these cells to "normal" (15–19). Similarly, in animals, solid and ascites tumor growth can be greatly reduced or even inhibited if the animals are treated with certain derivatives or the cyclic nucleotides (20–22).

It has been postulated that cyclic AMP and cyclic GMP may exert a portion of their biological actions upon metabolic systems which are in opposition (i.e., cellular differentiation versus cell division; see Refs. 23–31). Indeed, the in vivo cyclic GMP:cyclic AMP ratio was increased in liver and kidney neoplasia (12, 26; Table 1), in viral transformed cells (27–28), and in proliferating epidermal tissues (29). The levels of both of these cyclic nucleotides are influenced by hormones and numerous pharmacological agents. Therefore, regulation of the in vivo levels of the cyclic nucleotides may occur through alterations of synthesis, hydrolysis, intracellular binding, and cellular excretion. Studies with the hepatomas, proliferating cell systems of lymphocytes (24–26), epidermal tissues (29, 32), fibroblasts (27–28, 33), and plants (34), support the concept that these two cyclic nucleotides do promote opposing regulatory influences that may directly involve cellular growth processes. It is obvious that these regulatory influences are quite complex.

In addition to cyclic AMP and cyclic GMP, two other cyclic nucleotides have been recently identified. Cytidine 3',5'-monophosphate (cyclic CMP) has been observed in leukemia cells and in the urine of leukemia patients (35). It has been postulated to play a role in regulating the growth of these cells (36). The same labora-

Table 1
In Situ Levels of Cyclic AMP and Cyclic GMP in Normal and Neoplastic Rat Tissues*

Tissue	pmoles/mg protein Cyclic AMP	pmoles/mg protein Cyclic GMP	Cyclic GMP: Cyclic AMP Ratio	
Normal Liver	3.58	0.11	0.03	
Liver of Hepatoma Bearing Rats	3.50	0.09	0.03	0.04
Normal Kidney Cortex	8.01	0.32	0.04	
Slow Growing Tumors				
Hepatoma 16	8.30	0.26	0.03	
Hepatoma 9633F	5.11	0.39	0.08	
Hepatoma 20	4.00	0.19	0.05	0.06
Hepatoma 21	6.30	0.50	0.08	
Hepatoma 9633	6.64	0.51	0.08	
Intermediate Growing Tumors				
Hepatoma 7800	11.00	0.56	0.05	
Hepatoma 9121	9.94	0.38	0.04	
Hepatoma 7316B	11.11	0.45	0.04	0.08
Kidney Sarcoma (MK2)	5.10	0.70	0.14	
Kidney Adenoma (MK3)	4.50	0.55	0.12	
Fast Growing Tumors				
Hepatoma 7288ctc	8.41	2.50	0.30	
Hepatoma 9618A2	5.10	0.99	0.19	
Hepatoma 5123tc	8.80	0.88	0.10	0.71
Hepatoma 3924A	9.86	22.40	2.27	

*Compiled from reference 12 and our unpublished observations.

tory has also recently identified uridine 3′,5′-monophosphate (cyclic UMP) in rat liver and in the urine of leukemia patients (37). After confirmation of their existence is some tissues and extracellular fluids, the regulatory roles of these latter two cyclic nucleotides and their relationships to the neoplastic state remain to be elucidated.

ADENYLATE CYCLASE ACTIVITY

Adenylate cyclase activity is associated with various membrane structures of the cell, although there have been reports of "soluble" adenylate cyclase activity in Yoshida ascites hepatoma cells (38) and in rat testis (39). The membrane-bound adenylate cyclase system in mammalian cells are modulated by biogenic amines, polypeptide hormones, guanine nucleotides, halides, and prostaglandins. Thus adenylate cyclase activity is highly regulatory and must be considered an important aspect of the cell's regulatory control mechanisms. It is probable that these adenylate cyclase systems are composed of membrane elements which include specific receptors for each of the above listed modulators, enzymatic catalytic units which are capable of converting ATP to cyclic AMP, and some form of transducing elements which allow or disallow the interaction of modulators with the catalytic units (40–45).

Several laboratories have examined the adenylate cyclase systems in preneoplastic and neoplastic cells and tissues. Preneoplastic or hyperplastic liver tissue from rats which have been fed 3'-methyl-4-dimethylaminoazobenzene (46), ethionine (47), or acetylaminofluorene (48–49), showed modified responses to glucagon, epinephrine, and prostaglandins. SV40 transformed WI38 cells had lower basal adenylate cyclase activities when compared to non transformed WI-38 cells, but their adenylate cyclase was hyperresponsive to catecholamines, adenosine, and prostaglandins (50); while sarcoma virus transformed cells showed decreased adenylate cyclase activity and were nonresponsive to prostaglandin E_1 (51). One study found that HeLa and hepatoma (HTC) cells had adenylate cyclases stimulatable by glucagon, epinephrine, or prostaglandins; and that the cyclases from Chang's liver cells and fibroblast cell lines L929 and 3T3 responded to catecholamines and prostaglandins (52). Adenylate cyclases from normal adrenal cells responded to ACTH by producing cortisone while the adrenocortical carcinoma 494 did not synthesize cortisone upon ACTH stimulation (53). Yet this same adrenocortical carcinoma contained an adenylate cyclase which had multiple hormone responsiveness, including ACTH (54). The sensitivity of the adenylate cyclase system to neurotransmitters and divalent ions and the sensitivity of prostaglandin E_1 stimulated activity to GTP increased in malignant neuroblastoma cells which were induced to differentiate with cyclic AMP (55). Neoplastic thyroid nodules were reported to have increased basal activity and showed hyperresponsiveness to thyroid stimulating hormone (56–57). Liver tissue contained a membrane adenylate cyclase which was stimulated by glucagon, epinephrine, guanine nucleotides, fluoride, and prostaglandins; fast-growing hepatomas contained a mem-

Table 2
Stimulation of Adenylate Cyclase and Binding of Glucagon to Membranes from Liver and Hepatoma Tissue*
(Fl⁻ = fluoride; Epi = epinephrine; Gluc = glucagon; Gpp (NH) p = guanyl-5'-yl-imidodiphosphate)

	Enzymatic Activity (nmoles/10 min/protein)							^{125}I-Glucagon Binding (pmoles/mg protein)
	Basal	+Fl⁻	+Epi	+Gluc	+GPP (NH) p	PGA_1	PGE_1	
Normal Adult Liver	0.6	3.5	0.7	3.1	3.3	0.8	1.2	0.72
Slow Growing Hepatomas								
20	0.06	0.44	0.06	0.21	0.40	0.08	0.12	⟨0.01
21	0.08	0.98	0.11	0.35	0.84	0.10	0.15	0.15
Intermediate Growing Hepatomas								
9633F	0.3	2.0	0.3	0.4	0.6	0.4	0.5	0.06
Fast Growing Hepatomas								
3924A	0.1	2.8	0.1	0.2	2.3	0.1	0.2	0.01
7777	0.2	2.0	0.3	0.2	0.7	0.3	0.4	0.02
5123tc	0.5	4.9	0.6	1.2	2.3	0.3	0.5	0.04
9618A2	0.4	3.0	0.4	0.4	1.3	0.3	0.4	⟨0.01

*composited from references 63-65

brane adenylate cyclase which was lower in basal activity and which had decreased or no response to glucagon, epinephrine, and prostaglandins (58–64). Thus, there would appear to be a wide variation in the ability of tumor adenylate cyclase systems to respond to various hormones.

We have recently examined the hepatoma membrane bound adenylate cyclase in greater detail (63–64; Table 2). Using purified plasma membranes, which did not contain the epinephrine sensitive adenylate cyclase, the adenylate cyclase systems from adult liver, and slow, intermediate, and fast growing hepatomas were compared. The slow growing tumors and decreased basal adenylate cyclase activity, were responsive to fluoride, glucagon, Gpp(NH)p, and prostaglandins, but showed decreased glucagon binding. The intermediate growing hepatomas had decreased basal enzymatic activity, were responsive to fluoride, prostaglandins and Gpp(NH)p, were not responsive to glucagon, and had very low glucagon binding. The membranes from the fast growing tumors showed low to normal adenylate cyclase activities, responded to fluoride and Gpp(NH)p, were inhibited and/or stimulated by prostaglandins, did not respond to glucagon, and showed negligible binding of glucagon (Table 2). When the membranes from adult liver, from a slow-growing hepatoma, which had low basal activity but good response and binding to glucagon, and from a fast-growing hepatoma, which had intermediate basal activity and did not show response or binding to glucagon were compared at several temperatures, several pH's, and with various concentrations of cations and modulators, the following conclusions were made with regard to adenylate cyclase activity in malignant liver tumors. There was (1) decreased stimulation by glucagon, (2) decreased binding by glucagon, (3) widely variable stimulation by prostaglandins, (4) fully intact responses to both fluoride and Gpp(NH)p, (5) decreased and altered activity of the basal catalytic unit, and (6) altered component to component relationships within the membrane. It is not known whether these modified characteristics reflect changes in membrane composition and structure, or whether they reflect changes in the individual components of the membrane bound adenylate cyclase system.

GUANYLATE CYCLASE ACTIVITY

Guanylate cyclase activity is distributed throughout mammalian cells. It is associated with all membrane fractions and is also found to be soluble (66–73). The soluble and membrane bound (particulate) guanylate cyclases differ with respect to their activation and inhibition by Triton X-100, Ca^{2+}, ATP, maleate, p-chloromercuriphenyl sulfonic acid, and sodium azide; they also have different molecular weights (70–73). The basal activities and subcellular distribution of both forms of guanylate cyclase vary from tissue to tissue within the same animal (73). Ratios of particulate to soluble guanylate activity were altered in liver during development and regeneration (69). And data have been published indicating the possibility of a soluble macromolecular activator and soluble macromolecular inhibitor of guanylate cyclase activity (72, 74).

Table 3
Subcellular Distribution of Guanylate Cyclase Activities in Normal and Neoplastic Tissues*

	Cyclic GMP Formed (nmoles/min/g tissue)		Particulate: Soluble Ratio
	particulate	soluble	
Normal Adult Liver	0.6	2.5	0.24
Normal Adult Kidney	0.9	1.8	0.50
Slow Growing Tumors			
Hepatoma 20	1.2	0.5	2.40
Hepatoma 21	1.8	0.9	2.00
Intermediate Growing Tumors			
Hepatoma 9633F	1.3	0.5	2.60
Kidney Sarcoma (MK2)	0.6	0.5	1.20
Kidney Adenoma (MK3)	0.7	0.7	1.00
Fast Growing Tumors			
Hepatoma 3924A	1.3	0.4	3.25
Hepatoma 9618A2	0.8	0.6	1.33

*composited from references 76-77

Only two reports appear in the scientific literature which describe guanylate cyclase in neoplastic cells or tissues (69, 75). These studies, both reported in Morris transplantable hepatoma systems, showed decreased soluble and increased particulate guanylate cyclase activities in fast growing tumors.

We recently separated and characterized the guanylate cyclase activities from Morris liver and kidney tumors and compared them with the guanylate cyclase activities from normal adult liver and kidney tissues (76-77; Table 3). Total basal homogenate guanylate cyclase activities were lower in all tumors (with 10-fold varitation in growth rates) when compared to corresponding normal tissues. The soluble form of guanylate cyclase was predominant in the normal tissues, while the particulate form predominated in the tumor tissues. Comparisons of particulate enzymes from these normal and neoplastic tissues and comparison of soluble enzymes from these normal and neoplastic tissues indicated that the tumor guanylate cyclases were less able to utilize Mg^{2+} at low Mn^{2+} concentrations, were less stimulatable by Triton X-100, and apparently lacked the soluble macromolecular (azide-dependent) activator. Ca^{2+} activation of the soluble enzymes and inhibition of the particulate enzymes, as well as the greater susceptibility of the soluble enzymes to inhibition by ATP, was similar with guanylate cyclases from normal and neoplastic tissues. It is therefore possible that the observed elevated levels of cyclic GMP in these tumors may be related to the increased amount of the particulate form of guanylate cyclase, which is regulated differently by this group of modulators, than is the soluble form of the enzyme.

PHOSPODIESTERASE

Phosphodiesterases occur in most biologic systems which have cyclic nucleotides. Phosphodiesterase activity is both membrane associated and soluble in most mammalian cells (78). It exists in multiple molecular forms which differ in their substrate specificities (79–82), substrate affinities (82–84), heat and cation sensitivities (80, 85–86), chromatographic and electrophoretic mobilities (82, 87), senstivities to various inhibitors (78, 88–89), subcellular location (79, 90–91), and possibily in physiologic function (92–93). There would appear to be two major forms of phosphodiesterase activity in most tissues, high Km ($~10^{-4}M$) and low Km ($~10^{-6}M$) forms. However, recent studies support the view that both cyclic AMP and cyclic GMP phosphodiesterases exist in cells (78, 88–89, 93–94). There may be interconversion of the high and low Km enzymes; and macromolecular activators and inhibitors of phosphodiesterase activity have been described (95–100). In addition, the levels of phosphodiesterase activity are affected by hormones (101–104). Therefore, breakdown of cyclic AMP and cyclic GMP by mammalian cell phosphodiesterase hydrolysis is accomplished by a complex and interconcerted set of mechanisms.

Phosphodiesterase activity has been examined in cultured astrocytoma (104–105), and neuroblastoma (94, 105–107) cells, Krebs 2 ascites tumor cells (108), Walker carcinoma (109), thymic lymphoblasts (110), HeLa cells (111), Mouse L cells (111), hyperplastic liver nodules (112), Novikoff and Morris hepatomas (111, 113–116), and an adrenocortical carcinoma (117). Total cyclic AMP phosphodiesterase activity in whole homogenates was decreased in the hepatomas and in the adrenocortical carcinoma tissues when compared with normal liver and adrenal tissues, respectively. In the hepatoma tissues, there was a large decrease in the high Km form and a small increase in the low Km form of cyclic AMP phosphodiesterase. The normal adrenal and adrenal tumor contained only one form of the enzyme which hydrolized cyclic AMP and cyclic GMP at about the same rate. Cyclic GMP hydrolysis was not examined in any of the hepatoma studies. The authors are not aware of other reports on the cyclic GMP phosphodiesterase activity in solid tumors. However, one recent manuscript which examined phosphodiesterase activity in normal and viral transformed fibroblasts showed decreased capacity of the transformed cells to hydrolize cyclic GMP (118). Interpretations of the physiological role of the cyclic nucleotide phosphodiesterases in tumors must await comparative studies with both cyclic nucleotides and with heat stable modulating protein and Ca^{2+}.

PROTEIN KINASE

A large number of physiologic processes are modulated by the enzymatic phosphorylation of proteins. Some of these phosphorylation reactions require protein kinases that are activated by cyclic AMP and/or cyclic GMP, and some are independent of cyclic nucleotide concentrations (119–125). The postulated enzymatic model for the skeletal muscle protein kinase is

$$[R_2C_2 \cdot (MgATP)_2 + 2 \text{ cAMP} = R_2 \cdot (cAMP)_2 + 2C + 2 \text{ MgATP}]$$
$$\underset{inactive}{} \qquad\qquad\qquad \underset{active}{}$$

The native inactive enzyme is composed of two regulatory subunits (R) and two catalytic subunits (C). Cyclic AMP causes disassociation of the regulatory and catalytic subunits by binding to the regulatory subunits. The "free" catalytic subunits then enzymatically catalyze the phosphorylation of substrate proteins. MgATP and a heat stable protein modulator (122, 126–128) decrease the affinity of the catalytic subunits for the regulatory subunits. Proteins which are located in the plasma membranes, ribosomes, cytoplasm, and chromosomes serve as substrate proteins for protein kinase catalyzed phosphorylations (122–125, 129–137).

The independent and cyclic nucleotide dependent protein kinase systems have been examined in normal and viral transformed thymic fibroblasts (138), normal and malignant glioma cells (139), Ehrlich ascites tumor cells (140), HeLa cells (141), Hepatoma cells (142–145), Walker 256 carcinoma cells (146), adrenocortical carcinoma (147-148), and Morris hepatomas (149-150). The studies with Ehrlich ascites tumor cells and HeLa tumor cells both show a complete loss of cyclic AMP responsiveness protein kinase activities. Fractionation of cyclic AMP binding activity and protein kinase activity from rat liver and a hepatoma cell line (HTC) revealed that the HTC cells did not contain one of two cyclic AMP responsive protein kinase fractions which was found in the liver tissue. This deficiency was attributed to the absence of a specific cyclic AMP binding component (144). Purification and comparison of the cyclic AMP responsive protein kinase from rat adrenal tissue and from an adrenocortical carcinoma indicated that there was decreased responsiveness to cyclic AMP and altered protein substrate specificity in the tumor system (148). None of the above systems were examined for cyclic GMP responsive protein kinase activities. One recent report does compare independent, cyclic AMP responsive, and cyclic GMP responsive protein kinase activities in cytoplasmic and particulate extracts from normal rat liver and a fast growing Morris hepatoma (150). The latter study showed an increase in basal protein kinase activity, which was independent of cyclic nucleotides, and increased responsiveness to cyclic AMP and cyclic GMP in the tumor system. It is thus quite possible that neoplastic tissues may have modified protein kinase activities which could readily lead to altered regulatory controls.

SUMMARY

Considering the numerous publications on the cyclic nucleotide systems in various tissues, it is surprising that there are relatively few reports that have examined the cyclic nucleotide systems in solid tumors. Most of these latter reports are limited to liver, adrenal, or kidney tumors in laboratory animals. From these reports it is difficult to delineate specific and consistent trends. However, some generalizations, while risky, seem apparent. There does appear to be an increase in the ratio of cyclic GMP to cyclic AMP levels in cancer tissues. Such changes in the in vivo levels of these two cyclic nucleotides could result from alterations in synthesis, hydrolysis, or from cellular excretion. Synthesis of cyclic GMP may be enhanced because of an increase in membrane bound guanylate cyclase and/or an alteration of its properties and regulation. There is altered regulation of adenylate cyclase activity in tumors by various agents; the apparent trend is decreased activation of adenylate cyclase. Hydrolysis of cyclic nucleotides by the several phosphodiesterases indicates that these hydrolytic activities may be low in tumors. It is also quite likely that neoplastic tissues have a unique "set" of protein kinases which may directly influence numerous metabolic functions, from membrane transport to genomic expression. Obviously, much information is needed before one can begin to determine "essential" macromolecular lesions within the cyclic nucleotide systems of solid tumor tissues. Hopefully, the variability in the reported alterations of the cyclic nucleotide systems in different animal models of tumors and cell culture systems will prepare us for the even more complex anaylsis of data dervied from human tumors which to date is noticeably nonexistent.

REFERENCES

1. Ball, J. H., Kaminsky, N. I., Hardman, J. G., Broadus, A. F., Sutherland, E. W., and Liddle, G. W.: Effects of catecholamines and adrenergic blocking agents on plasma and urinary cyclic nucleotides. Man. J. Clin. Invest. *51*, 2124-2129, 1972.
2. Bernstein, R. A., Linarelli, L., Facktor, M. A., Friday, G. A., Drash, A., and Fireman, P.: Decreased urinary cyclic 3',5'-adenosine monophosphate after epinephrine in asthmatic patients. J. Allergy Clin. Immunol. *49*, 86-97, 1972.
3. Chase, L. R., Melson, G. L., and Aurbach, G. D.: Pseudohypoparathyroidism: defective excretion of 3',5'-AMP in response to parathyroid hormone. J. Clin. Invest. *48*, 1832-1844, 1969.
4. Fichman, M., and Brooker, G.: Cyclic AMP and the antidiuretic effect on chloropropamide in diabetes insipidus. Clin. Res. *18*, 121-134, 1970.
5. Kaminsky, N. I., Broadus, A. E., Hardman, J. G., Jones, D. J., Ball, J. H., Sutherland, E. W., and Liddle, G. W.: Effects of parathyroid hormone on plasma and urinary adenosine 3',5'-monophosphate in man. J. Clin. Invest. *49*, 2387-2394, 1970.
6. Kanamori, T., Kuzuya, H., and Nagatsu, T.: Excretion of cyclic GMP and cyclic AMP into human parotid saliva. J. Dental Res. *53*, 760, 1974.
7. Murad, F.: Clinical studies and applications of cyclic nucleotides. Adv. Cyclic Nucleotide Res. *3*, 355-383, 1973.
8. Murad, F., Moss, W., Johanson, A. J., and Selden, R. F.: Urinary excretion of adenosine 3',5'-monophosphate in normal children and those with cystic fibrosis. J. Clin. End. Met. *40*, 552-558, 1975.
9. Murad, F., and Pak, C. Y. C.: Urinary excretion of adenosine 3',5'-monophosphate and guanosine 3',5'-monophosphate. N. Engl. J. Med. *286*, 1382-1387, 1972.
10. Taylor, A. L., Davis, B. B., Pawlson, L. G., Josimovich, J. B., and Mintz, D. H.: Factors influencing the urinary excretion of 3',5'-adenosine mono-

phosphate in humans. J. Clin. Endocrinol. Metab. *30*, 316-323, 1970.
11. Wray, H. L., Corrigan, D. F., Bruton, J., Schaaf, M., and Earle, J. M.: Elevated urinary cyclic GMP in patients with Cushing's syndrome. Proceedings of the 57th Annual Meeting of the Endocrine Society 362, 1975.
12. Thomas, E. W., Murad, F., Looney, W. B., and Morris, H. P.: Adenosine 3',5'-monophosphate and guanosine 3',5'-monophosphate concentrations in Morris hepatomas of different growth rates. Biochem. Biophys. Acta *297*, 564-567, 1973.
13. Murad, F., Kimura, H., Hopkins, H. A., Looney, W. B., and Kovacks, C. J.: Increased urinary excretion of cyclic guanosine monophosphate in rats bearing Morris hepatoma 3924A. Science *190*, 58-60, 1975.
14. Criss, W. E., and Murad, F.: Urinary excretion of cyclic 3',5'-GMP and Cyclic 3',5'-AMP in rats bearing transplantable liver and kidney tumors. Cancer Res. *36*, 1714-1721, 1976.
15. Hsie, A. W., and Puck, T. T.: Morphological transformation of chinese hamster cells by dibutyryl adenosine cyclic 3',5'-monophosphate and testosterone. Proc. Natl. Acad. Sci., USA *68*, 358-361, 1971.
16. Johnson, G. S., Friedman, R. M., and Pastan, I.: Restoration of several morphological characteristics of fibroblasts in sarcoma cells treated with cyclic AMP and its derivatives. Proc. Natl. Acad. Sci., USA *68*, 425-429, 1971.
17. Johnson, G. S. and Pastan, I.: Role of 3',5'-adenosine monophosphate in regulation of morphology and growth of transformed and normal fibroblasts. J. Natl. Cancer Inst. *48*, 1377-1378, 1972.
18. Van Wijk, R., Wicks, W. D., and Clay, K.: Effects of derivatives of cyclic 3',5'-adenosine monophosphate on the growth, morphology, and gene expression of hepatoma cells in culture. Cancer Res. *32*, 1905-1911, 1972.
19. Macintyre, E. H., Wintersgill, C. J., Perkins, J. P., and Vatter, A. E.: The responses in culture of human tumor astrocytes and neuroblasts to $N^6,O^{2'}$-dibutyryl adenosine 3',5' monophosphoric acid. J. Cell Sci. *2*, 639-667, 1972.
20. Cho-Chung, Y. S.: *In vivo* inhibition of tumor growth by cyclic adenosine 3',5'-monophosphate derivatives. Cancer Res. *34*, 3492-3496, 1974.
21. Cho-Chung, Y. S.: The role of cyclic AMP in neoplastic cell growth and regression. Biochem. Biophys. Res. Comm. *60*, 528-534, 1974.
22. Cotton, F. A., Gillen, R. G., Gohil, R. N., Hazen, E. E., Kirchner, C. R., Nagyvary, J., Rouse, J. P., Stanislowski, A. G., Stevens, J. D., and Tucker, P. W.: Tumor inhibiting properties of neutral P-O-Ethyl of adenosine 3',5'-monophosphate in correlation with its crystal and molecular structure. Proc. Natl. Acad. Sci., USA *72*, 1335-1339, 1975.
23. Goldberg, N. D., Haddox, M. K., Hartle, D. K., and Hadden, J. W.: The biological role of cyclic 3',5'-guanosine monophosphate. In: Pharmacology and the Future of Man, 5th Int. Congress of Pharmacology, *5*, 146-169 (Karger, Basel: 1973).
24. Goldberg, N. D., Haddox, M. K., Estensen, R., White, J. G., Lopez, C., and Hadden, J. W.: Evidence for a dualism between cyclic GMP and cyclic AMP in the regulation of cell proliferation and other cellular processes. In: Cyclic AMP, Cell Growth and the Immune Response, W. Braun, L. Lichtenstein, and C. Parker (eds.), pp. 247-262, (Springer-Verlag, New York: 1974).

25. Goldberg, N. D., Haddox, M. K., Nicol, S. E., Glass, D. B., Sanford, C. H., Kuehl, F. A., and Estensen, R.: Biological regulation through opposing influences of cyclic AMP and cyclic GMP: The Ying:Yang hypothesis. Adv. Cyclic Nucleotide Res. 5, 307-338, 1975.
26. Goldberg, M. L., Burke, G. C., and Morris, H. P.: Cyclic AMP and cyclic GMP content and binding in malignancy. Biochem. Biophys. Res. Comm. 62, 320-327, 1975.
27. Rudland, P. S., Seeley, M., and Seifert, W.: Cyclic GMP and cyclic AMP levels in normal and transformed fibroblasts. Nature 251, 417-419, 1974.
28. Moens, W., Vokaer, A., and Kram, R.: Cyclic AMP and cyclic GMP concentrations in serum and density restricted fibroblast cultures. Proc. Natl. Acad. Sci., USA 72, 1063-1067, 1975.
29. Voorhees, J. J., Stawiski, M., Duell, E. A., Haddox, M. K., and Goldberg, N. D.: Increased cyclic GMP and decreased cyclic AMP levels in the hyperplastic, abnormally differentiated epidermis of psoriasis. Life Sciences 13, 639-653, 1973.
30. Kram, R., and Tomkins, G. M.: Pleiotypic control by cyclic AMP: interaction with cyclic GMP and possible role of microtubles. Proc. Natl. Acad. Sci., USA 70, 1659-1663, 1973.
31. Watson, J.: The influence of intracellular levels of cyclic nucleotides on cell proliferation and the induction of antibody synthesis. J. Exp. Med. 141, 97-111, 1975.
32. Voorhees, J., Kelsey, W., Stawiski, M., Smith, E., and Duell, E.: Increased cyclic GMP and decreased cyclic AMP in rapidly proliferating epithelium of psoriasis. In: The Role of Cyclic Nucleotides in Carcinogenesis, J. Schultz and H. G. Gratzner (eds.), pp. 325-373 (Academic Press, N.Y.: 1973).
33. Seifert, W. E., and Rudland, P. S.: Possible involvement of cyclic GMP in growth control of cultured mouse cells. Nature 248, 138-140, 1974.
34. Haddox, M. K., Stephenson, J. H., and Goldberg, N. D.: Cyclic GMP in meristematic and elongating regions of bean roots. Fed. Proc. 33, 522, 1974.
35. Bloch, A.: Cytidine 3',5'-monophosphate (cyclic CMP). I. Isolation from extracts of leukemia L-1210 cells. Biochem. Biophys. Res. Comm. 58, 652-659, 1974.
36. Bloch, A., Dutschman, G., and Maue, R.: Cytidine 3',5'-monophosphate (cyclic CMP). II. Initiation of leukemia L-1210 cell growth *in vitro*. Biochem. Biophys. Res. Comm. 59, 955-1002, 1974.
37. Bloch, A.: Uridine 3',5'-monophosphate (cyclic UMP). I. Isolation from rat liver extracts. Biochem Biophys. Res. Comm. 64, 210-218, 1975.
38. Tomasi, V., Rethy, A., and Trevisani, A.: Soluble and membrane bound adenylate cyclase activity in Yoshida ascites hepatoma. Life Sci. 12, 145-150, 1973.
39. Braun, T., and Dods, R. F.: Development of a Mn^{2+}-sensitive, "soluble" adenylate cyclase in rat testis. Proc. Natl. Acad. Sci., USA 72, 1097-1101, 1975.
40. Oka, H., Kaneko, T., Yamashita, K., Suzuki, S., and Oda, T.: The glucagon and fluoride sensitive adenylate cyclase in plasma membranes of rat liver. Endocrinol. Japon. 20, 263-270, 1973.
41. Robison, G. A., Butcher, R. W., and Sutherland, E. W.: Adenyl cyclase

as an adrenergic receptor. Ann. N.Y. Acad. Sci. *139*, 703-723, 1967.
42. Robison, G. A., Butcher, R. W., and Sutherland, E. W.: In: *Fundamental Concepts in Drug-Receptor Interactions*, J. F. Danielli, J. F. Moran, and D. J. Triggle (eds.), pp. 59-110 (Academic Press, N.Y.: 1968).
43. Rodbell, M., Lin, M. C., Salomon, Y., Londos, C., Harwood, J. P., Martin, B. R., Rendell, M., and Berman, M.: Role of adenine and guanine nucleotides in the activity and response of the adenylate cyclase systems to hormones: evidence for multisite transition states. Acta Endocrinol. Suppl. *191*, 11-37, 1974.
44. Harwood, J. P., Löw, H., and Rodbell, M.: Stimulatory and inhibitory effects of guanyl nucleotides on fat cell adenylate cyclase. J. Biol. Chem. *248*, 6239-6245, 1973.
45. DeRubertis, F., Zenser, T. V., and Curnow, R. T.: Inhibition of glucagon —mediated increases in hepatic cyclic adenosine 3',5'-monophosphate by prostaglandin E_1 and E_2. Endocrinol. *95*, 93-100, 1974.
46. Boyd, H., Louis, C. J., and Martin, T. J.: Activity and hormone responsiveness of adenyl cyclase during induction of tumors in rat liver with 3'-methyl-4-dimethylaminoazobenzene. Cancer Res. *34*, 1720-1725, 1974.
47. Chayoth, R., Epstein, S. M., and Field, J. B.: Glucagon and E_1 stimulation of cyclic adenosine 3',5'-monophosphate levels and adenylate cyclase activity in benign hyperplastic nodules and malignant hepatomas of ethionine treated rats. Cancer Res. *33*, 1970-1974, 1973.
48. Christoffersen, T., and Oye, I.: Alterations in hormone responsiveness of hepatic adenylate cyclase during ontogenesis and oncogenesis. Acta Endocrinol. Suppl. 191, *77*, 67-71, 1974.
49. Christoffersen, T., Moreland, J., Osnes, J. B., and Elgjo, K.: Hepatic adenyl cyclase: alterations in hormone response during treatment with a chemical carcinogen. Biochim. Biophys. Acta *279*, 363-366, 1972.
50. Kelly, L. A., Hall, M. S., and Butcher, R. W.: Cyclic adenosine 3',5'-monophosphate metabolism in normal and SV40 transformed WI-38 cells. J. Biol. Chem. *249*, 5182-5187, 1974.
51. Anderson, W. B., Gallo, M., and Pastan, I.: Adenylate cyclase activity in fibroblasts transformed by Kirsten or Moloney Sarcoma Viruses. J. Biol. Chem. *249*, 7041-7048, 1974.
52. Makman, M. H.: Conditions leading to enhanced response to glucagon, epinephrine, or prostaglandins by adenylate cyclase of normal and malignant cultured cells. Proc. Natl. Acad. Sci., USA *68*, 2127-2130, 1971.
53. Brush, J. S., Sutliff, L. S., and Sharma, R. K.: Metabolic regulation and adenyl cyclase activity of adrenocortical carcinoma cultured cells. Cancer Res. *34*, 1495-1502, 1974.
54. Schoor, I., Ratham, P., Saxena, B. B., and Ney, R. L.: Multiple specific hormone receptors in the adenylate cyclase of an adrenocortical carcinoma. J. Biol. Chem. *246*, 5806-5811, 1971.
55. Prasad, K. N., Gilmer, K. N., Sahu, S. K., and Baker, G.: Effect of neurotransmitters, guanosine triphosphate, and divalent ions on the regulation of adenylate cyclase activity in malignant and adenosine cyclic 3',5'-monophosphate-induced "differentiated" neuroblastoma cells. Cancer Res. *35*, 77-81, 1975.

56. DeRubertis, F., Yamashita, K., Dekker, A., Larsen, P. R., and Field, J. B.: Effects of thyroid stimulating hormone on adenylate cyclase activity and intermediary metabolism of "cold" thyroid nodules and normal human thyroid tissue. J. Clin. Invest. *51*, 1109-1117, 1972.
57. Field, J. B., Larsen, P. R., Yamashita, K., Mashiter, K., and Dekker, A.: Demonstration of iodide transport defect but normal iodide organification in nonfunctioning nodules of human thyroid. J. Clin. Invest. *52*, 1973-2404, 1973.
58. Brown, H. D., Chattopadhyay, S. K., Morris, H. P., and Pennington, S. N.: Adenyl cyclase activity in Morris hepatoma 7777, 7794A, and 9618A. Cancer Res. *30*, 123-126, 1970.
59. Allen, D. O., Munshower, J., Morris, H. P., and Weber, G.: Regulation of adenyl cyclase in hepatomas of different growth rates. Cancer Res. *31*, 557, 1971.
60. Butcher, F. R., Scott, D. F., Potter, V. R., and Morris, H. P.: Effect of prostaglandins on adenylate cyclase activities in membranes from liver and transplatnable hepatomas. Cancer Res. *32*, 2135-2140, 1972.
61. Hickie, R. A., Walker, C. M., Croll, G. A.: Decreased basal cyclic adenosine 3',5'-monophosphate levels in Morris hepatoma 5123tc(h). Biochem. Biophys. Res. Comm. *59*, 167-173, 1974.
62. Hickie, R. A., Jan, S-H., and Datta, A.: Comparative adenylate cyclase activities in homogenate and plasma membrane fractions of Morris hepatoma 5123tc(h). Cancer Res. *35*, 596-600, 1975.
63. Criss, W. E., and Morris, H. P.: Regulation of the adenylate cyclase system in transplantable hepatomas. Cancer Res. *36*, 1740-1747, 1976.
64. Pradhan, T. K., Criss, W. E., and Morris, H. P.: Effect of prostaglandins on adenylate cyclase activities in membranes from liver and transplantable hepatomas. Cancer Biochem. Biophys. (In press).
65. Criss, W. E., and Prodhan, T. K.: Characteristics of temperature dependency on the modulation of adenylate cyclase from liver hepatomas. European J. Cancer (In press).
66. Hardman, J. G., and Sutherland, E. W.: Guanyl cyclase, an enzyme catalyzing the formation of guanosine 3',5'-monophosphate from guanosine triphosphate. J. Biol. Chem. *244*, 6363-6370, 1969.
67. Ishikawa, E. M., Ishikawa, S., Davis, J. W., and Sutherland, E. W.: Determination of guanosine 3',5'-monophosphate in tissues and of guanyl cyclase. J. Biol. Chem. *244*, 6371-6376, 1969.
68. Kimura, H., Murad, F.: Evidence for two different forms of guanylate cyclase in rat heart. J. Biol. Chem. *249*, 6910-6919, 1974.
69. Kimura, H., Murad, F.: Increased particulate and decreased soluble guanylate cyclase activity in regenerating liver, fetal liver, and hepatoma 3924A. Proc. Natl. Acad. Sci., USA *72*, 1965-1969, 1975.
70. Kimura, H., and Murad, F.: Two forms of guanylate cyclase in mammalian tissues and possible mechanisms for their regulation. Metabolism *24*, 439-445, 1975.
71. Kimura, H., and Murad, F.: Localization of particulate guanylate cyclase in plasma membranes and microsomes of rat liver. J. Biol. Chem. *250*, 4810-4817, 1975.

72. Kimura, H., and Murad, F.: Activation of guanylate cyclase from rat liver and other tissues by sodium azide. J. Biol. Chem. *250*, 8016-8022, 1975.
73. Kimura, H., and Murad, F.: Subcellular localization of guanylate cyclase. Life Sci. *17*, 837-844, 1975.
74. Mittal, C. K., Kimura, H., and Murad, F.: Requirement for a macromolecular factor for sodium azide activation on guanylate cyclase. J. Cyclic Nucleotide Res. *1*, 261-269, 1975.
75. Northup, S. J., Barthel, J. S., Brown, H. D., Chattopadhyay, S. K., and Morris, H. P.: Guanyl cyclase activity in Morris hepatoma 7787, 7795, 7800, and 9618A2. Missouri Med. *69*, 934-937, 1972.
76. Criss, W. E., Murad, F., Kimura, H., and Morris, H. P.: Properties of guanylate cyclase activity in adult rat liver and in a series of Morris hepatomas. Cancer Res. (In press).
77. Criss, W. E., Murad, F., and Kimura, H.: Properties of guanylate cyclase from rat kidney cortex and transplantable kidney tumors. J. Cyclic Nucleotide Res. *2*, 11-20, 1976.
78. Amer, M. S., and Kreighbaum, W. E.: Cyclic nucleotide phosphodiesterases: properties, activators, inhibitors, structure to activity relationships, and possible role in drug development. J. Pharm. Sci. *64*, 1-37, 1975.
79. Thompson, W. J., and Appleman, M. M.: Multiple cyclic nucleotide phosphodiesterase activities from rat brain. Biochemistry *10*, 311-316, 1971.
80. Kakiuchi, S.: Cyclic 3',5'-Nucleotide Phosphodiesterase of rat brain and other tissues: regulation of activity by Ca^{2+} and the modulator protein. Pharmacology and Future of Man *5*, 192-206, 1973.
81. Beavo, J. A., Hardman, J. G., and Sutherland, E. W.: Hydrolysis of cyclic guanosine and adenosine 3',5'-monophosphates by rat and bovine tissues. J. Biol. Chem. *245*, 5649-5655, 1970.
82. Thompson, W. J., and Appleman, M. M.: Characterization of cyclic nucleotide phosphodiesterases of rat tissues. J. Biol. Chem. *246*, 3145-3150, 1971.
83. Song, S-Y, and Cheung, W. Y.: Cyclic 3',5'-nucleotide phosphodiesterase properties of the enzyme of human blood platelets. Biochim. Biophys. Acta *242*, 593-605, 1971.
84. Peterkofsky, A., and Gazder, G.: Glucose and the metabolism of adenosine 3',5'-cyclic monophosphate in E. Coli. Proc. Natl. Acad. Sci., USA *68*, 2794-2798, 1971.
85. Murray, A. W., Spiszman, M., and Atkinson, D. E.: Adenosine 3',5'-monophosphate phosphodiesterase in the growth medium of Physarum polycephalum. Science *171*, 496-498, 1971.
86. Rutten, W. J., Schoot, B. M., Depont, J. J. H. H. M., and Bonting, S. L.: Adenosine 3',5'-monophosphate phosphodiesterase in the rat pancreas. Biochim. Biophys. Acta *315*, 384-393, 1973.
87. Ramanathan, S., and Chou, S. C.: Cyclic nucleotide phosphodiesterase from Tetrahymena. Comp. Biochem. Physiol. *46B*, 93-97, 1973.
88. Amer, M. S.: Cyclic AMP and gastric secretion. Amer. J. Dig. Dis. *17*, 945-953, 1972.
89. Sheppard, H., Wiggan, G., and Tsien, W. H.: Structure-activity relationships fro inhibitors of phosphodiesterases from erythrocytes and other

tissues. Adv. Cyclic Nucleo. Res. *1*, 103-112, 1971.
90. Robison, G. A., Butcher, R. W., and Sutherland, E. W.: *Cyclic AMP*, Academic Press, New York, 1970.
91. Schmidt, S. Y., and Lolley, R. N.: Cyclic nucleotide phosphodiesterase— an early defect in the inherited retinal degeneration of C3H mice. J. Cell Biol. *57*, 117-123, 1973.
92. Farber, D. B., and Lolley, R. N.: Proteins in the degenerative retina of C3H mice: deficiency of a cyclic nucleotide phosphodiesterase and opsin. J. Neurochem. *21*, 817-828, 1973.
93. Schultz, G., Hardman, J. G., Schultz, J., Davis, J. W., and Sutherland, E. W.: A new enzymatic assay for guanosine 3′,5′-cyclic monophosphate and its application to the ductis deferens of the rat. Proc. Natl. Acad. Sci. USA, *70*, 1721-1725, 1973.
94. Prasad, K. N., Becker, G., and Tripathy, K.: Differences and similarities between guanosine 3′,5′-cyclic monophosphate phosphodiesterase and adenosine 3′,5′-cyclic monophosphate phosphodiesterase activities in neuroblastoma cells in culture (38893). Proc. Soc. Exp. Biol. Med. *149*, 757-762, 1975.
95. Cheung, W. Y.: Cyclic 3′,5′-nucleotide phosphodiesterase: evidence for and properties of a protein activator. J. Biol. Chem. *246*, 2859-2869, 1971.
96. Teo, T. S., and Wang, J. H.: Mechanism of activation of a cyclic adenosine 3′,5′-monophosphate phosphodiesterase from bovine heart by calcium ion: adentification of the protein activator as a Ca^{2+} binding protein. J. Biol. Chem. *248*, 5950-5955, 1973.
97. Miki, N., and Yoshida, H.: Purification and properties of cyclic AMP phosphodiesterase from rat brain. Biochim. Biophys. Acta *268*, 166-174, 1972.
98. Kakiuchi, S., Yamazaki, R., Teshima, Y., Uenishi, K., and Miyamoto, E.: Multiple cyclic nucleotide phosphodiesterase activities from rat tissues and occurrence of a calcium-plus-magnesium-ion-dependent phosphodiesterase and its protein activator. Biochem. J. *146*, 109-120, 1975.
99. Kakiuchi, S., Yamazaki, R., Teshima, Y., Uenishi, K., and Miyamoto, E.: Ca^{2+}/Mg^{2+}-dependent cyclic nucleotide phosphodiesterase and its activator protein. Adv. Cyclic Nucleotide Res. *5*, 163-178, 1975.
100. Stancel, G. M., Leunz, K. M. T., and Gorski, J.: Estrogen receptors in the rat uterus. Multiple forms produced by concentration-dependent aggregation. Biochemistry *12*, 2130-2135, 1973.
101. Appleman, M. M., Thompson, W. J., and Russell, T. R.: Cyclic nucleotide phosphodiesterases. Adv. Cyclic Nucleo. Res. *3*, 65-98, 1973.
102. Berridge, M. J.: Interaction of cyclic nucleotides and calcium in the control of cellular activity. Adv. Cyclic Nucleo. Res. *6*, 1-98, 1975.
103. Thompson, W. J., Little, S. A., and Williams, R. H.: Effect of insulin and growth hormone on the rat liver cyclic nucleotide phosphodiesterase. Biochemistry *12*, 1889-1894, 1973.
104. Uzunov, P., Shein, H. M., and Weiss, B.: Cyclic AMP Phosphodiesterase in cloned astrocytoma cells: norepinephrine induces a specific enzyme form. Science *180*, 304-306, 1973.
105. Uzunov, P., Shein, H. M., and Weiss, B.: Multiple forms of cyclic 3′,5′-AMP phosphodiesterase of rat cerebrum and cloned astrocytoma and

neuroblastoma cells. Neuropharm. *13*, 337-391, 1974.
106. Prasad, K. N., and Kumar, S.: Cyclic 3',5'-AMP phosphodiesterase activity during cyclic AMP induced differentiation of neuroblastoma cells in culture. Proc. Soc. Exptl. Biol. Med. *142*, 406-410, 1973.
107. Kumar, S., Beckerm, G., and Prasad, K. N.: Cyclic adenosine 3',5'-mnonphosphate phosphodiesterase activity in malignant and cyclic adenosine 3',5'-monophosphate-induced differentiated neuroblastoma cells. Cancer Res. *35*, 82-87, 1975.
108. Burdon, R. H., and Pearce, C. A.: Enzymic modification of chromosomal macromolecules. Biochim. Biophys. Acta *246*, 561-571, 1971.
109. Tisdale, M. J.: Inhibition of cyclic adenosine 3',5'-monophosphate phosphodiesterase from Walker carcinoma by ascorbic and dehydroascorbic acids. Biochem. Biophys. Res. Comm. *62*, 887-881, 1975.
110. Whitfield, J. F., Rixon, R. H., MacManus, J. P., and Balk, S. D.: Calcium, cyclic adenosine 3',5'-monophosphate, and the control of cell prolifieration: a review. In Vitro *8*, 257-278, 1973.
111. Schröder, J., and Plagemann, P. G. W.: Cyclic 3',5'-nucleotide phosphodiesterases of Novikoff raf hepatoma, mouse L, and HeLa cells growing in suspension culture. Cancer Res. *32*, 1082-1087, 1972.
112. Chayoth, R., Epstein, S. M., and Field, J. B.: Glucagon and prostaglanoin E_1 stimulation of cyclic adenosine 3',5'-monophosphate levels and adenylate cyclase activity in benign hyperplastic nodules and malignant hepatomas in ethionine treated rats. Cancer Res. *33*, 1970-1974, 1973.
113. Rhoads, A. R., Morris, H. P., and West, W. L.: Cyclic 3',5'-nucleotide monophosphate phosphodiesterase activity in hepatomas of different growth rates. Cancer Res. *32*, 2651-2655, 1972.
114. Clark, J. F., Morris, H. P., and Weber, G.: Cyclic adenosine 3',5'-monophosphate phosphodiesterase activity in normal, differentiating, regenerating, and neoplastic liver. Cancer. Res. *33*, 356-361, 1973.
115. Hickie, R. A., Walker, C. M., and Datta, A.: Increased activity of low Km cyclic adenosine 3',5'-monophosphate phosphodiesterase in plasma membranes of Morris hepatoma 5123tc(h). Cancer Res. *35*, 601-605, 1975.
116. Criss, W. E., and Morris, H. P.: Cyclic AMP phosphodiesterase activity in three Morris hepatomas. Enzyme *20*, 65-70, 1975.
117. Sharma, R. K.: Studies on adrenocortical carcinoma of rat cyclic nucleotide phosphodiesterase activities. Cancer Res. *32*, 1734-1736, 1972.
118. Lynch, T. J., Tallant, F. A., and Cheung, W. J.: Marked reductions of cyclic GMP phosphodiesterase activity in virally transformed fibroblasts. Biochem. Biophys. Res. Comm. *65*, 1115-1122, 1975.
119. Beavo, J. A., Bechtel, P. J., and Krebs, E. G.: Mechanisms of control for cAMP-dependent protein kinase from skeletal muscle. In: Advances in Enzyme Regulation, G. I. Drummond, P. Greengard, and G. A. Robison (eds.), *5*, 241-251.
120. Miyamoto, E., Petzgold, G. L., Kuo, J. F., and Greengard, P.: Disassociation and activation of adenosine 3',5'-monophosphate-dependent and guanosine 3',5'-monophosphate-dependent protein kinases by cyclic nucleotides and by substrate proteins. J. Biol. Chem. *248*, 179-189, 1973.
121. Rubin, C. S., Erlichman, J., and Rosen, O. M.: Molecular forms and subunit composition of a cyclic adenosine 3',5'-monophosphate-dependent pro-

tein kinase purified from bovine heart muscle. J. Biol. Chem. 247, 36-44, 1972.
122. Walsh, D. A., and Ashby, C. D.: Protein kinases: aspects of their regulation and diversity. Rec. Prog. Horm. Res. 29, 329-359, 1973.
123. Nishiyama, K., Katakami, H., Yamamura, H., Takai, Y., Shimomura, R., and Nishizuka, Y.: Functional specificity of guanosine 3',5'-monophosphate-dependent and adenosine 3',5'-monophosphate-dependent protein kinasas from silkworm. J. Biol. Chem. 250, 1297-1300, 1975.
124. Takai, Y., Nishiyama, K., Yamamura, H., Nishizuka, Y.: Guanosine 3',5'-monophosphate-dependent protein kinase from bovine cerebellum. J. Biol. Chem. 250, 4690-4695, 1975.
125. Takeda, M., Matsumura, S., and Nakaya, Y.: Nuclear phosphoprotein kinases from rat liver. J. Biochem. 75, 743-751, 1974.
126. Ashby, C. D., and Walsh, D. A.: Characterization of the interaction of a protein inhibitor with adenosine 3',5'-monophosphate dependent protein kinase. J. Biol. Chem. 248, 1255-1261, 1973.
127. Haddox, M. K., Newton, N. E., Hartle, D. K., and Goldberg, N. D.: ATP (Mg^{2+}) induced inhibition of cyclic AMP reactivity with a skeletal muscle protein kinase. Biochem. Biophys. Res. Comm. 47, 653-661, 1972.
128. Donnelly, T. E., Kuo, J. F., Reyes, P. L., Lui, Y. P., and Greengard, P.: Protein kinase modulator from lobster tail muscle. J. Biol. Chem. 248, 190-198, 1973.
129. Matsumura, S., and Nishizuka, Y.: Phosphorylation of endogenous hepatic proteins by adenosine 3',5'-monophosphate protein kinase. J. Biochem. 76, 29-38, 1974.
130. Langan, T. A.: Histone phosphorylation: stimulation by adenosine 3',5'-monophosphate. Science 162, 579-580, 1968.
131. Stein, G. S., Spelsberg, T. C., and Kleinsmith, L. J.: Nonhistone chromosomal proteins and gene regulation. Science 183, 817-824, 1974.
132. Ueda, T., Maeno, H., and Greengard, P.: Regulation of endogenous phosphorylation of specific proteins in synaptic membrane fractions from rat brain by adenosine 3',5'-monophosphate. J. Biol. Chem. 248, 8295-8305, 1973.
133. Walton, G. M., Gill, G. N., Abrass, I. B., and Garren, L. D.: Phosphorylation of ribosome associated protein by an adenosine 3',5'-cyclic monophosphate-dependent protein kinase. Proc. Natl. Acad. Sci. USA 68, 880-884, 1971.
134. Rubin, C. S., and Rosen, O. M.: The role of cyclic AMP in the phosphorylation of proteins in human erythrocyte membranes. Biochem. Biophys. Res. Comm. 50, 421-429, 1973.
135. Kemp, B. E., Bylund, D. B., Huang, T-S, and Krebs, E. G.: Substrate specificity of the cyclic AMP-dependent protein kinase. Proc. Natl. Acad. Sci. USA, 72, 3448-3452, 1975.
136. Chang, K-J, Marcus, N. A., and Cuatrecasas, P.: Cyclic adenosine monophosphate dependent phosphorylation of specific fat cell membrane proteins by an endogenous membrane bound protein kinase. J. Biol. Chem. 249, 6854-6865, 1974.
137. Ljungström, O., Berglund, L., Hjelmquist, G., Humble, E., and Engstrom, L.: Cyclic 3',5'-AMP stimulated and nonstimulated phosphorylations of

protein fractions from rat liver cell sap on incubation with (γ-^{32}P) ATP. Upsala J. Med. Sci. *79*, 129-137, 1974.
138. Troy, F. A., Vijay, I. K., and Kawakami, T. G.: Cyclic 3′,5′-AMP-dependent protein kinase levels in normal and feline sarcoma virus transformed cells. Biochem. Biophys. Res. Comm. *52*, 150-163, 1973.
139. Agren, G., and Ronquist, G.: (^{32}P) phosphoryl transfer by endogenous protein kinase at the glia and glioma cell surface in culture into extrinsic acceptor proteins. Acta Physiol. Scand. *92*, 430-432, 1974.
140. Ronquist, G., and Agren, G.: (^{32}P) phosphoryl transfer by endogenous protein kinase at the Ehrlich cell surface into extrinsic acceptor proteins. Upsala J. Med. Sci. *79*, 138-142, 1974.
141. Blanchard, J. M., Ducamp, Ch., and Jeanteur, Ph.: Endogenous protein kinase activity in nuclear RNP particles from HeLa cells. Nature *253*, 467-468, 1975.
142. Olson, M. O. J., Orrick, L. R., Jones, C., and Bush, H.: Phosphorylation of acid soluble nucleolar proteins of Novikoff hepatoma ascites cells *in vivo*. J. Biol Chem. *249*, 2823-2827, 1974.
143. Granner, D. K.: Protein kinase: altered regulation in a hepatoma cell line deficient in adenosine 3′,5′-cyclic monophosphate-binding protein. Biochem. Biophys. Res. Comm. *46*, 1516-1526, 1972.
144. Mackenzie, C. W., and Stellwagen, R. H.: Differences between liver and hepatoma cells in their complements of adenosine 3′,5′-monophosphate-binding proteins and protein kinases. J. Biol. Chem. *249*, 5755-5762, 1974.
145. Mackenzie, C. W., and Stellwagen, R. H.: Heterofeneity and unusually high affinity in the interactions of adenosine 3′,5′-monophosphate with specific binding proteins from liver and hepatoma cells. J. Biol. Chem. *249*, 5763-5771, 1974.
146. Smith, D. L., Chen, C-C, Bruegger, B. B., Holtz, S. L., Halpern, R. M., and Smith, R. A.: Characterization of protein kinases forming acid-labile histone phosphates in Walker-256, carcinosarcoma cell nuclei. Biochemistry *13*, 3780-3784, 1974.
147. Sharma, R. K., and Brush, J. S.: Metabolic regulation of steroidogenesis in isolated adrenal and adrenocortical carcinoma cells of rat. The incorporation of (20S)-20 [7-^{3}H] hydroxycholesterol into deoxycorticosterone and corticosterone. Arch. Biochem. Biophys. *156*, 560-562, 1973.
148. Sharma, R. K.: Regulation of steroidogenesis in adrenal carcinoma, in Modified Cellular and Molecular Controls in Neoplasia, W. E. Criss, T. Ono, and J. Sabine, eds., Raven Press, New York (In press).
149. Criss, W. E., and Morris, H. O.: Protein kinase activity in Morris hepatomas. Biochem. Biophys. Res. Comm. *54*, 380-386, 1973.
150. Shoji, M., Huguley, C. M., Groth, D. P., Morris, H. P., and Kuo, J. F.: Cyclic AMP-dependent and cyclic GMP-dependent protein kinasas in fast growing Morris hepatomas. International Res. Comm. System: Medical Science (suppl. cancer) *3* (10) 2, 1975.

Index

Acetaminophen, 112-113, 119, 136-137
Acetazolamide, 394
Acetylcholine, see Cholinergic agents
Adenine nucleotides, see Adenosine,
 ADP, ATP, Cyclic AMP
  gastrointestinal motility, 253
  psoriasis, 422
Adenosine, 338
Adenylate cyclase,
  asthma, 204-206
  brain, 327-328, 334
  cell proliferation, 72-73
  cholera toxin, 266-275
  ciliary processes, 392
  ethanol, 369
  fetal tissues, 74
  histamine, 199, 238-239
  intestine, 250-251
  iodide, 21
  localization, 249, 268-269
  lungs, 194, 199, 201, 204, 238-239
  narcotics, 363-364, 368
  pancreas, 249-256
  platelets, 298, 300-301, 310
  prostaglandins, 199
  psoriasis, 418-419
  retina, 380-381, 397
  shock, 182-183
  smooth muscle, 200, 201
  thyroid, 21-23, 25-28
  tumors, 431-434
  vasopressin, 98-100
ADP, 296-297, 302-304
Adrenal steroids, see Corticosteroids
Affective disorders, 346-352
Alzheimer's disease, 343-345
Aminophylline, see Theophylline
Amphotericin B, 114-115, 120
Anaphylaxis, 207-208, 214, 216-217
Antidepressants, 329-349
Apomorphine, 365-367, 380
Aqueous homor,
  formation, 390-392
  outflow, 392-401
Arachidonic acid, 298, 308-309, 313, 420
Aspirin, 298, 305, 313-314
Asthma, 203-219
ATP, 335, 383

Basal ganglia, 339-342
Betazole, 240-243
Bile,
  cyclic nucleotide levels 231
  secretion 232-237

449

Blindness, 386-390
Blood brain barrier, 331
Bordetella pertussis, 207-208

Calcium,
   cyclic AMP excretion, 11-12
   disorders, 1-16
   pancreas, 245-246
   platelet function, 297, 310-312
   vasopressin, 99, 114-115
Cancer, see Tumors
Carbamazepine, 112-113, 121-122, 137
Catecholamines,
   Alzheimer's disease, 344
   aqueous humor, 392, 395-397
   cerebral infarction, 332, 334-335
   cholera toxin, 269-270
   cyclic AMP, 194-197, 200, 204-206, 208, 211-212, 216-217, 329, 332-333
   cyclic GMP, 196-197, 200-201
   gastrointestinal motility, 253-254
   histamine, 211-212, 216-217
   levels, 40-41
   parkinsonism, 340-343
   phosphodiesterase, 195-196, 205, 207
   platelet aggregation, 297, 299-302
   psoriasis, 418-419
   retina, 380-381
   seizure threshold, 339
   sensitivity, 25-28, 194-196, 200-201, 209, 211-214, 369
   SRS-A, 217
   thyroid hormone, 21, 23-24
   tumors, 432-433
   vasopressin, 105-108, 112-113, 121
Cell proliferation, 71-73, 411-416
Cerebral,
   atrophy, 345
   blood flow, 336
   edema, 335
   ischemia, 330-335
Cerebrospinal fluid, 329-330, 332-333
Chloride diarrhea, 252
Clorpromazine, 112-113, 123
Chlorpropamide, 14-15, 112-113, 116-120, 135
Cholecystokinin, 245-247, 256
Cholera toxin,
   adenylate cyclase, 212-213, 266-275
   ECF-A release, 217
   mechanism of action, 263-264, 275
   structure, 265-266
Cholinergic agents,
   cyclic AMP, 196
   cyclic GMP, 196-197, 239-240
   ECF-A release, 217
   histamine release, 214-215
   SRS-A release, 214-215, 217
Clofibrate, 112-113, 122, 137
Clonidine, 361
Colchicine, 114-115, 121, 314
Collagen, 297, 306-307
Cornea, 401-402
Corticosteroids,
   shock, 189-190
   vasopressin, 108-109, 112-113
Cyclic AMP, see also Cyclic nucleotides
   affective disorders, 346-352
   behavioral effects, 334
   catecholamines, 195-197, 200, 204-205, 208, 211-212, 216-227
   cholinergic agents, 196
   cornea, 401-402
   ECF-A release, 217
   excretion,
      affective disorders, 347-348
      asthma, 208-209
      physical activity, 351
      stroke, 330
   histamine, 201, 207, 210-214
   hypertension,
      essential, 171-175
      pheochromocytoma, 175
      renovascular, 175-177
   intraocular pressure, 398-401
   level,
      aqueous humor, 396-398
      arteriovenous differences, 331, 335
      brain, 328, 333-334
      cerebrospinal fluid, 329, 337, 340-342, 344-346
      gingiva, 287-290
      plasma, 330-331, 337
      saliva, 284-287
   pituitary hormones, 70, 95-98
   prostaglandins, 198, 201, 212
   psoriasis, 417-419
   renal effects, 123-124
   smooth muscle, 200-201
   SRS-A release, 217-218

stroke, 338
vasomotor control, 335-336
Cyclic CMP 72, 414-415, 430
Cyclic GMP, see also Cyclic nucleotides
  catecholamines, 196
  cholinergic agents, 196, 208
  ECF-A release, 217
  histamine, 207-214
  intraocular pressure, 399-400
  prostaglandins, 198
  psoriasis, 419-420
  smooth muscle, 200, 201
  SRS-A release, 217-218
  thyroid gland, 24, 44-45
  vasopressin, 99, 107-108
Cyclic nucleotides, see also Cyclic AMP, Cyclic GMP,
  cardiovascular effects, 187-188
  cell proliferation, 71-73, 413-416
  compartmentalization, 230
  diabetes mellitus,
    glucagon, 49
    insulin, 47-48
  gastrointestinal motility, 253-256
  implantation, 70-71
  lactation, 71
  levels,
    amniotic fluid, 76, 79
    bile, 231, 236-237
    brain, 365, 367, 370-373
    gastric juice, 231, 240
    intestine, 231
    liver, 231
    pancreas, 231
    stomach, 231, 240
    tumors, 430-431
  methodology,
    plasma, 4-5, 170
    skin, 416-417
    urine, 4-5
  platelets, 298-299
  psoriasis, 416-422
  retina, 380-390
  second messengers, 1-3
  secretion,
    bile, 232-235
    gastric, 237-244
    insulin, 46-47
    intestinal, 247-253
    pancreatic, 244-248

  shock, 182-190
  source, 5, 7, 84-86
  variability,
    age, 8
    circadian, 5-6
    menstrual cycle, 75-76
Cyclic UMP, 431
Cyclophosphamide, 112-113, 122, 137
Cystic fibrosis, 8

Dementia, 343-345
Denervation supersensitivity, 395, 397-398
Depression, 346-352
Diabetes insipidus, 107, 116-119, 138-141
  hypothalamic, 124
  nephrogenic, 124-129
Diabetes mellitus, 45-51
Diarrhea, 249-253
Diazapam, 372
Diphenylhydantoin, 112-113, 122-123
Dipyridamole, 314-315
L-DOPA, 341-342, 347
Dopamine, see Catecholamines
Down's syndrome, 284-286

Eaton-Lambert syndrome, 346
ECF-A release, 217
Endoperoxides, 299, 304, 308-309
Enterotoxins, 251
Epilepsy, 339
Epinephrine, see Catecholamines
Essential tremor, 343
Ethacrynic acid, 114-115, 134
Ethanol, 112-113, 121-122, 368-373

Fetal tissues, 72-75
Fluoride, see Sodium fluoride
Furosemide, 114-115, 134

Gamma-amino butyric acid, 341, 343-344, 371
Gastric
  inhibitory peptide, 250
  juice, 231, 237-244
Gingiva, 287-290
Glaucoma, 394-395
Glucagon,
  bile secretion, 232, 234-236

452 INDEX

cholera toxin, 269-270
cyclic nucleotides, 49-51
mechanism of action, 49
secretion, 48-49
tumors, 432-434
Glycerol, 335
Glycolysis, 334, 351
Graves' disease, 22-23
GTP analogs, 270-272-274
Guanethidine, 25
Guanylate cyclase,
  localization, 249
  lung, 196
  retina, 381, 385
  tumors, 434-436

Haloperidol, 329
Halothane, 112-113, 121
Heparin, 385
Histamine,
  adenylate cyclase, 199, 238
  brain, 329
  cholinergic agents, 214
  cyclic AMP, 201-202, 207, 210-214, 238
  cyclic GMP, 207
  gastric secretion, 238
  prostaglandins, 214
  receptors, 207
  release, 210-215, 217
  smooth muscle, 201
Huntington's disease, 241, 343
Hybrid cells, 368
Hyperparathyroidism, 10-12, 32
Hypertension,
  essential, 170-175
  pregnancy, 78-84
  renovascular, 175-177
Hyperthyroidism,
  catecholamine sensitivity, 25-28, 110
  cyclic nucleotides,
    excretion, 30-35
    plasma levels, 35-44
    tissue levels, 36, 38, 40-41
  polyuria, 110-111
Hypothermia, 338
Hypoparathyroidism, 10
Hypothyroidism,
  catecholamine sensitivity, 27-28

cyclic nucleotides, 110
  excretion, 29, 32-35
  plasma levels, 36

Indomethacin, 112-113, 120, 140-141
Insulin,
  biosynthesis and secretion, 46-47
  mechanism of action, 47-48
Intestinal secretion, 247
Intraocular pressure, 398-401
Iodide, 21
Isoproterenol, see Catecholamines

Lithium, 347
  cyclic AMP excretion, 347
  intoxication, 131-132
  vasopressin, 100-102, 114-115, 120
Long-acting thyroid stimulator, 21-22

Magnesium, 99, 114-115
Menstrual cycle, 75-76
Metabolic inhibitors, 111-115
Methodology,
  plasma, 4-5, 170
  skin, 416-417
  urine, 4-5
Methoxyfluorane,
  renal effects, 133
  vasopressin, 114-115, 120-121
Migraine, 336-338
Morphine, 112-113, 121, 363-368
Myasthenia gravis, 346
Myocardial infarction, 186-187

Nephrolithiasis, 10-11
Nephropathy, 129-131
Neuromuscular transmission, 346
Norepinephrine, see Catecholamines

Osteoporosis, 10-11

Pancreatic secretion, 244-248
Papaverine, 314, 336, 346, 382, 422
Parathyroid hormone, 9-15, 42-44
Parkinson's syndrome, 340-343
Pentagastrin, 240-243
Peptic ulcer, 240-242
Pernicious anemia, 240-242
Picrotoxin, 371

Pilocarpine, 394, 399
Phenacetin,
    nephritis, 132
    vasopressin, 114-115, 119-120
Phenformin, 112-113, 119, 136
Phentolamine, 21, 23, 27
Pheochromocytoma, 175
Phosphodiesterase,
    asthma, 205
    brain, 327-328, 330, 336
    catecholamines, 195, 205, 207
    cell proliferation, 415
    gingiva, 287-290
    kidney, 118
    localization, 384
    platelets, 300-301, 314
    psoriasis, 419
    purification, 385, 387-388
    retina, 381-390
    shock, 189
    smooth muscle, 200
    stomach, 239
    thyroid, 21, 26-27
    tumors, 436-437
Phosphoribosyl pyrophosphate synthetase, 414
Platelets,
    affective disorders, 351-352
    cyclic nucleotides, 288-289
    mechanism of function, 296-297
Polydipsia, 124
Polyuric syndromes, 124-133
Potassium, 99-100, 114-115
Pregnancy, 69-87
Probenecid, 329, 340, 342, 344-345
Propoxyphene, 114-115, 119-120
Propranolol,
    essential tremor, 343
    hyperthyroidism, 38, 41
    thyroid function, 21, 23, 25
    withdrawal, 362
Prostaglandins
    catecholamines, 217
    cell proliferation, 415
    diabetes insipidus, 128-129
    gastrointestinal,
        motility, 255-256
        secretion, 241, 243-244, 247, 250
    histamine, 213-214

    indomethacin, 215, 216
    morphine, 368
    platelets, 297-298, 301-302, 307-310, 312
    pregnancy, 83
    psoriasis, 420-421
    smooth muscle, 201
    thyroid function, 21
    tumors, 433-434
    vasopressin, 102-105, 112-115, 121
Protein kinase, 28, 195, 437-438
Pseudohypoparathyroidism, 12-15, 29
Psoriasis, 407-411, 416-422

Receptors,
    adrenergic, 194
    dopaminergic, 340-343
Reserpine, 25
Retina, 25
Rhodopsin, 386

Saliva, 284-287
Sarcoidosis, 11
Secretin, 323-237, 244-247
Seizures, 334
Sensitization, 204-208, 211-219
Serotonin, 329, 331-332, 334-335, 344
Shock,
    cardiogenic, 186-187
    hemorrhagic, 182-185
    septic, 185-186
Sickle cell anemia, 132
Skin anatomy, 407-408
Smooth muscle,
    bronchial, 199-203
    gastrointestinal, 253-256
Sodium dehydrocholate, 232, 234-236
Sodium fluoride, 21, 102, 114-115, 120, 432-433
SRS-A, 217-219
Stroke, see Cerebral ischemia
Subarachnoid hemorrhage, 335-336
Sulfinpyrazone, 314

Tetracyclines, 114-115, 120, 132-133
Theophylline,
    aqueous humor, 396
    brain trauma, 338
    cornea, 401

diuretic effect, 134-135
Eaton-Lambert syndrome, 346
involuntary movements, 342
pancreatic secretion, 246
platelet function, 314
pseudohypoparathyroidism, 14-15
retina, 382
vasopressin, 97, 112-113
Thiazide diuretics, 114-116, 134-136, 138-140
Thrombin, 297, 304-305
  -sensitive protein, 304-305
Thrombopathia, 312-313
Thromboxanes, 308-309
Thyroid hormone,
  adenylate cyclase, 22-23, 27, 28
  adipose tissue, 26-28
  bone growth, 28-29
  brain, 28
  cardiovascular system, 25-26
  cyclic nucleotide excretion, 30
  TSH, 20-24
  vasopressin, 109-113
Tolbutamide, 112-113, 116-119, 135
Trauma, 182-185, 338-339
Tumors, 286-287, 429-438

Vasoactive intestinal peptide, 245-247, 250, 252
Vasopressin,
  adrenal steroids, 108-109
  analgesics, 119-120
  catecholamines, 105-108
  cyclic AMP, 95-98
  diuretics, 114-116
  drug interactions, 112-115
  ionic modifiers, 99-102
  metabolic inhibitors, 111-114
  prostaglandins, 102-105
  release, 93-95, 112-113, 121-123
  sulfonylurea drugs, 116-119
  thyroid hormone, 109-111
Vincristine, 112-113, 122, 137

Wound healing, 401-402

Zollinger-Ellison disease, 240-242